Modeling and Parameter Extraction Techniques of Silicon-Based Radio Frequency Devices

Modeling and Parameter Extraction Techniques of Silicon-Based Radio Frequency Devices

Ao Zhang
Nantong University, China

Jianjun Gao
East China Normal University, China

中国工信出版集团

Published by

World Scientific Publishing Co. Pte. Ltd.
5 Toh Tuck Link, Singapore 596224
USA office: 27 Warren Street, Suite 401-402, Hackensack, NJ 07601
UK office: 57 Shelton Street, Covent Garden, London WC2H 9HE

Library of Congress Cataloging-in-Publication Data
Names: Zhang, Ao, 1995– author. | Gao, Jianjun, 1968– author.
Title: Modeling and parameter extraction techniques of silicon-based radio frequency devices / Ao Zhang, Nantong University, China, Jianjun Gao, East China Normal University, China.
Description: Hackensack, NJ : World Scientific Publishing Co. Pte. Ltd., [2023] | Includes bibliographical references and index.
Identifiers: LCCN 2022036005 | ISBN 9789811255359 (hardcover) | ISBN 9789811255366 (ebook for institutions) | ISBN 9789811255373 (ebook for individuals)
Subjects: LCSH: Radio frequency integrated circuits--Simulation methods.
Classification: LCC TK7874.78 .Z49 2023 | DDC 621.3841/2--dc23/eng/20221011
LC record available at https://lccn.loc.gov/2022036005

British Library Cataloguing-in-Publication Data
A catalogue record for this book is available from the British Library.

For any available supplementary material, please visit
https://www.worldscientific.com/worldscibooks/10.1142/12805#t=suppl

Desk Editors: Nimal Koliyat/Steven Patt

Typeset by Stallion Press
Email: enquiries@stallionpress.com

Printed in Singapore

Preface

This textbook is written for the beginning user of silicon-based radio frequency integrated circuits design. Our purposes are as follows:

- to introduce the basic concepts of silicon-based radio frequency devices,
- to provide state-of-the-art modeling and measurement technologies,
- to provide equivalent circuit parameter extraction methods for passive and active devices.

With the rapid growth of the demand for low-power, low-cost, and high-integration wireless communication systems, the development of on-chip passive and active devices for radio frequency integrated circuits (RFICs) has emerged as a critical issue recently. The accurate equivalent circuit model is a supposition for the device performance analysis in the design of microwave circuits and the characterization of the device technological process. State-of-the-art computer-aided design software relies heavily on models of real devices. Thence, it is crucial to be able to accurately predict the performance of silicon radio frequency circuits in order to shorten the design cycles and achieve first-time success. Our primary objective with this book is to bridge the gap between silicon-based passive and active device modeling and integrated circuit design by utilizing computer-aided design tools. Even for those who do not have a good microwave background, the readers can readily understand the contents of the book.

The presentation of this book assumes only a basic course in electronic circuits as a prerequisite.

This book is intended to serve as a reference book for practicing engineers and technicians working in the areas of RF microwave integrated circuit design. This book should also be useful as a textbook for microwave active and passive devices circuit courses designed for senior undergraduate and first-year graduate students. At the same time, students can benefit from it if they are assigned problems requiring reading the original research papers.

About the Authors

Ao Zhang was born in Nanjing, Jiangsu province, China, in 1995. She received the B.S. degree at Nanjing University of Posts and Telecommunications, Nanjing, China, in 2017 and the Ph.D degree at East China Normal University, Shanghai, China, in 2021. In 2019, she visited Carleton University, Ottawa, ON, Canada, as a research associate working on HBT modeling using artificial neural network techniques. Since 2022, she has been a lecturer at Nantong University, Nantong, Jiangsu province, China. Her research focuses on modeling and on-wafer millimeter-wave measurements of active and passive devices.

Jianjun Gao was born in Hebei province, P.R.China in 1968. He received the B.Eng. and Ph.D. degrees from the Tsinghua University, in 1991 and 1999, respectively, and M.Eng. degree from Hebei semiconductor research institute, in 1994.

From 1999 to 2001, he was a Post-Doctoral Research Fellow at the Microelectronics R&D Center, Chinese Academy of Sciences developing PHEMT optical modulator driver. In 2001, he joined the School of Electrical and Electronic Engineering, Nanyang Technological University (NTU), Singapore, as a Research

Fellow in semiconductor device modeling and on wafer measurement. In 2003, he joined the Institute for High-Frequency and Semiconductor System Technologies, Berlin University of Technology, Germany, as a research associate working on the InP HBT modeling and circuit design for high-speed optical communication. In 2004, he joined the Electronics Engineering Department, Carleton University, Canada, as Post-doctor Fellow working on the semiconductor neural network modeling technique. From 2004 to 2007, he was a Full Professor of radio engineering department at the Southeast University, Nanjing, China. Since 2007, he has been a Full professor of the School of Information Science and Technology, East China Normal University, Shanghai, China. He authored *RF and Microwave Modeling and Measurement Techniques for Field Effect Transistors* (SciTech Publishing, 2009), *Optoelectronic Integrated Circuit Design and Device Modeling* (Wiley, 2010), and *Heterojunction Bipolar Transistor for Circuit Design — Microwave Modeling and Parameter Extraction* (Wiley, 2015). His main areas of research are characterization, modeling and on wafer measurement of microwave semiconductor devices, optoelectronics device, and high-speed integrated circuit for radio frequency and optical communication.

Acknowledgments

I would like to thank the collaboration with Prof. Sun Ling of Nantong University, Prof. Wang Hong of Nanyang Technical University (Singapore), and Dr. Andreas Werthof of Infineon Technologies for their cooperation.

I would also like to thank our team members: Dr. Chen Jiali, Dr. Chen Bo, Dr. Wang Huang, Dr. Yu Panpan, Dr. Zhang Yixin, Dr. Yan Lingling, Dr. Chen Ran, and Dr. Zhou Ying.

I would also like to thank our families for their great support, patience, and understanding provided throughout the period of writing.

This work was supported in part by the National Natural Science Foundation of China under Grant 62201293 and 62034003, in part by the Talent Plan of Nantong University.

Contents

Nomenclature

nm	Nanometer, one-billionth of a meter ($= 10^{-9}$ m)
μm	Micrometer, one-millionth of a meter ($= 10^{-6}$ m)
ps	Picosecond, one-thousandth of a billionth of a second ($= 10^{-12}$ s)
MHz	Megahertz, 1 million vibrations per second ($= 10^6$ Hz)
GHz	Gigahertz, 1 billion vibrations per second ($= 10^9$ Hz)
mW	Milliwatt, one-thousandth of a watt ($= 10^{-3}$ W)
q	Electron charge ($= 1.6 \times 10^{-19}$ C)
k	Boltzmann's constant ($= 1.38 \times 10^{-23}$ J/k)
fF	Femto Farad, one-billionth of a Farad ($= 10^{-15}$ F)
Gb/s	Gigabits per second, 1 billion bits per second ($= 10^9$ bits per second)
pF	Pico Farad, one-thousandth of a billionth of a Farad ($= 10^{-12}$ F)
pH	Pico Henry
3D	Three-dimension
BiCMOS	Bipolar complementary metal oxide semiconductor
BJT	Bipolar junction transistor
CMRR	Common Mode Rejection Ratio
CAD	Computer-aided design
CMOS	Complementary metal oxide semiconductor
DUT	Device under test
EDA	Electronic Design software automation
IMD	Inter-metal dielectric
MEMS	Micro Electro Mechanical System
MOSFET	Metal-oxide semiconductor field effect transistor

NMS	Noise measurement system
PDK	Process design kits
RF	Radio frequency
RFCs	Radio frequency coils
RFIC	Radio frequency integrated circuit
PCB	Printed circuit board
SEM	Scanning electron microscope
SPICE	Simulation program with integrated circuit emphasis
UWB	Ultra wideband

Chapter 1

Introduction

1.1 Overview of Silicon-Based Devices

With the rapid growth of the demand for low-power, low-cost, and high-integration wireless communication systems, the development of on-chip passive and active devices for radio frequency integrated circuits (RFICs) has emerged as a critical issue recently. RFICs such as low noise amplifiers, voltage-controlled oscillators, mixers, and power amplifiers rely on a number of passive components including capacitors, resistors, inductors, transformers, and transmission lines. They are primarily used for impedance matching, resonance circuits, filters, and bias circuitry. The area scaling of the RFICs is limited by passive devices. Various semiconductor materials can be used as substrates to fabricate integrated circuits, such as GaAs, InP, GaN, and silicon. Since silicon is easy to purify in nature, has low cost, has stable properties, and is nontoxic and harmless, silicon-based integrated circuits become the mainstream of RFICs. Silicon-based semiconductor active devices, with their low-cost, high-volume production, have improved frequency response significantly as the channel length is made smaller and up to sub-10 nm.

Figure 1.1 illustrates the commonly used semiconductor devices in the RFIC design. In terms of the operation mechanism, microwave and radio frequency semiconductors can be categorized into field effect transistors and bipolar transistors. Metal-oxide semiconductor field effect transistor (MOSFET) and bipolar junction transistor (BJT) are the two key active components and play important roles in amplifier design. The BJTs can be combined with MOSFETs

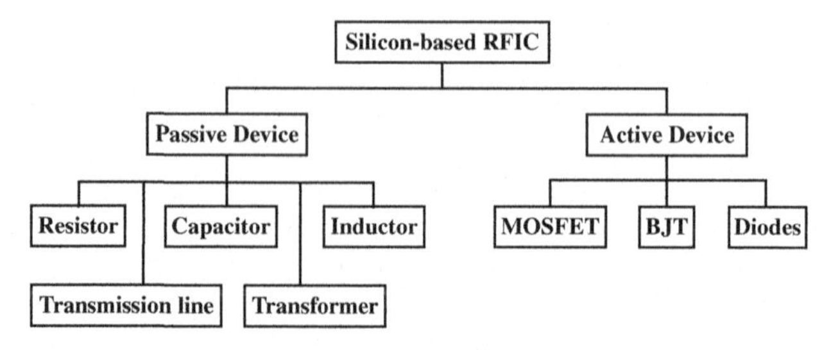

Figure 1.1. Passive and active devices in RFICs.

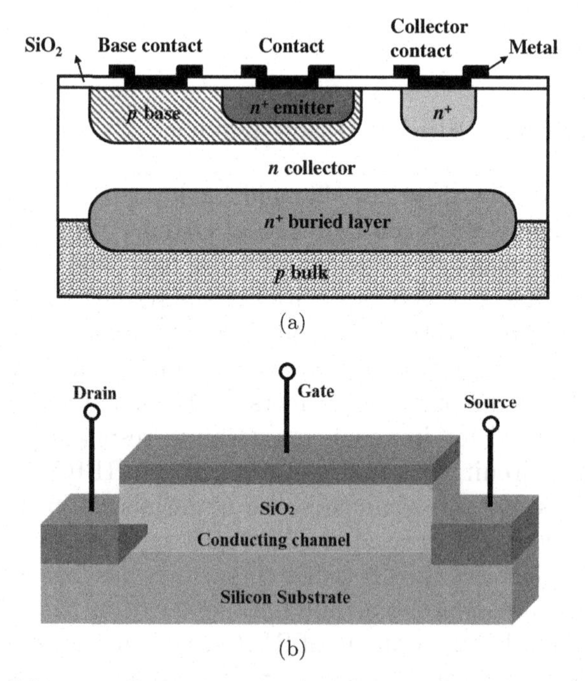

Figure 1.2. Schematic illustration of generic BJT and MOSFET: (a) Schematic view of *npn* BJT; (b) Schematic view of MOSFET.

to create innovative circuits that take advantage of the high input impedance and low power operation of MOSFETs and the very high frequency operation and high current driving capability of BJTs. The resulting technology is known as the BiCMOS process.

Figure 1.2(a) exhibits a more realistic cross-section and cubic diagram of an *npn* BJT. Note that the collector virtually surrounds the emitter region, thus making it difficult for the electrons injected into

Table 1.1. Comparison of FET and bipolar transistor.

Device	MOSFET	BJT
Physical structure	Planar	Vertical
Junction	Schottky	PN
Terminal	4	3
Key dimension	Gate length	Base thickness
Turn-on	Threshold voltage	Base current
Noise	Channel current noise	Shot noise

the thin base to escape being collected. Moreover, the schematic view of MOSFET is presented in Figure 1.2(b).

The comparison of some device parameters for both MOSFET and bipolar transistor devices is tabulated in Table 1.1. It can be clearly seen that the MOSFET has a planar structure, while the BJT has a vertical structure. A terminal is connected to each of the three semiconductor regions of the BJT, with the terminals labeled emitter, base, and collector. The transistor consists of two *pn* junctions, the emitter-base junction and the collector-base junction, which are sufficiently close together that they interact with each other. The MOSFET can be viewed as a unipolar device where only electrons are involved in the movement of carriers. The gate voltage determines the drain current by controlling the channel. Metal is deposited on top of the oxide layer to form the gate electrode of the device. Metal contacts are also made to the source region, the drain region, and the substrate, also known as the body. The physical dimension limitation set the ultimate device speed performance. Fundamentally, the shorter gate length in MOSFET can reduce the carrier transport time and narrower base as well as the thinner collector in the bipolar transistor can decrease the carrier transit time.

Owing to the technological breakthrough of the process, the gate length of MOSFET became shorter and shorter, and MOSFET replaced BJT and became the mainstream process device of silicon-based integrated circuits. The variation curve of device characteristics with years is sketched in Figure 1.3. Obviously, the cutoff frequency and maximum oscillation frequency rise rapidly with the decrease in the gate length and have been widely used in the terahertz band circuit design. Since the carrier transit time is inversely proportional to the channel length, the characteristic frequency of the device becomes higher and higher with the shortening of the gate

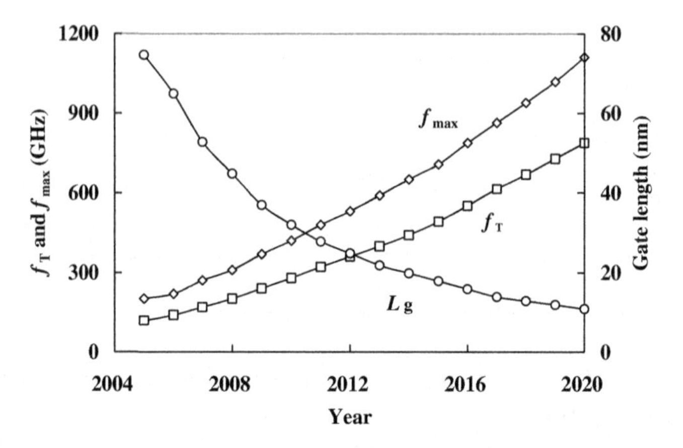

Figure 1.3. Progress of MOSFET with years.

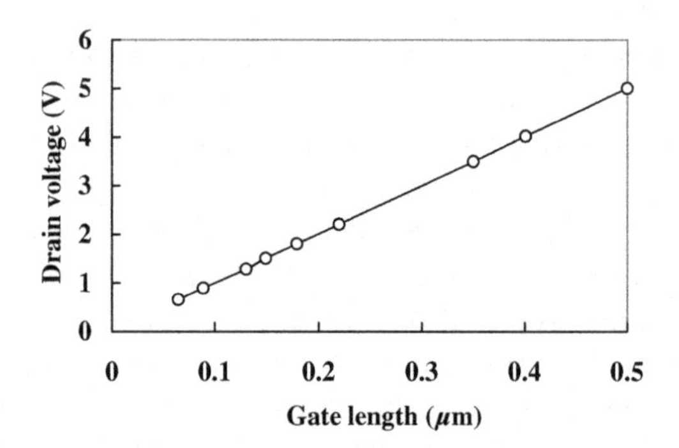

Figure 1.4. Drain voltage versus device gate length.

length for the device. The tremendous success of CMOS technology is due to the scalability of MOSFET transistors. The drain voltage versus device gate length is illustrated in Figure 1.4. As seen, the maximum drain voltage is directly proportional to the device gate length. The maximum voltage that can be handled for the device with a 0.5 μm gate length is 5V roughly, while the device with a 0.1 μm gate length device can handle the voltage of about 1V [1–3].

1.2 What is Modeling

Semiconductor device modeling is the process of characterizing the physical characteristics of the device with mathematical formulas.

The combination of device model and circuit simulation software is to bridge the gap between device process and circuit design and make the design and process manufacturing as consistent as possible. The accuracy of the device model (also called process design kits, PDK) is an important factor for circuit design engineers. The simulation results will be closed to the final chip tape-out test results by utilizing the accurate model. With the increase in the circuit scale, the requirements for the accuracy of the device models are getting higher and higher, and new device models need to be established to guide the circuit design.

Before computers, researchers mainly relied on analytical formulas for manual calculations. With the popularization of computers, the analytical formulas of devices and circuits can be completed by computers. This software is called computer-aided design (CAD) software or electronic design software automation (EDA). The advantages of computer-aided design software are as follows [4–6],

(1) With the increase in the integration of semiconductor devices, the number of transistors integrated on a chip has increased significantly, and the operating frequency becomes higher and up to millimeter-wave. The influence of various parasites cannot be neglected. Obviously, the hand calculation for analysis and design of integrated circuits has been unable to meet the requirement.
(2) The performance of the integrated circuits can be obtained from the computer simulation, for example, the gain and noise figure of the amplifiers. Using the Monte Carlo method, process tolerances also can be considered. Therefore, the EDA tools can be used to improve the chip yield, shorten the design cycle, reduce production costs, and improve market competitiveness.
(3) In order to reduce the production cost and keep low consumption, the integrated circuits need to be successfully tape-out at one time. This requires accurate simulation of integrated circuits to help designers predict circuit behavior.

There are mainly two types of computer-aided design tools for RF microwave devices and circuits: one is semiconductor device simulation software and the other is microwave circuit design software. Semiconductor device simulation software refers to analyzing the physical structure of the device, solving the corresponding

Poisson equation and current continuity equation, etc., and finally obtaining the output characteristics of the device such as DC and AC characteristics to guide device design and production. These models account for process-related parameters (such as geometry, recess depth, material parameters, and doping profile). The difficulty in applying physical device models to microwave CAD simulators is the large execution times required.

The popular RF microwave circuit simulation software can be categorized into frequency-domain analysis software and time-domain analysis software. The most commonly used time-domain analysis software is Berkeley SPICE (Simulation program with integrated circuit emphasis). Frequency-domain analysis and time-domain analysis can be converted into each other through Fourier transform. The accuracy and calculation speed of the device model directly determine the calculation speed of the circuit simulation [4–7]. The CAD tools need to be improved until the simulated and measured RF performance of the component being designed is in good agreement. This will permit the design to be completed, simulated, and fully tested by engineers working at computer workstations before fabrication is implemented.

Although the performance for semiconductor devices can be obtained by utilizing two-dimensional nonlinear differential equations which describe the electron transport in the channel. Nevertheless, this kind of method requires specialized software, it is not suited to the inclusion of package parasitics, and device-circuit interactions are not readily taken into account. Another alternative approach is to use circuit analysis based on device circuit models. The semiconductor device modeling concept is that the complex active devices are represented as two-port circuits which include the basic circuit elements, such as resistances, inductances, capacitances, and controlled sources (see Figure 1.5). From the equivalent circuit model, circuit designers can readily understand the operation mechanism of the complex active devices.

Semiconductor device modeling is to seek an equivalent circuit model so that it can represent the physical device for circuit design and analysis. The model should meet the following conditions:

- accurately predict linear and nonlinear characteristics, etc.,
- keep good accuracy within a very wide frequency band,

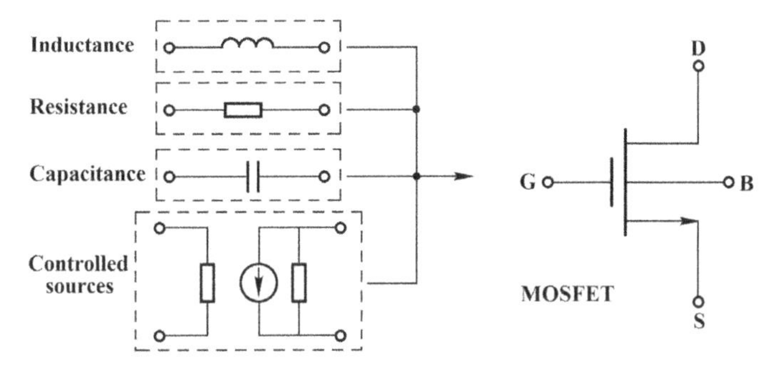

Figure 1.5. Semiconductor device modeling concept.

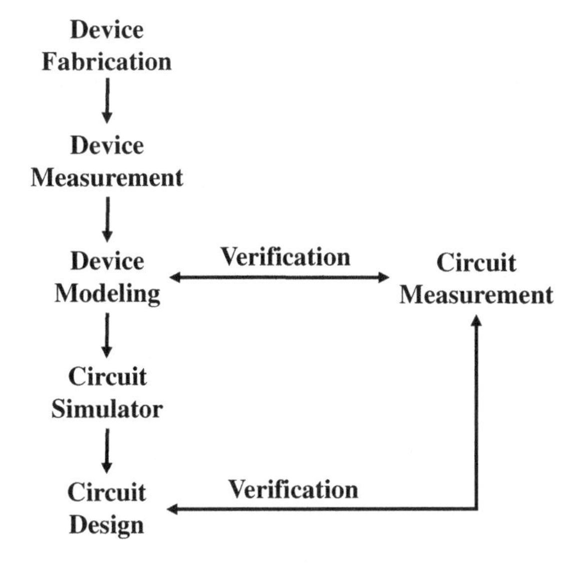

Figure 1.6. Relationship between modeling and measurement.

- model parameters are as few as possible,
- model parameters must be easy to extract.

The process of model development is depicted in Figure 1.6; microwave and RF measurement techniques are the basis of characterization of the microwave and RF devices and circuits. The procedure can be carried out as follows:

- device fabrication and performance measurement,
- modeling and parameter extraction,

- incorporate equivalent circuit models into CAD software,
- model verification.

After the model is embedded in the computer-aided software, the simulation data can be compared with the measurement results. We need to continue to modify the model and re-extract the model parameters until the error value is within the allowable range.

1.3 Organization of This Book

This book consists of six chapters, focusing on the modeling methods and parameter extraction techniques of passive and active devices used in silicon-based RFICs. Passive components include on-chip spiral inductors and transformers; active devices are MOSFETs.

The second chapter mainly introduces the basic structure and characteristics of commonly used on-chip spiral inductors, including spiral inductors under standard CMOS technology and new inductors, and proposes modeling and model parameter extraction methods of on-chip spiral inductors. Finally, the influence of the main physical structure parameters of the on-chip spiral inductor on the inductance and quality factor is discussed.

In the design of radio frequency integrated circuits, the on-chip transformer can complete impedance matching, feedback, single-ended to double-ended conversion, and AC coupling between stages of the circuit. Chapter 3 mainly introduces the typical structure of the on-chip transformer, the equivalent circuit model of the on-chip transformer, and the corresponding parameter extraction method.

In Chapter 4, we introduce the physical structure and operation concept of MOSFET devices. The small signal modeling and parameter extraction method are described, especially determination methods for pad capacitances, feedline inductances, extrinsic resistances, and intrinsic elements. The scalable rules for intrinsic elements are also given. The sensitivity analysis and uncertainty estimation for intrinsic elements and substrate parasitics related to the S-parameters are also described.

The influence of guard-ring on the DC and high-frequency performance of deep-submicrometer MOSFET device is investigated in

Chapter 5. A deep submicrometer large signal model for silicon-based MOSFET device with DC/AC dispersion model is discussed.

Based on the noise correlation matrix, a scalable noise and small-signal model for deep-submicrometer metal–oxide semiconductor field-effect transistors, which consist of multiple elementary cells, are presented in Chapter 6. It allows exact modeling of all noise and small-signal model parameters from the elementary cell to the large-size device. The scalable rules for noise and small-signal model parameters are given in detail.

References

[1] W. Haensch, E. J. Nowak, R. H. Dennard, *et al.*, "Silicon CMOS devices beyond scaling," *IBM Journal of Research and Development*, 50(4/5): 339–361, 2006.

[2] D. A. Antoniadis, I. Aberg, C. N. Chleirigh, *et al.*, "Continuous MOSFET performance increase with device scaling: The role of strain and channel material innovations," *IBM Journal of Research and Development*, 50(4/5): 363–376, 2006.

[3] P. H. Aaen, J. A. Pla, and J. Wood, *Modeling and Characterization of RF and Microwave Power FETs*. UK: Cambridge University Press, 2007.

[4] T. Ytterdal, Y. Cheng, and T. A. Fjeldly, *Device Modeling for Analog and RF CMOS Circuit Design*. England: John Wiley & Sons, Ltd., 2003.

[5] J. Gao, *RF and Microwave Modeling and Measurement Techniquesfor Field Effect Transistors*. Raleigh, NC: SciTech Publishing, Inc., 2010.

[6] J. Gao, *Optoelectronic Integrated Circuit Design and Device Modeling*. Singapore: John Wiley & Sons, 2010.

[7] W. Liu, *MOSFET Models for SPICE Simulation, Including BSIM3v3 and BSIM4*. USA: John Wiley & Sons, 2001.

Chapter 2

On-Chip Spiral Inductor

Continuous growth in wireless-communication systems has stimulated research in low-cost, low-power, and high-performance CMOS RF integrated-circuit (IC) components for system-on-chip solutions. The surging demand for silicon-based radio frequency integrated circuits has raised tremendous interest in on-chip passive components. On-chip passive devices include capacitors, resistors, inductors, transformers, and transmission lines. The roles of passive devices are impedance matching, filtering of signals, and power supply. All radio frequency integrated circuits such as low noise amplifiers, oscillators, mixers, and power amplifiers all rely on these passive components. High-quality on-chip inductors are essential to the monolithic integration of both RFICs and monolithic microwave integrated circuits (MMICs) [1].

The researchers found that the layout of RF integrated circuits is constrained by the area of passive devices rather than active devices. In other words, since the layout area of the on-chip inductor is not determined by the characteristic size of the CMOS process, the on-chip inductor becomes the most area-consuming element. Hence, the on-chip spiral inductors become very important in integrated circuit and layout design. It is worth noting that lumped-element design utilizing inductors is a key technique for reducing MMIC chip area resulting in more chips per wafer and thus lower cost. Inductors used as circuit components, matching networks, and biasing chokes play a significant role in the realization of compact integrated chips. Manufacturing high-performance integrated inductors has always been one of the important challenges in realizing monolithic integrated circuits. With the increasing demand for high-performance on-chip spiral inductors in microwave integrated circuits, an ideal lumped-element equivalent circuit supplemented with a high-precision modeling technique is crucial to inductor optimization and circuit simulation.

This chapter mainly introduces the physical structure and basic characteristics of on-chip inductors commonly used in RF microwave integrated circuits. The corresponding modeling and parameter extraction methods of on-chip spiral inductors are given. The influence of the main physical structure parameters of the on-chip spiral inductor on the inductance and quality factor is discussed.

2.1 Physical Structure

A circuit element built to possess inductance is called an inductor, which has different behaviors from resistors and capacitors. Unlike capacitors, inductors are not readily available in standard CMOS technology. As a result, some design tricks have to be used, which usually limit the performance of the inductor. In its simplest form, an inductor is simply a coil of wire; a coil is generally formed by winding a straight wire on a cylindrical former. Inductors are used in many places. In radios, they are part of the tuning circuit that you adjust when you select a station. RF coils (RFCs) can be used to short-circuit the device to DC voltage conditions. In the low-frequency ranges, there are two types of inductors: one is an air-core coil inductance with a small inductance and the other is a magnetic core inductance with a large inductance (as shown in Figure 2.1). The main feature of the low-frequency inductors (also called coil inductors) is that it has three-dimension (3D) structures, which require a large space volume. It can be directly soldered on the surface of the printed circuit board (PCB) [2].

Compared with low-frequency applications, the shape, structure, and fabrication of inductors in RF integrated circuits have been changed. Comparison of low-frequency inductors and RF microwave monolithic inductors is tabulated in Table 2.1. As seen from Table 2.1, compared with the low frequency inductors, the quality factor of RF microwave monolithic inductors is very low, which brings difficulties to the design of low-noise amplifiers. On-chip

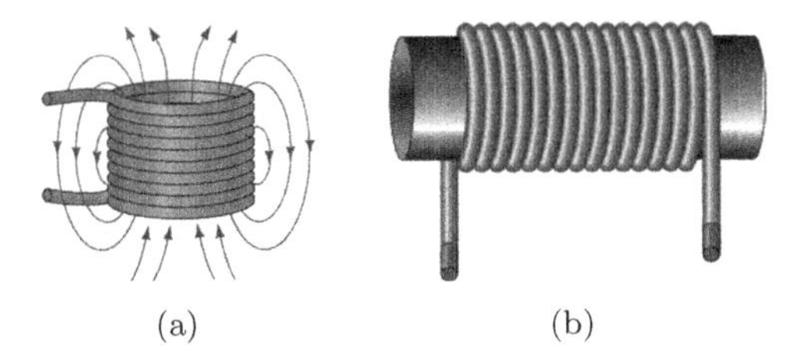

Figure 2.1. Air-core inductors (a) and magnetic core inductor (b).

Table 2.1. Comparison of low-frequency inductors and RF microwave inductors.

Inductor	Low frequency	RF microwave
Structure	3D	Planar
Volume	Large	Small
Value	μH	nH
Quality factor	Several Hundred	Dozens
Fabrication	Surface mount	Multi-layer

inductors have two common features: (1) multi-turns for saving chip area; and (2) an overwhelmingly wider metal trace width than spacing in order to reduce the resistive losses [3].

The schematic of the low noise amplifier with on-chip octagonal spiral inductors peaking is depicted in Figure 2.2 [4]. Base and collector peaking inductors have been used. Base inductance peaking involves placing an inductor at the base of the transistor. A collector peaking inductance is used to increase the high-frequency gain.

This emitter inductance acts as series feedback and shifts the input impedance to lower values. On account of the larger values of inductances required for circuit design, three inductors occupy about half of the chip area.

The schematic of the low noise amplifier with feedback inductances and chip photograph is illustrated in Figure 2.3 [5]. The methodology takes advantage of the Miller effect for ultra-wideband input impedance matching and the inductive shunt–shunt feedback technique for bandwidth extension by pole and zero cancellation. The profile of the on-chip inductor is a square spiral structure. The chip layout with the circular on-chip inductor is sketched in Figure 2.4.

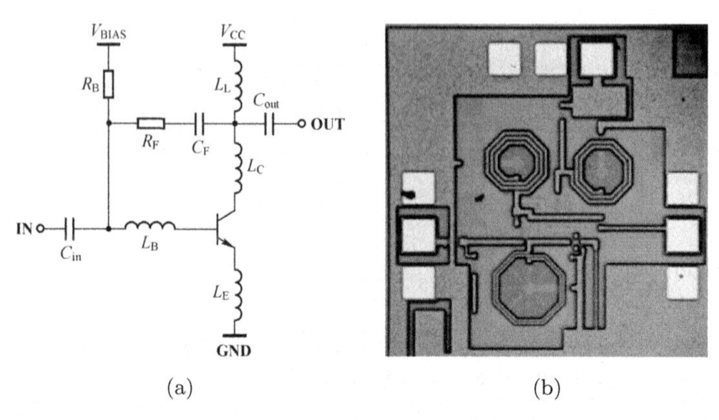

Figure 2.2. Schematic of the low noise amplifier (a) and chip photograph (b) [4].

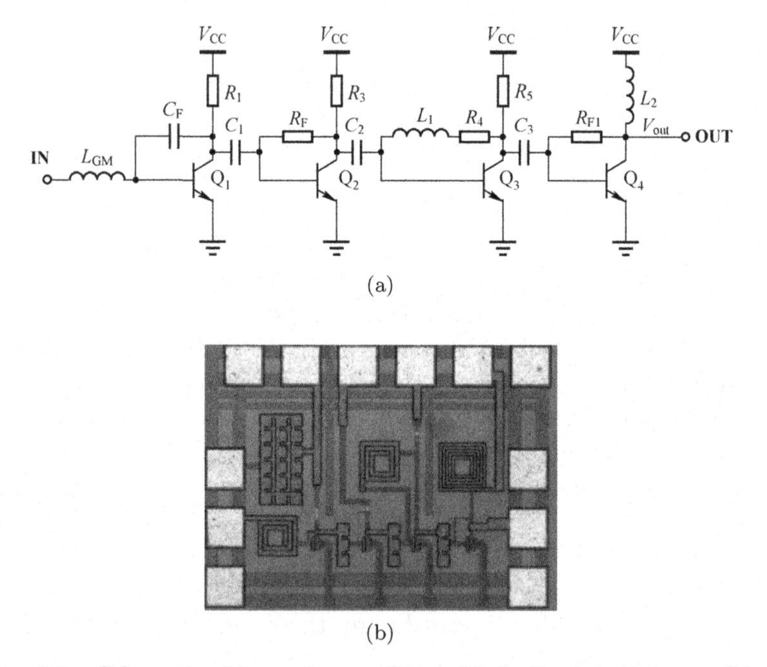

Figure 2.3. Schematic of low noise amplifier with feedback inductances (a) and chip photograph (b) [5].

Planar spiral inductors mainly include asymmetric (single-ended) and symmetrical structures. The structure of the asymmetric planar spiral inductor is illustrated in Figure 2.5. Figures 2.5(a)–2.5(d) are

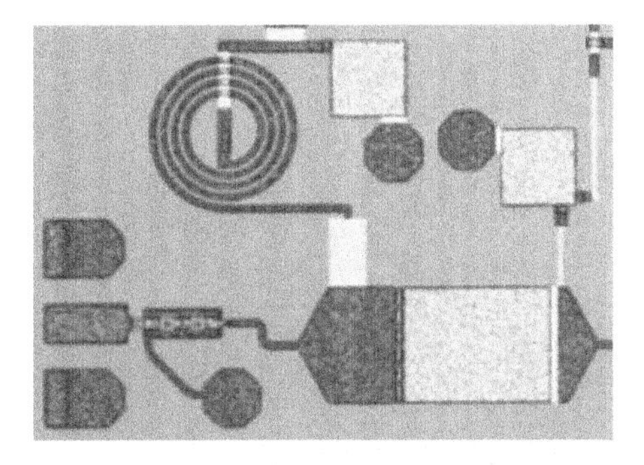

Figure 2.4. Chip layout with the circular on-chip inductor.

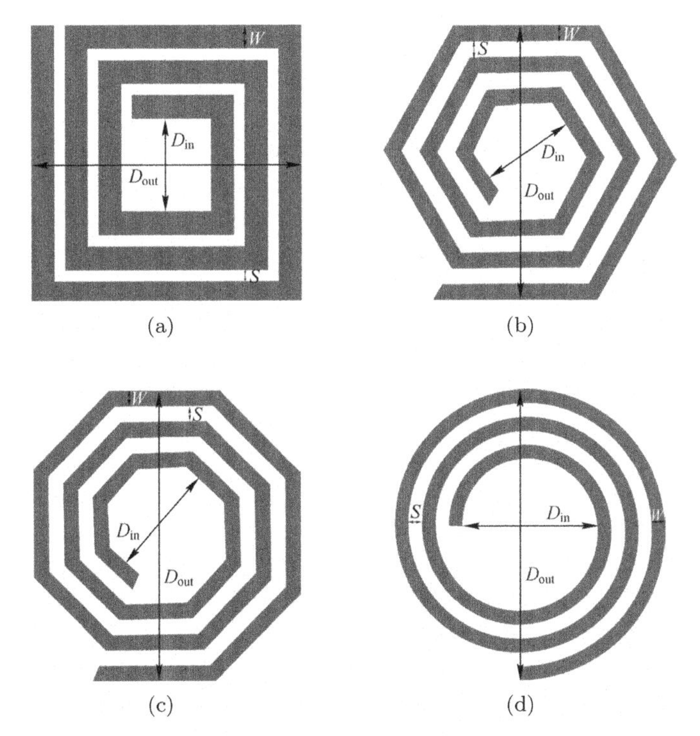

Figure 2.5. On-chip inductor realizations: (a) square; (b) hexagonal; (c) octagonal; (d) circular [6].

quadrilateral, hexagonal, octagonal, and circular structures, respectively. Among these, hexagonal and octagonal inductors are utilized widely. For a given shape, an inductor is completely specified by the

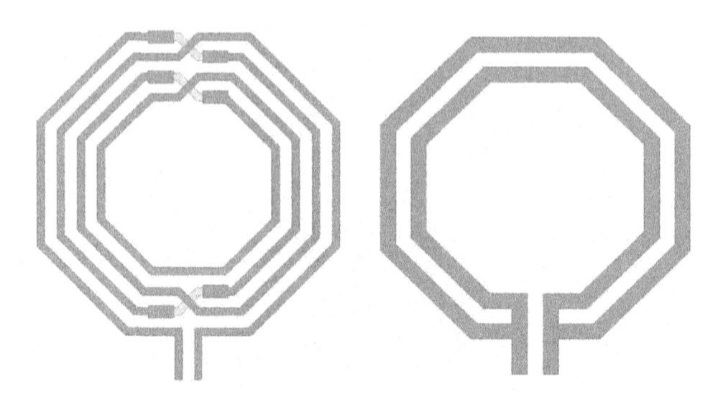

Figure 2.6. Symmetrical planar spiral inductor.

number of turns n, the turn width w, the turn spacing s, and any one of the following: the outer diameter d_{out} or the inner diameter d_{in}.

In general, the performance of the inductors improves with the increase in the number of sides, thence the most perfect structure is circular. Once the geometric parameters are fixed, the self-resonant frequency of the circular structure is higher than that of the quadrilateral structure, and the ohmic loss is smaller. Nevertheless, from a tape-out point of view, it is usually difficult to fabricate the mask into a circular structure; the more sides mean the more difficulty in fabrication.

The structure of the symmetrical planar spiral inductor is depicted in Figure 2.6. The layout is almost completely symmetrical except for the cross-region, hence, the input and output ports of the inductor have the same characteristics.

By leading a tap at the midpoint, the symmetrical inductor can be used as a center-tapped differential inductor. A multi-turn octagonal differential inductor is constructed with the top layer metal for winding and the second and third layers from the top for underpass and crossing. Since the symmetrical inductance can be regarded as winding two identical spiral inductors together, it has a higher self-inductance value due to the magnetic field coupling between two subinductors.

The on-chip inductors need to be connected with the upper and lower layers of metal through a via hole for a given process, as shown in Figure 2.7. The additional capacitive coupling between two layers

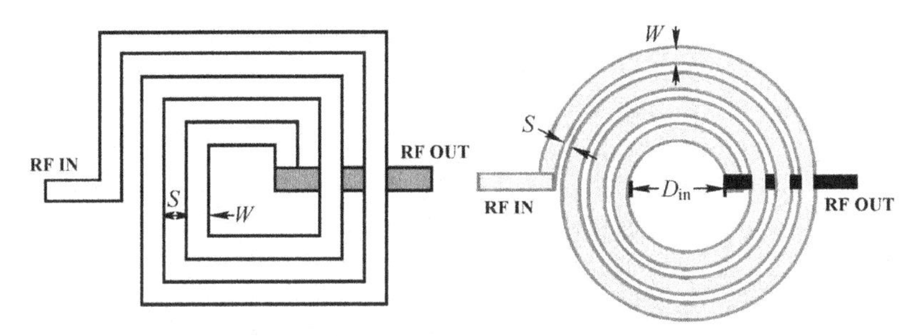

Figure 2.7. Terminal connection of on-chip inductor.

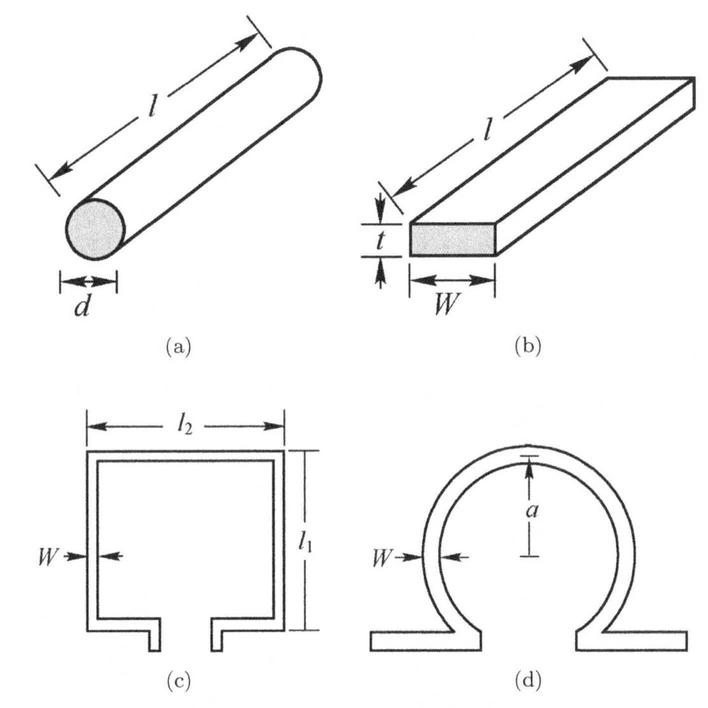

Figure 2.8. Small inductor configurations: (a) cylinder; (b) strip section; (c) half rectangular loop; (d) half circular loop.

of metal wires should be considered. If the inductance is very small, for example, below 100 pH, it can be realized by microstrip transmission line, half square, or circular spiral structures (as illustrated in Figure 2.8). In this case, two layers of metal are not necessary.

2.2 Basic Characteristics of the Planar Spiral Inductor

Spiral inductors implemented in silicon processes suffer from several power dissipation mechanisms, leading to poor inductor quality factors. The performance of on-chip inductors is greatly affected by the thin metal layers and lossy silicon substrate in very large-scale integration (VLSI) technology.

For the on-chip inductors in silicon technology, there are four major electromagnetic effects:

- skin effect in the metal lines,
- proximity effect between the metal lines,
- dielectric capacitive coupling effect,
- lossy substrate effect.

Both skin and proximity effects become more severe with increased frequency and are also dependent on the geometry of the device. The quality factor of the inductor is limited by the series resistance of the metal traces. At high frequencies, the skin effect and other magnetic field effects cause a nonuniform current distribution in the inductor. Thence, the analysis of electromagnetic interaction between metal coils and between coils and substrate is very important. Figure 2.9 depicts the cross-sectional view of the conventional on-chip spiral inductor, as well as the corresponding electromagnetic effect and loss mechanism [7,8].

The loss of on-chip spiral inductor mainly includes two parts: metal loss and substrate loss. Metal loss refers to the loss caused by the metal coil itself and the interaction between the metal coils. The skin effect is a tendency for high-frequency currents to flow mostly near the outer surface of a solid electrical conductor. The effect becomes more and more apparent as the frequency increases, and the result is a current flow that tends to reside at the outer perimeter with increasing frequency. That means the resistance of metal (metal loss) will be increased with the increase in frequency. A first-order approximation of the skin effect of a straight conductor suggests that the resistance of the conductor increases proportionally to the square root of the frequency, whereas the inductance remains almost unchanged. The proximity effect occurs between the adjacent metal wires to the spiral inductor, as illustrated in Figure 2.10. The magnetic field interaction of the two metal wires forms an eddy current

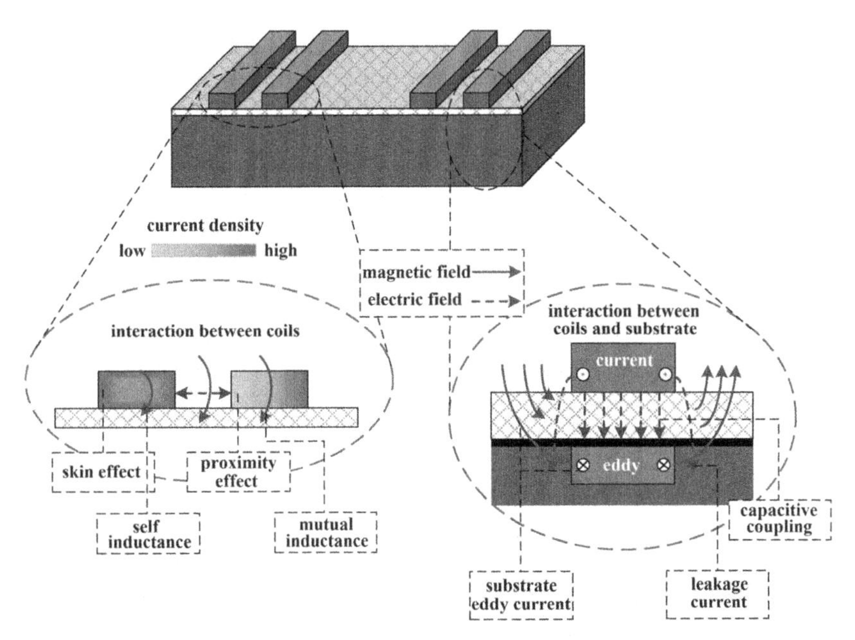

Figure 2.9. Electromagnetic effect and loss mechanism of the on-chip spiral inductor.

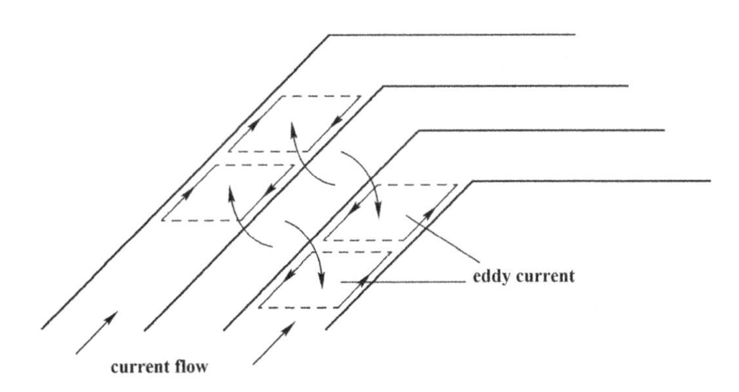

Figure 2.10. The proximity effect between the adjacent metal wires.

in the conductor, and the current distribution has been changed and the loss becomes large.

Substrate loss mainly includes the following:

- ohmic loss of substrate caused by eddy current of the substrate,
- loss caused by the leakage current of the coil in the substrate,

- the capacitance formed by metal wire and multi-layer structure can conduct displacement current and results in loss through the substrate.

In order to reduce the substrate loss, the substrate with high resistivity will be used in the design of on-chip inductance. For example, in the CMOS process with the lower resistivity of the substrate, the quality factor of the inductor is quite low due to the eddy current dominating the loss of the substrate. In contrast, for the BiCMOS process, the high resistivity of the substrate has been used to eliminate the eddy current loss, hence the quality factor can be improved a lot.

2.3 Equivalent Circuit Model of the Spiral Inductor

In order to help researchers understand the physical characteristics of devices, it is particularly important to develop corresponding equivalent circuit models [9,10]. In the high-frequency characterization of microwave devices, small-signal models are often used to parameterize complicated behaviors with relatively simple equations. A small-signal model is preferably designed so that the model parameters represent something physical in the devices. This can provide important information to optimize the test structure's layout and to perform the simulation of the complete structure using an equivalent circuit.

Due to nonuniform current distribution in the conductor, line resistance and inductance show a strong dependence on frequency. The frequency-independent ladder circuit representation is displayed in Figure 2.11 [11]. A resistance R_0 and inductance L_0 in series are used to represent the behavior of DC going through the metal wire. In order to represent the skin effect, a branch (resistance R_1 and inductance L_1 in series) is added in parallel to resistance R_0. By adding the mutual inductance L_m between L_0 and L_1, the magnetic field generated by neighboring lines further changes the current distribution, resulting in a higher current density at the edges of the metal wires.

Figure 2.12(a) illustrates the cross-section of the conventional π-type equivalent circuit model of the spiral inductor on the

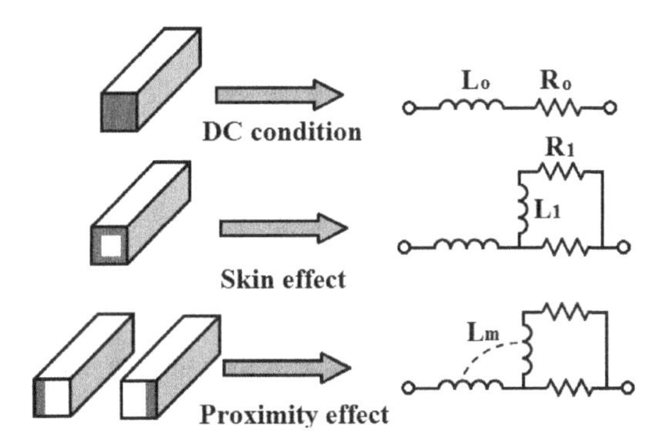

Figure 2.11. Frequency-independent ladder circuit representation [11].

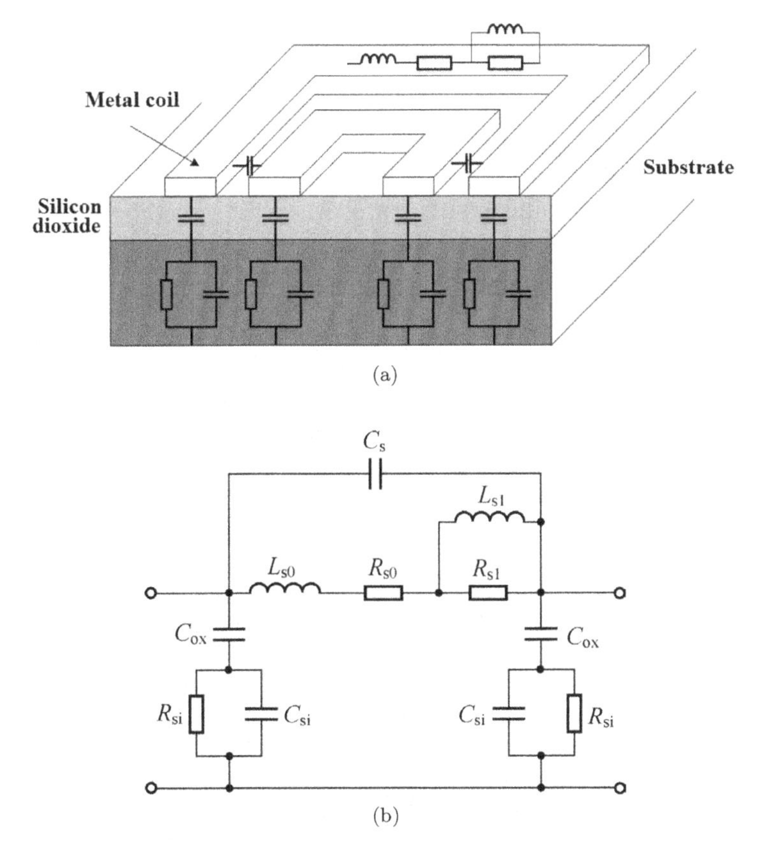

Figure 2.12. Cross-section of the conventional π-type equivalent circuit model (a) and simplified model (b).

silicon substrate. This model belongs to the distributed model; the coupling capacitance between each coil and coupling between coil and substrate are considered. To reduce the burden of parameter extraction, Figure 2.12(b) depicts the simplified π-type equivalent circuit model, which consists of the following two parts:

(1) Series branch: inductances L_{s0} and L_{s1}, resistances R_{s0} and R_{s1}, and capacitance C_s,
(2) Parallel branch: capacitances C_{ox} and C_{si} and resistance R_{si},

where inductance L_{s0} and resistance R_{s0} represent the self-inductance and intrinsic loss, and inductance L_{s1} and resistance R_{s1} are used to model the skin effect. The capacitance C_s represents the capacitive coupling between the input and output ports. The capacitance C_{ox} represents the oxide capacitances between the metal segments and Si substrate. The substrate resistance R_{si} and capacitance C_{si} model the ohmic loss in the conductive silicon substrate.

Figure 2.13 displays the frequency-dependent characteristics of spiral inductance on silicon substrate. According to the tendency of the intrinsic inductance value, the frequency can be divided into the following three different bands:

I region: The intrinsic inductance is roughly independent of the frequency, therefore this band is the operating region of the inductor.

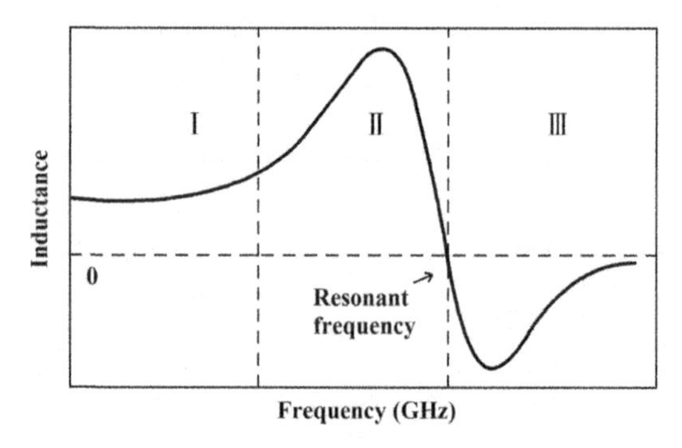

Figure 2.13. Frequency-dependent characteristics of spiral inductance on the silicon substrate.

II region: With the increase of frequency, the intrinsic inductance increases slowly and then decreases rapidly and becomes zero at the frequency resonance point. The frequency point at which the intrinsic inductance becomes zero for the first time is called the first self-resonance point.
III region: After the first self-resonance point, the inductance value is negative, which also means that the inductance has become a capacitor and the quality factor is zero.

The application frequency range of the simplified π equivalent circuit model is up to the first self-resonant frequency point. Such a single π model has obvious disadvantages, that is, it does not consider the proximity effect between coils at high frequencies. Based on the ladder circuit representation, the symmetrical double π equivalent circuit model is built to capture the distributed characteristics of the on-chip spiral inductor. With all frequency-independent elements, the double π ladder circuit accounts for all major physical phenomena occurring in spiral inductors [9,11]. The double π equivalent circuit model is exhibited in Figure 2.14(a). The double π model can be regarded as a cascade of two single π subnetworks. For spiral inductors with symmetrical layout, their subnetworks are identical. The double π model is more convenient in fitting the characteristics of the two ports for the inductor, which is particularly important when the asymmetry of the inductor is significant.

Compared with single π and double π models, T-type model has a simpler topology and considerable fitting bandwidth [10]. Figure 2.14(b) depicts the T-type equivalent circuit model, in order to characterize the conductor loss of inductor coil caused by substrate loss loop; a resistance R_p is used to improve the accuracy of quality factor.

Table 2.2 summarizes the three equivalent circuit models mentioned above: the single π model has the simplest circuit structure but does not include the effects of various parasitic effects. The double π model has the highest accuracy, but the circuit structure is complex and the parameter extraction is difficult. The topology of T-type model is in the middle, and the accuracy needs to be improved.

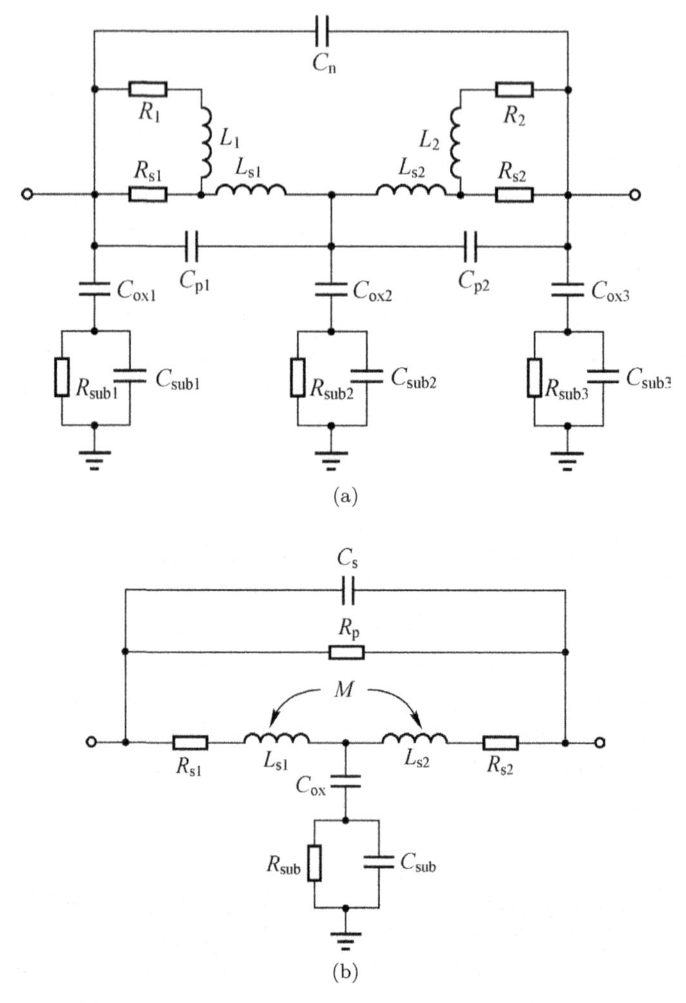

Figure 2.14. (a) Double π equivalent circuit model and (b) T-type equivalent circuit model.

Table 2.2. Comparison of three equivalent circuit models.

Model	Topology	Accuracy	Parameter extraction
Single π	Simple	Middle	Easy
Double π	Complicated	High	Difficult
T-type	Simple	Middle	Easy

2.4 Octagonal Spiral On-Chip Inductor

A parameter-extraction approach for the spiral on-chip inductors, which combines the analytical approach and the empirical optimization procedure, is developed in this paper. The model parameters determined from the analytical expressions are considered as an initial guess of a subsequent optimization procedure leading to the final model parameters. Scaling rules for the inductances of spiral on-chip inductors have been developed and are very useful for predicting the performance of the on-chip inductors with the larger size. Good agreement is obtained between simulated and measured results for the spiral inductors with different sizes on silicon in the frequency range from 500 MHz to 40 GHz [12–14].

On-chip inductors are the most area-consuming devices, the layout area of which is not determined by the feature size of the CMOS process, and therefore the RF modeling of the on-chip spiral inductor has gained great importance in the circuit and layout design. Since the accuracy of the model is really dependent on the model parameters, an accurate procedure for the extraction of model parameters is extremely important for optimizing device performance of on-chip inductors based on silicon material system. Optimization methods have been usually used for the determination of these parameters. In general, these mathematical treatments can greatly improve the model accuracy, but they often require considerable computing resources, as well as proper initial guesses to obtain accurately converged solutions. Moreover, it is difficult to acquire the related physical conceptions of the effects of model parameters on the inductor characteristics from these mathematics-based methodologies, while the analytical methods allow us to extract the equivalent circuit model parameters in a straightforward manner. However, it is noted that there are slight slopes of the extracted elements against frequency, making it difficult to decide which values are the best to use; further optimization is needed, thence.

A minimum of two metal layers is needed to build the basic spiral coil and an underpass contact to return the inner terminal of the coil to the outside. The most common realizations use the topmost metal layer for the main part of the inductor and provide a connection to the center of the spiral with a cross under implemented with some lower level of metal. These conventions arise from quite practical

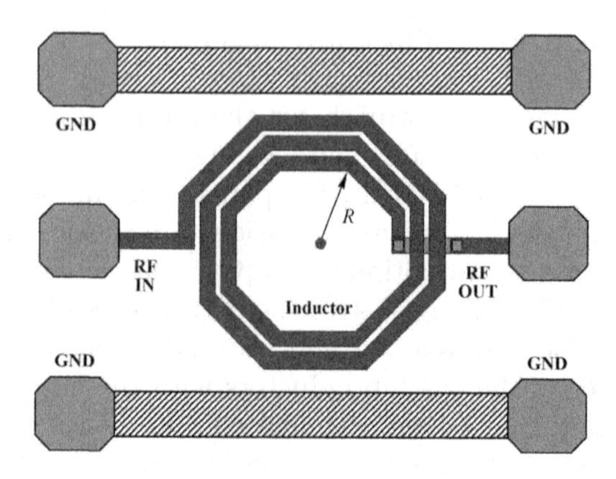

Figure 2.15. Top view of the octagonal spiral inductor.

considerations: the topmost metal layers in an integrated circuit are usually the thickest and thus generally the lowest in resistance. The top view of multi-loops' spiral inductor is illustrated in Figure 2.15, and the lateral structure of an inductor is defined by the major physical dimensions, such as the number of turns, the trace width, the trace thickness, line spacing, and the inner diameter.

2.4.1 *Open and short test structure*

The pad parasitics can be determined by measuring the open test structure, which consists of only the pads. Measurements of the open test structure are modeled as a RC π-type network of capacitance. The open test structure layout with the corresponding equivalent circuit model is exhibited in Figure 2.16. $C_{\text{ox}i}$ and $C_{\text{ox}o}$ represent the oxide capacitances between the metal segments and Si substrate. $C_{\text{sub}i}$, $C_{\text{sub}o}$ and $R_{\text{sub}i}$, $R_{\text{sub}o}$ are the capacitances and resistances of the Si substrate, respectively. These model parameters are determined by measuring an open structure that consisted of only the pads.

In the low-frequency ranges, the oxide capacitances and substrate resistances can be determined as follows:

$$C_{\text{ox}i} = -\frac{1}{\omega\,\text{Im}\left(\frac{1}{Y_{11}^{o}+Y_{12}^{o}}\right)} \tag{2.1}$$

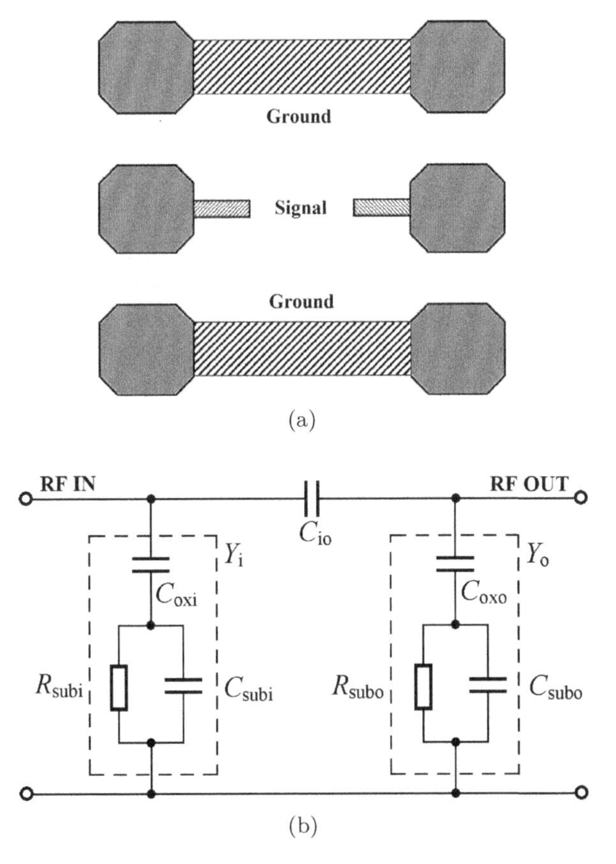

Figure 2.16. Open test structure (a) and equivalent circuit model (b).

$$C_{\rm oxo} = -\frac{1}{\omega\,{\rm Im}\left(\frac{1}{Y_{22}^o+Y_{12}^o}\right)} \tag{2.2}$$

$$C_{io} = -\frac{{\rm Im}(Y_{12}^o)}{\omega} \tag{2.3}$$

$$R_{{\rm sub}i} = {\rm Re}\left(\frac{1}{Y_{11}^o + Y_{12}^o}\right) \tag{2.4}$$

$$R_{{\rm sub}o} = {\rm Re}\left(\frac{1}{Y_{22}^o + Y_{12}^o}\right) \tag{2.5}$$

In the high-frequency ranges, the substrate capacitance ($C_{\text{sub}i}$ and $C_{\text{sub}o}$) can be determined as follows:

$$C_{\text{sub}i} = \frac{1}{\omega}\,\text{Im}\left\{1\bigg/\left[\frac{1}{Y_{11}^{o}+Y_{12}^{o}}\right)-\frac{1}{j\omega C_{\text{ox}i}}\right]\right\} \tag{2.6}$$

$$C_{\text{sub}o} = \frac{1}{\omega}\,\text{Im}\left\{1\bigg/\left[\frac{1}{Y_{22}^{o}+Y_{12}^{o}}\right)-\frac{1}{j\omega C_{\text{ox}o}}\right]\right\} \tag{2.7}$$

The superscript o denotes the open test structure. This method also can be considered as an initial guess of a subsequent optimization procedure leading to the final model parameters.

The parasitic device-connection impedances can be determined by measuring a test pattern, which consists of the pads, the device feeds, and a short replacing the inductor. The short test structure is modeled as a T-network of resistors and inductors in series. The short test structure and corresponding equivalent circuit model are exhibited in Figure 2.17, where L_i, L_o, and L_s are the feedline inductances, R_i, R_o, and R_s are the losses of the feedlines. The extrinsic inductances and feedline losses can be directly determined by Z parameters of the short test structure:

$$L_s = \frac{1}{\omega} I_m(Z_{12}^{s}) = \frac{1}{\omega} I_m(Z_{21}^{s}) \tag{2.8}$$

$$L_i = \frac{1}{\omega}(Z_{11}^{s} - Z_{12}^{s}) \tag{2.9}$$

$$L_o = \frac{1}{\omega}(Z_{22}^{s} - Z_{21}^{s}) \tag{2.10}$$

$$R_i = R_e(Z_{11}^{s} - Z_{12}^{s}) \tag{2.11}$$

$$R_o = R_e(Z_{22}^{s} - Z_{12}^{s}) \tag{2.12}$$

$$R_s = R_e(Z_{12}^{s}) = R_e(Z_{21}^{s}) \tag{2.13}$$

The superscript s denotes the short test structure.

2.4.2 *Intrinsic equivalent circuit model*

The equivalent circuit model for an octagonal spiral on-chip inductor is depicted in Figure 2.18, the basic cell (dashed box) is represented by intrinsic inductance L_0, R_1 is used to represent the loss, and the

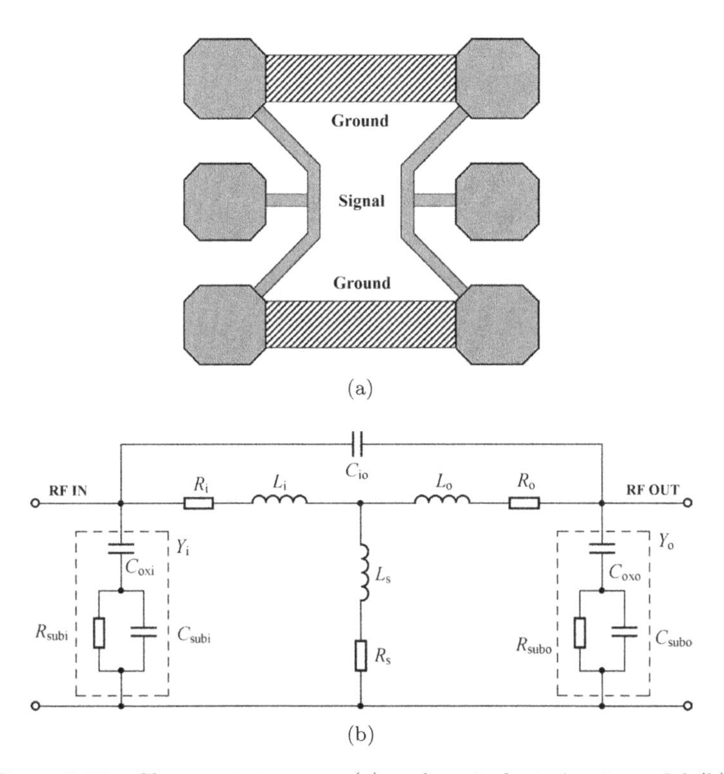

Figure 2.17. Short test structure (a) and equivalent circuit model (b).

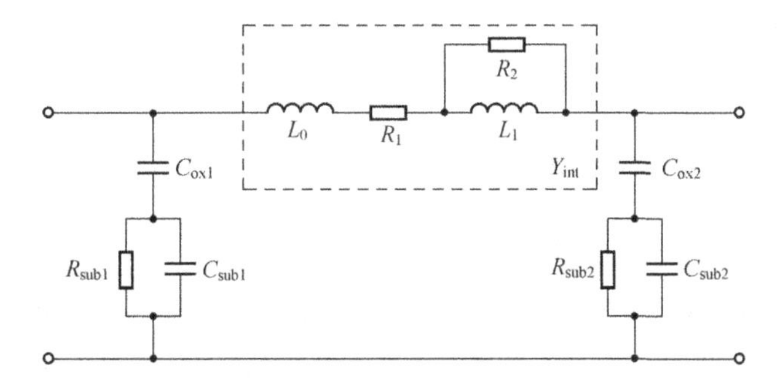

Figure 2.18. Equivalent circuit model for octagonal spiral on-chip inductor.

resistance R_2 in parallel with the L_1 are used to model the skin effect. The Y parameter of a basic cell can be expressed as follows:

$$1/Y_{\text{int}} = R_1 + j\omega L_0 + \frac{j\omega R_2 L_1}{R_2 + j\omega L_1} \tag{2.14}$$

At low frequencies, the sum of L_0 and L_1 can be obtained from the imaginary part of $1/Y_{\text{int}}$:

$$L_t = L_0 + L_1 = \frac{\text{Im}(1/Y_{\text{int}})}{\omega} \tag{2.15}$$

R_1 can be determined from the real part of $1/Y_{\text{int}}$:

$$R_1 = \text{Re}(1/Y_{\text{int}}) \tag{2.16}$$

Subtraction of $R_1 + j\omega(L_0 + L_1)$, we have

$$1/Y'_{\text{int}} = 1/Y_{\text{int}} - R_1 - j\omega(L_0 + L_1) = \frac{\omega^2 L_1^2}{R_2 + j\omega L_1} \tag{2.17}$$

The resistance R and inductance L_1 can be determined as follows:

$$R_2 = \frac{|1/Y'_{\text{int}}|}{\omega^2 k^2}\sqrt{1 + \omega^2 k^2} \tag{2.18}$$

$$L_1 = kR_2 \tag{2.19}$$

where

$$k = \frac{L_1}{R_2} = -\frac{Im(1/Y'_{\text{int}})}{\omega Re(1/Y'_{\text{int}})}$$

Since the intrinsic basic cell is dominated in the whole model, the elements caused by the substrate effect (C_{ox1}, C_{ox2}, C_{sub1}, C_{sub2} R_{sub1}, R_{sub2}) cannot be determined utilizing direct-extraction method accurately, and optimization is needed.

2.4.3 *Layout design*

To illustrate the above model and parameter extraction method, the extracted model parameters for a group of on-chip inductors have been developed, grown, and fabricated by using 0.13 μm RF CMOS process. In this work, on-chip inductors are designed with 0.5 $\sim$ 2.5 turns, the inner radius is 20 $\sim$ 50 μm, line spacing is 2 μm, and trace width is 8 μm. The S-parameter measurements for model extraction and verification were made up to 40 GHz using the Agilent E8363C network analyzer. Figures 2.19–2.22 show the layout of an octagonal inductor with the inner diameter from 20 μm to 50 μm. Besides, the on-wafer measurements for an octagonal inductor with pads are illustrated in Figure 2.23.

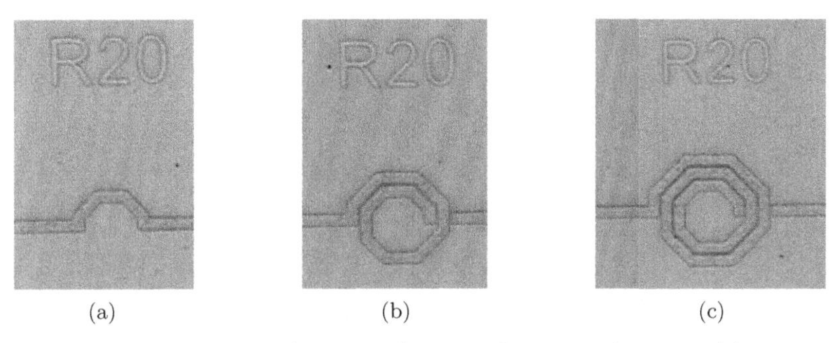

Figure 2.19. Octagonal inductor with inner diameter of 20 μm: (a) $n = 0.5$; (b) $n = 1.5$; (c) $n = 2.5$.

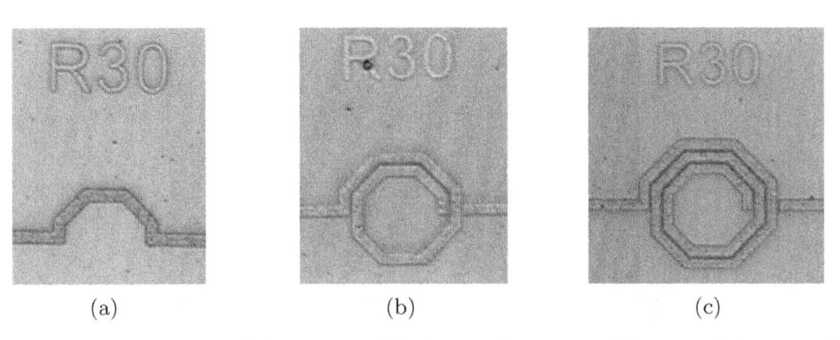

Figure 2.20. Octagonal inductor with inner diameter of 30 μm: (a) $n = 0.5$; (b) $n = 1.5$; (c) $n = 2.5$.

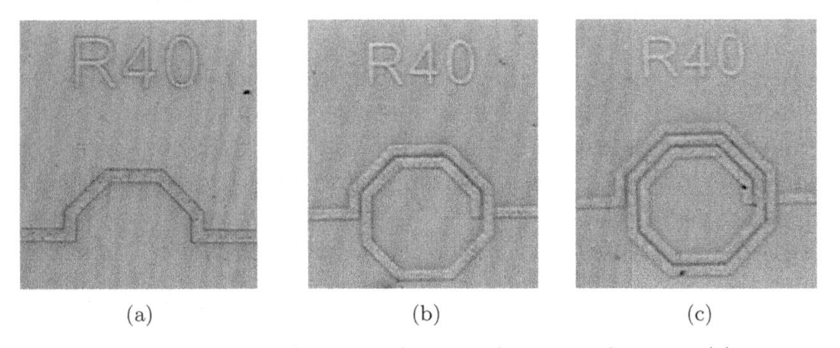

Figure 2.21. Octagonal inductor with inner diameter of 40 μm: (a) $n = 0.5$; (b) $n = 1.5$; (c) $n = 2.5$.

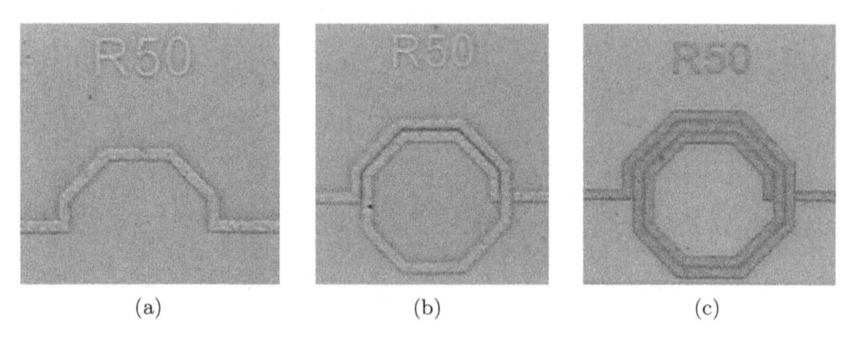

Figure 2.22. Octagonal inductor with inner diameter of 50 μm: (a) $n = 0.5$; (b) $n = 1.5$; (c) $n = 2.5$.

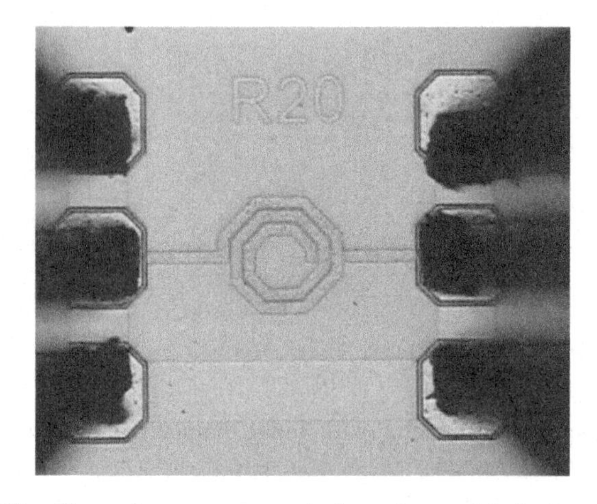

Figure 2.23. On-wafer measurements for octagonal inductor with pads.

2.4.4 *Model parameter extraction*

The extracted results of the pad capacitances and resistances are shown in Figures 2.24 and 2.25, respectively. Rather constant values are observed over a wide frequency range. The extracted results of the feedline inductances are given in Figure 2.26. It can be obviously seen that L_g and L_d are very close and larger than L_s. It is noted that there are slight slopes of the extracted elements against frequency, thence the extracted model parameters of open-short test structure can be regarded as the initial values for optimizing procedure using commercial circuit simulation software. The final parasitic parameters are summarized in Table 2.3. The comparison between modeled

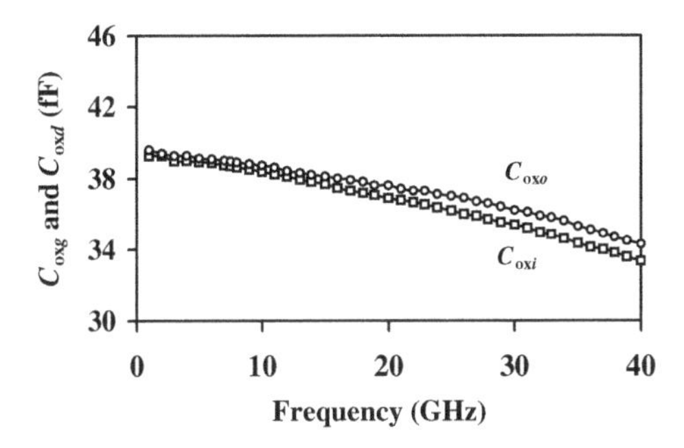

Figure 2.24. Extracted pad capacitances versus frequency.

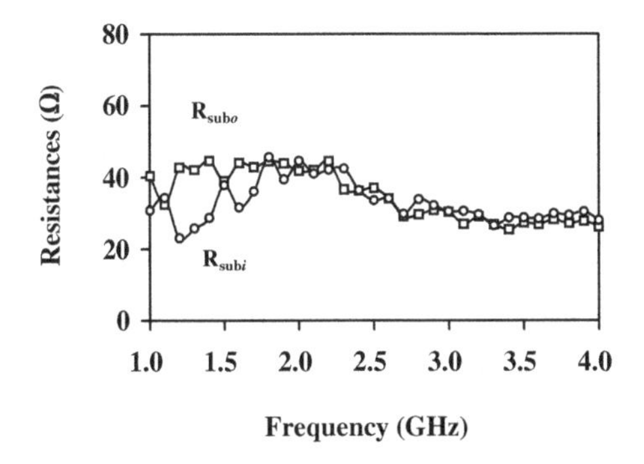

Figure 2.25. Extracted substrate resistances.

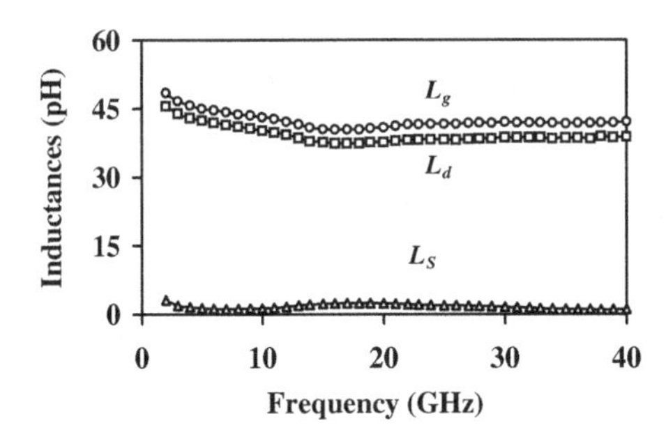

Figure 2.26. Extracted feedline inductances versus frequency.

and measured S parameters for the open-short test structure is illustrated in Figure 2.27. The squares indicate the measured values and the lines the modeled ones. Good agreement between modeled and measured S parameters over the whole frequency range is obtained with an error that is less than 5% for the open-short test structure.

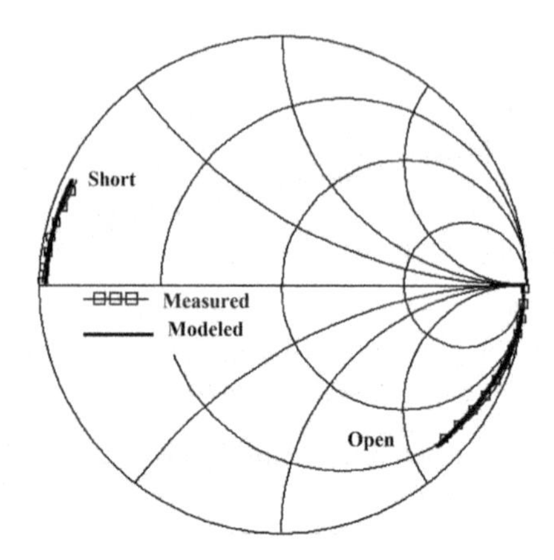

Figure 2.27. Comparison between the modeled and measured S parameters for the open-short test structure. The squares indicate the measured values and the lines the modeled ones.

Table 2.3. Parasitic parameters.

Elements	Parameters	Values	Units
PAD	$C_{\mathrm{ox}i}$	39	fF
	$C_{\mathrm{ox}o}$	39.5	fF
	C_{io}	0.4	fF
	$R_{\mathrm{sub}i}$	35	Ω
	$R_{\mathrm{sub}o}$	32	Ω
	$C_{\mathrm{sub}i}$	130	fF
	$C_{\mathrm{sub}o}$	200	fF
Feedline inductances	L_i	40	pH
	L_o	38	pH
	L_s	1.2	pH

The extracted total inductance $L_0 + L_1$ versus frequency is given in Figure 2.28. Furthermore, the extracted inductance k and resistance R_2 versus frequency are illustrated in Figure 2.29. The model parameters determined from the analytical expressions are considered as an initial guess of a subsequent optimization procedure leading to the final model parameters. Tables 2.4–2.6 tabulate the model parameters of a group of octagonal spiral on-chip inductors. In the

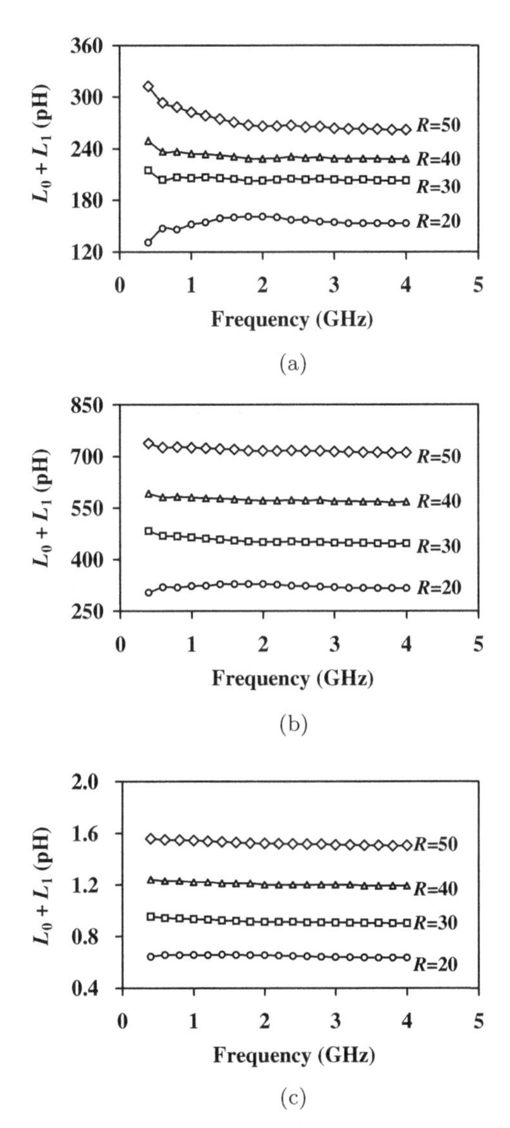

Figure 2.28. Extracted total inductance $L_0 + L_1$ versus frequency: (a) $n = 0.5$; (b) $n = 1.5$; (c) $n = 2.5$.

frequency range from 500 MHz to 40 GHz, the comparison between modeled and measured S parameters for the spiral inductors with different sizes on silicon are shown in Figures 2.30–2.32. Good agreement is achieved between the simulated and measured results.

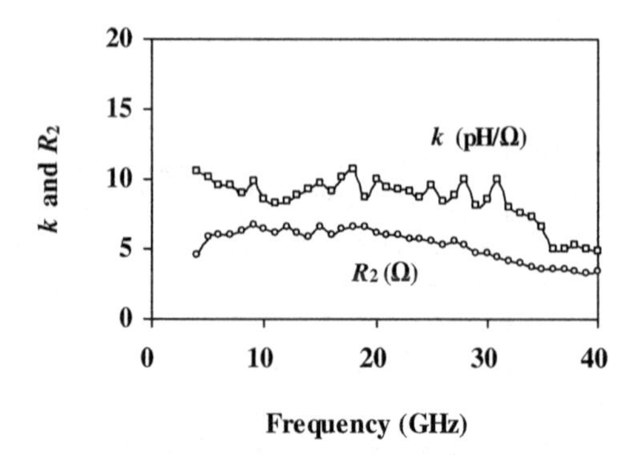

Figure 2.29. Extracted inductance k and resistance R_2 versus frequency.

Table 2.4. Model parameters for the inductor of $n = 0.5$ with different radii.

$n = 0.5$	$R = 20\ \mu$m	$R = 30\ \mu$m	$R = 40\ \mu$m	$R = 50\ \mu$m
L_0 (nH)	120	160	180	220
L_1 (nH)	40	50	50	60
R_1 (Ω)	0.2	0.43	0.4	0.5
R_2 (Ω)	5	5	6	6
C_{ox1} (fF)	8	10	12	14
C_{ox2} (fF)	8	10	12	14
C_{sub1} (fF)	2	10	30	50
C_{sub2} (fF)	2	10	30	50
R_{sub1} (Ω)	400	120	100	60
R_{sub2} (Ω)	400	120	100	60

Table 2.5. Model parameters for the inductor of $n = 1.5$ with different radii.

$n = 1.5$	$R = 20\ \mu$m	$R = 30\ \mu$m	$R = 40\ \mu$m	$R = 50\ \mu$m
L_0 (nH)	280	390	510	650
L_1 (nH)	50	50	60	70
R_1 (Ω)	0.3	0.4	0.5	0.6
R_2 (Ω)	5	5	6	6
C_{ox1} (fF)	10	14	16	19
C_{ox2} (fF)	10	15	17	20
C_{sub1} (fF)	5	30	50	70
C_{sub2} (fF)	5	30	50	70
R_{sub1} (Ω)	400	120	100	60
R_{sub2} (Ω)	400	120	100	60

Table 2.6. Model parameters for the inductor of $n = 2.5$ with different radii.

$n = 2.5$	$R = 20\ \mu$m	$R = 30\ \mu$m	$R = 40\ \mu$m	$R = 50\ \mu$m
L_0 (nH)	630	860	1140	1420
L_1 (nH)	30	40	60	80
R_1 (Ω)	0.4	0.5	0.6	0.7
R_2 (Ω)	3	4	5	6
C_{ox1} (fF)	15	19	23	26
C_{ox2} (fF)	15	20	23	26
C_{sub1} (fF)	10	80	120	150
C_{sub2} (fF)	10	80	120	150
R_{sub1} (Ω)	400	120	80	60
R_{sub2} (Ω)	400	120	80	60

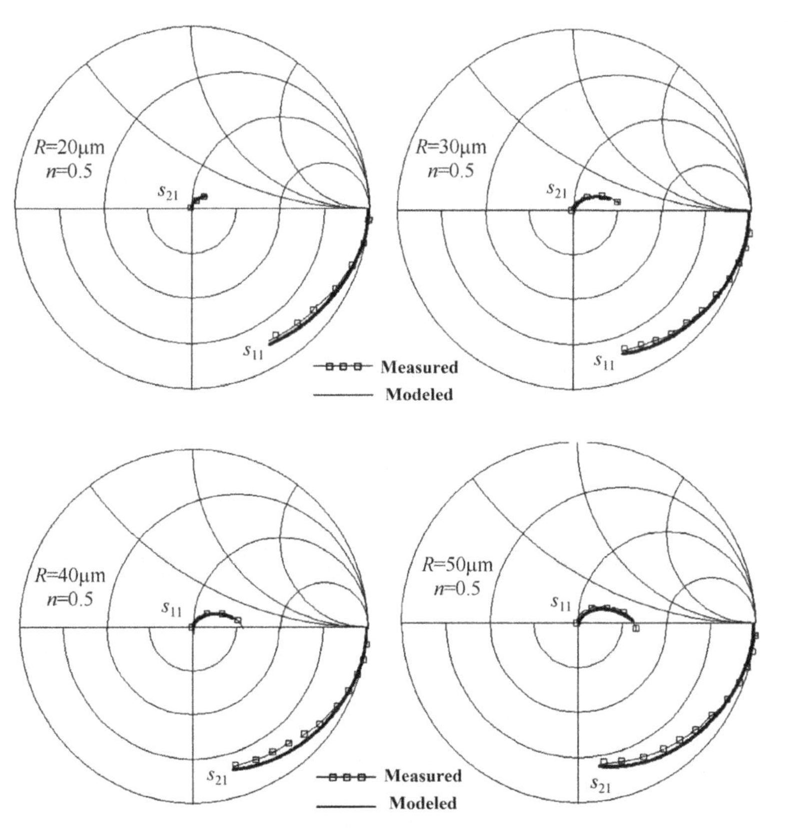

Figure 2.30. Comparison between the modeled and measured S parameters of four octagonal spiral inductors with 20–50 μm and $n = 0.5$.

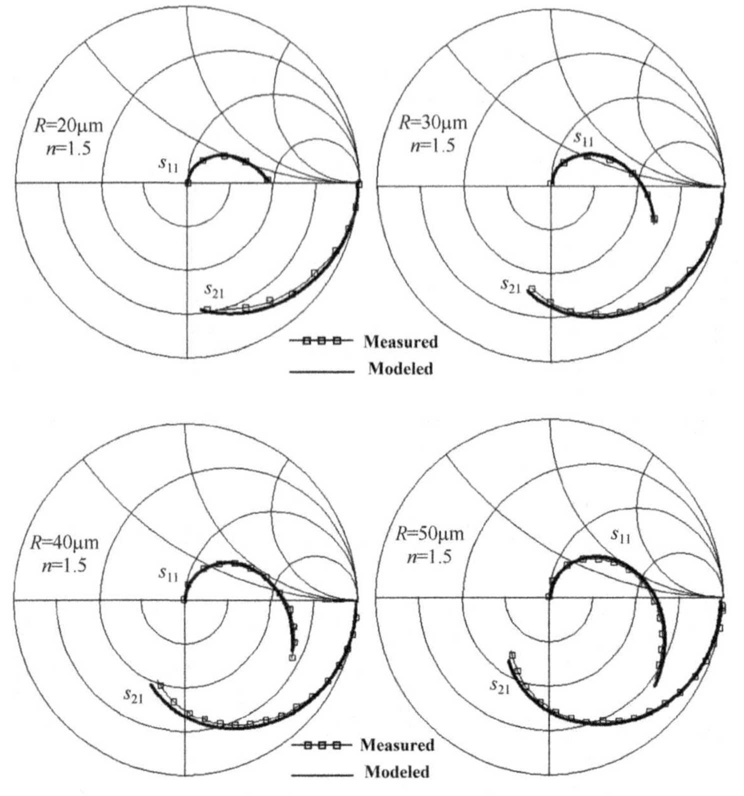

Figure 2.31. Comparison between the modeled and measured S parameters of four octagonal spiral inductors with 20–50 μm and $n = 1.5$.

The smallest inductor ($R = 20$ and $n = 0.5$) can be regarded as a criterion L_c; the scaling rule for the inductances of spiral on-chip inductors with a larger size can be expressed as follows:

$$L_0 + L_1 = \frac{R}{20}\left(1 + \frac{R - 20}{40}\right) kL_c \tag{2.20}$$

with

$$k = \begin{cases} 1 & \text{for } n = 0.5 \\ 2n - 1 & \text{for } n \geq 1 \end{cases}$$

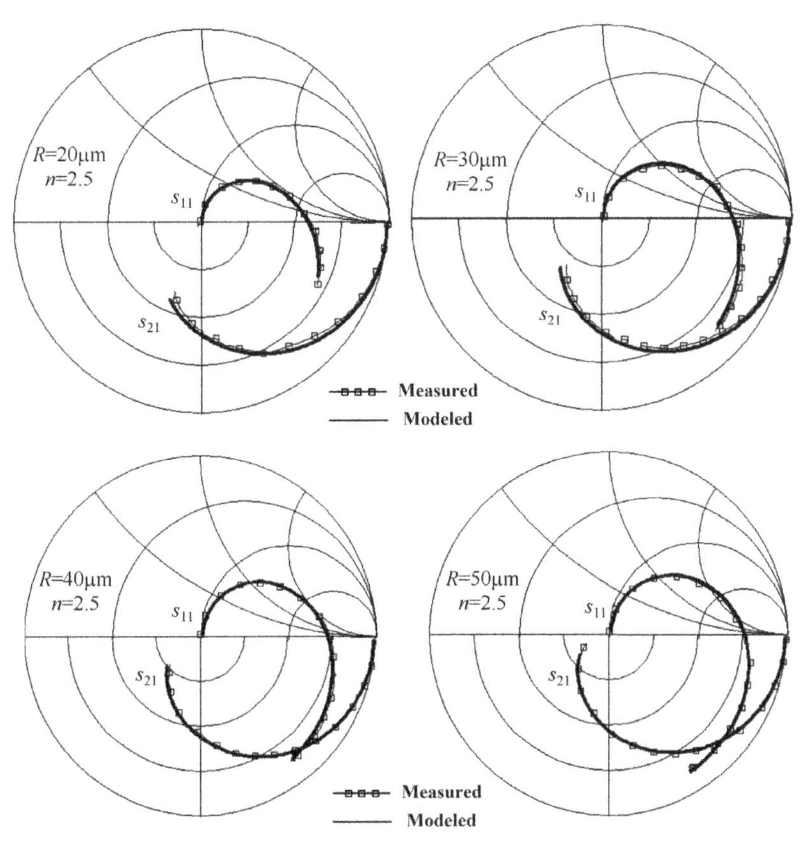

Figure 2.32. Comparison between the modeled and measured S parameters of four octagonal spiral inductors with 20–50 μm and $n = 2.5$.

2.5 3D MEMS Inductor

Fundamentally, the geometry of an on-chip planar spiral inductor is very different from the conventional discrete solenoid inductor with a magnetic core, which leads to low magnetic flux density and large footprint. The metal layer thickness and line width are limited by process, which translates into large series resistance. Typically, the inductance density of planar spirals for RFICs is only about $100\,\text{nH/mm}^2$. The spiral inductors implemented in the standard CMOS process suffer from poor quality factors due to the lossy property of the CMOS substrate and the thin metal layers. Therefore, realizing high quality factor on-chip spiral inductors in the standard CMOS process is one of the major challenges. To minimize the substrate loss, the substrate should have high resistivity or even be completely etched away by micromachining techniques. The three-dimensional (3-D) multi-layer on-chip inductor technology formed on the silicon substrate is an attractive solution. Stacked and miniature 3-D inductors use the multiple metal layers to achieve the required inductances in the small area [13–15].

2.5.1 *Physical structure of 3D inductor*

The on-chip inductors studied in this work consist of six metal layers fabricated using 0.18 μm standard CMOS technology. The typical schematic layout and I/O pads for the on-chip inductor are illustrated in Figure 2.33. The stacked inductor consists of series-connected spiral inductors in different metal layers. Normally, every spiral inductor in the different metal layers has the same or different turns, and the wires wind downward from the top metal layer to the bottom one. To implement the stacked-spiral inductors, a post-CMOS backend process module (CMOS+) was developed, as shown in Figure 2.34 [17]. First, vertically stacked inductors were pre-fabricated in 0.18 μm 6-metal aluminum low-κ inter-metal dielectric (IMD) RF CMOS. Then, the passivation and IMD in the coil region were etched by deep reactive ion etching (DRIE) to expose the metal spiral traces by utilizing post-CMOS processes as exhibited in Figure 2.35 [16].

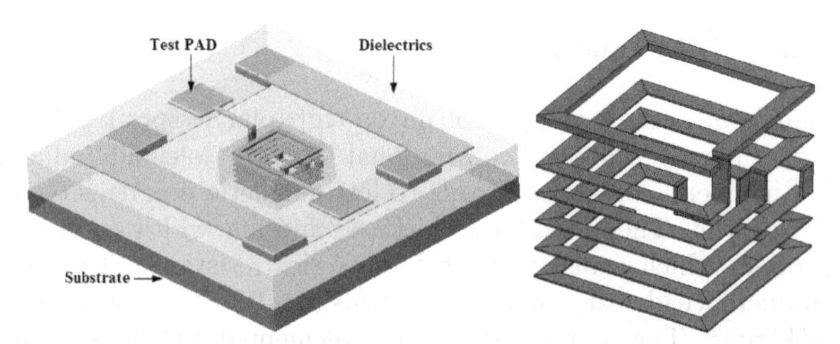

Figure 2.33. Schematic layout and I/O pads for the on-chip inductor with six metal layers.

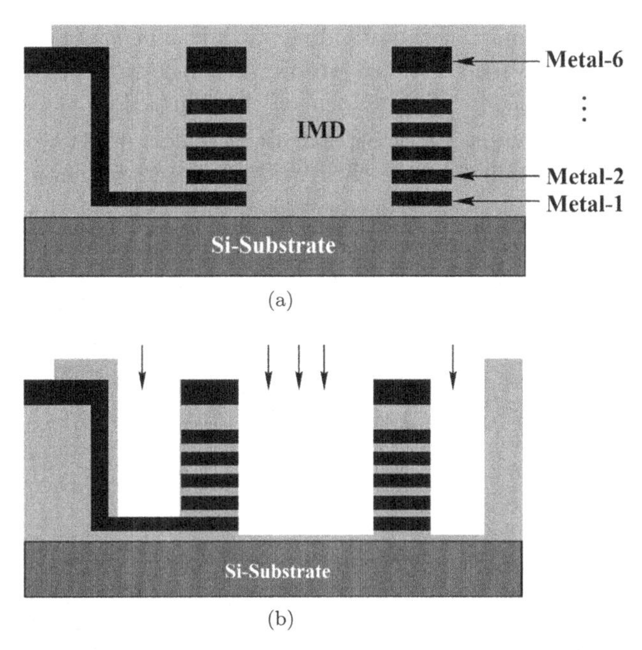

Figure 2.34. Process flow of multi-layer inductor fabrication: (a) six metal layer inductor fabrication using 0.18 μm CMOS process; (b) coil region releasing.

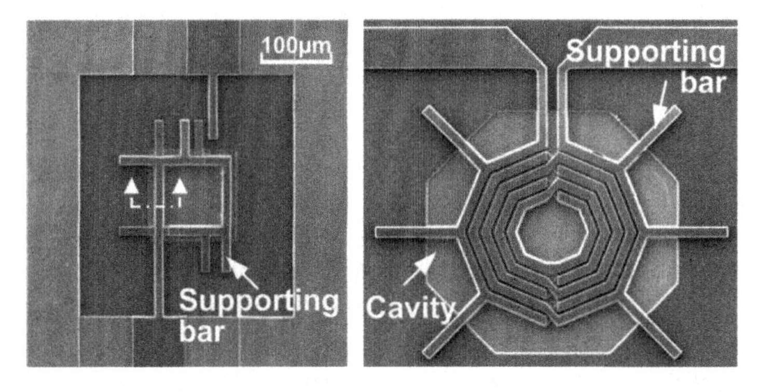

Figure 2.35. SEM photos for the fabricated magnetic-cored inductors [16].

2.5.2 *3D inductor equivalent circuit model*

Accurate modeling and model parameters extraction techniques for multi-layer on-chip inductors are extremely important for optimizing the device performance and understanding the device physical mechanisms. Normally, the equivalent circuit models of planar spiral inductors have been used to predict the 3D multi-layer on-chip inductors in the low-frequency ranges. Nevertheless, these models are based on the planar structures which consist of only one metal layer and are not suitable for the vertical structures which consist of multiple metal layers. Furthermore, these models have been found to have unsatisfactory accuracy in the high-frequency ranges for 3D multi-layer on-chip inductors. The main reason is that the distribution effect of the via-hole with feedline has not been taken into account. In other words, these models have to be improved, hence.

In order to overcome the limitations of previous literature, an improved equivalent circuit model for 3D multi-layer on-chip inductors is developed. A semi-analytical method for extracting the model parameters is investigated. This method has the following advantages:

(1) Single-stacked spiral inductor is regarded as an elementary cell, and the whole model consists of multiple elementary cells.
(2) The distribution effect of the via-hole which is used to connect the lowest mental to the pad has been taken into account.
(3) The initial values of the model parameters can be determined from S-parameter on-wafer measurement without any open-short test structures.

Figure 2.36 illustrates the proposed equivalent circuit model for 3D multi-layer on-chip inductor which consists of multiple elementary cells, where C_{io} represents the isolation capacitance between input and output pads and can be neglected since the value of C_{io} is very small (less than 1 fF normally) and does not affect the frequency response. C_{ox1} and C_{ox2} represent the oxide capacitances between the metal segments and Si substrate. C_{sub1}, C_{sub2} and R_{sub1}, R_{sub2} are the capacitances and resistances of the Si substrate, respectively. The distribution effects of via-hole and feedline are modeled by lumped inductance L_{via}. The basic elementary cell is represented by intrinsic inductance L_C^j in series with the resistance R_C^j, and capacitance C_C^j represents the coupling capacitance between the metal segments. j is the ordinal number of the elementary cells. It is noted that the model parameter of each elementary cell may be different as each spiral inductor in different metal layers probably has different turns.

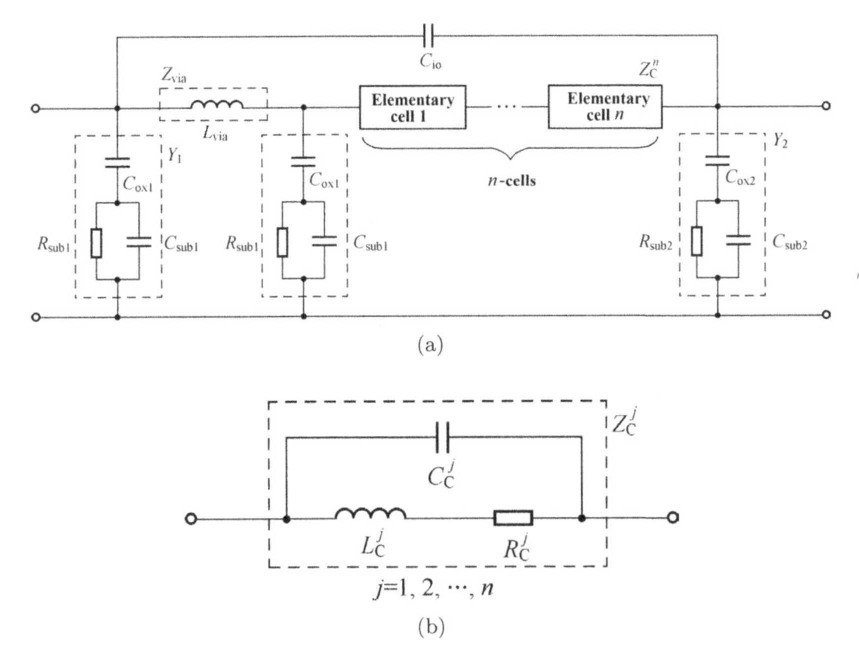

Figure 2.36. Equivalent circuit model for 3D multi-layer on-chip inductor: (a) outer part; (b) elementary cell.

It can be observed that the topology of the proposed model is asymmetric due to the distribution effect of the via-hole with feedline which is used to connect the test pad to the lowest mental layer at the input port. This equivalent circuit model can be divided into two parts, i.e., the outer part contains just pad parasitics and via-hole inductance and the inner part contains n elementary cells (n is the number of metal layers).

If all single-stacked spiral inductors have the same turns, we have

$$L_C = L_C^1 = L_C^2 = \cdots = L_C^n \tag{2.21}$$

$$R_C = R_C^1 = R_C^2 = \cdots = R_C^n \tag{2.22}$$

$$C_C = C_C^1 = C_C^2 = \cdots = C_C^n \tag{2.23}$$

Thence, the equivalent circuit model for 3D multi-layer on-chip inductor is exhibited in Figure 2.36 and the simplified equivalent circuit model is shown in Figure 2.37, where the total inductance, resistance, and capacitance of the n-elementary cells are nL_C, nR_C, and C_C/n ($j = 1, 2, \ldots, n$).

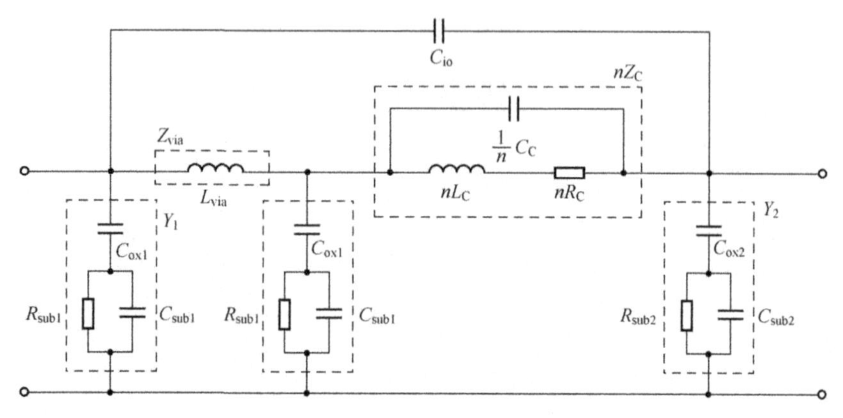

Figure 2.37. Simplified equivalent circuit model for 3D multi-layer on-chip inductor with same structure on each metal layer.

The short circuit Y parameters of the equivalent circuit can be expressed as follows:

$$Y_{11} = Y_1 + \frac{Y_1 + 1/nZ_C}{1 + (Y_1 + 1/nZ_C)Z_{\text{via}}} \tag{2.24}$$

$$Y_{12} = Y_{21} = \frac{-1/nZ_C}{1 + (Y_1 + 1/nZ_C)Z_{\text{via}}} \tag{2.25}$$

$$Y_{22} = \frac{Y_2 + 1/nZ_C}{1 + (Y_1 + 1/nZ_C)Z_{\text{via}}} \tag{2.26}$$

where

$$Y_1 = \frac{1}{\dfrac{1}{j\omega C_{ox1}} + \dfrac{R_{\text{sub1}}}{1 + j\omega R_{\text{sub1}} C_{\text{sub1}}}}$$

$$Y_2 = \frac{1}{\dfrac{1}{j\omega C_{ox2}} + \dfrac{R_{\text{sub2}}}{1 + j\omega R_{\text{sub2}} C_{\text{sub2}}}}$$

$$Z_C = \frac{1}{j\omega C_C + \dfrac{1}{R_C + j\omega L_C}}$$

2.5.3 *Determination of model parameters for 3D inductor*

The resistance of elementary cell R_C and the sum of via-hole inductance with feedline L_{via} and total inductance of n elementary cells $L_{\text{via}} + nL_C$ can be determined directly from the real part and imaginary part of $-1/Y_{21}$ in the low-frequency ranges, respectively:

$$nR_C = \mathrm{Re}\left(-\frac{1}{Y_{21}}\right) \tag{2.27}$$

$$L_{\text{via}} + nL_C = \frac{1}{\omega}\,\mathrm{Im}\left(-\frac{1}{Y_{12}}\right) \tag{2.28}$$

where n is the number of elementary cells and ω is the angular frequency.

The oxide capacitances C_{ox1} and C_{ox2} can be estimated from imaginary parts of $-1/(Y_{11} + Y_{12})$ and $-1/(Y_{22} + Y_{12})$ in the low-frequency ranges:

$$C_{\text{ox1}} \approx -\frac{1}{\omega\,\mathrm{Im}\left(\frac{2}{Y_{11}+Y_{12}}\right)} \tag{2.29}$$

$$C_{\text{ox2}} \approx -\frac{1}{\omega\,\mathrm{Im}\left(\frac{1}{Y_{22}+Y_{12}}\right)} \tag{2.30}$$

The substrate resistances R_{sub1}, R_{sub2} as well as substrate capacitances C_{sub1}, C_{sub2} can be estimated from real parts and imaginary parts of $-1/(Y_{11} + Y_{12})$ and $-1/(Y_{22} + Y_{12})$ in the high-frequency ranges:

$$R_{\text{sub1}} \approx 2\,\mathrm{Re}\left(\frac{1}{Y_{11} + Y_{12}}\right) \tag{2.31}$$

$$R_{\text{sub2}} \approx \mathrm{Re}\left(\frac{1}{Y_{22} + Y_{12}}\right) \tag{2.32}$$

$$C_{\text{sub1}} \approx \mathrm{Im}\left(\frac{1}{\frac{2\omega}{Y_{11}+Y_{12}} + j\frac{1}{C_{\text{ox1}}}}\right) \tag{2.33}$$

$$C_{\text{sub2}} \approx \mathrm{Im}\left(\frac{1}{\frac{\omega}{Y_{22}+Y_{12}} + j\frac{1}{C_{\text{ox2}}}}\right) \tag{2.34}$$

The coupling capacitance C_C can be estimated from the formula for an ideal parallel plate capacitor:

$$C_C \approx \varepsilon \frac{W_C L_C}{D} \tag{2.35}$$

where ε and D are the permittivity and thickness of the dielectric material between two adjacent metal layers. W_C and L_C are the width and length of the metal segment for single-stacked inductor, respectively.

The via-hole is fabricated by utilizing cylindrical metal with several μm highness, thence the total inductance of via-hole and feedline L_{via} can be estimated from the analytical formula of microstrip structure:

$$L_{\text{via}} \approx \frac{Z_o L}{c/\sqrt{\varepsilon}} \tag{2.36}$$

where c is the velocity of light in free space, Z_o is the characteristic impedance of the feedline, and L is the total length of feedline and cylindrical metal.

Figure 2.38 illustrates the extracted inductance $L_{\text{via}} + nL_C$ and resistance R_C in the low-frequency ranges. The magnitude variations of $L_{\text{via}} + nL_C$ and R_C are very small and almost negligible and can be considered to be constants. The extracted substrate resistances R_{sub1} and R_{sub2} are given in Figure 2.39. As observed, the substrate resistances at the input and out ports are very close. Table 2.7 tabulates the extracted model parameters utilizing the analytical method mentioned above, and the optimized values are also given. The final data are very close to the analytical values to verify the validity of the proposed parameter extraction method. Figure 2.40 compares the measured and modeled S parameters for six metal layers on-chip inductor in the frequency range of 10 MHz–20 GHz. The modeled S parameters agree very well with the measured ones to validate the accuracy of the proposed model. The proposed model is also compared with the conventional model, and Figure 2.41 indicates the comparison of accuracy between the conventional model and proposed model. It can be obviously seen that the accuracy (absolute error) of the proposed model is better than the conventional model, especially at high frequencies (above 5 GHz) for S_{11}.

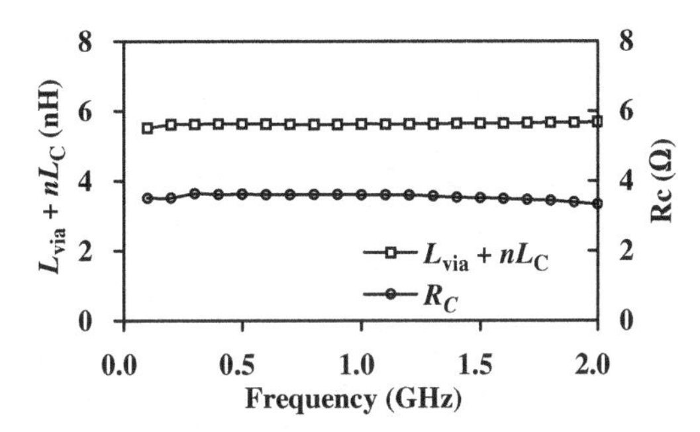

Figure 2.38. Extracted inductance $L_{via} + nL_C$ and resistance R_C versus frequency.

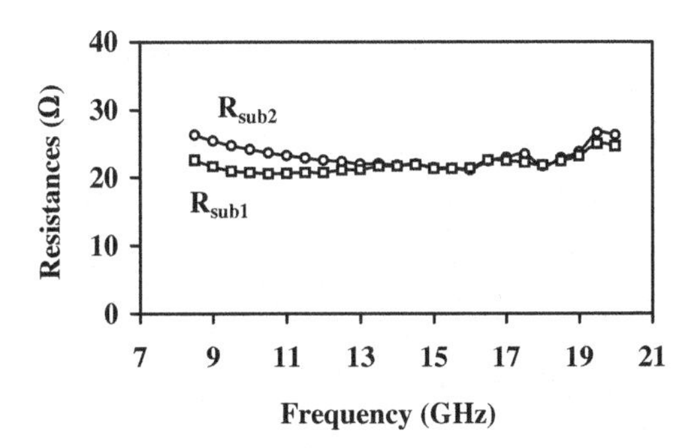

Figure 2.39. Extracted substrate resistances versus frequency.

Table 2.7. Extracted model parameters of 3D inductor.

Parameters	Values (analytical)	Values (optimized)
L_{via} (nH)	0.2	0.35
L_C (nH)	0.9	0.87
C_C (fF)	30	35
R_C (Ω)	3.5	3.5
C_{ox1} (pF)	0.12	0.12
C_{ox2} (pF)	0.40	0.36
R_{sub1} (Ω)	21	18
R_{sub2} (Ω)	22	25
C_{sub1} (pF)	0.0	0.05
C_{sub2} (pF)	0.2	0.22

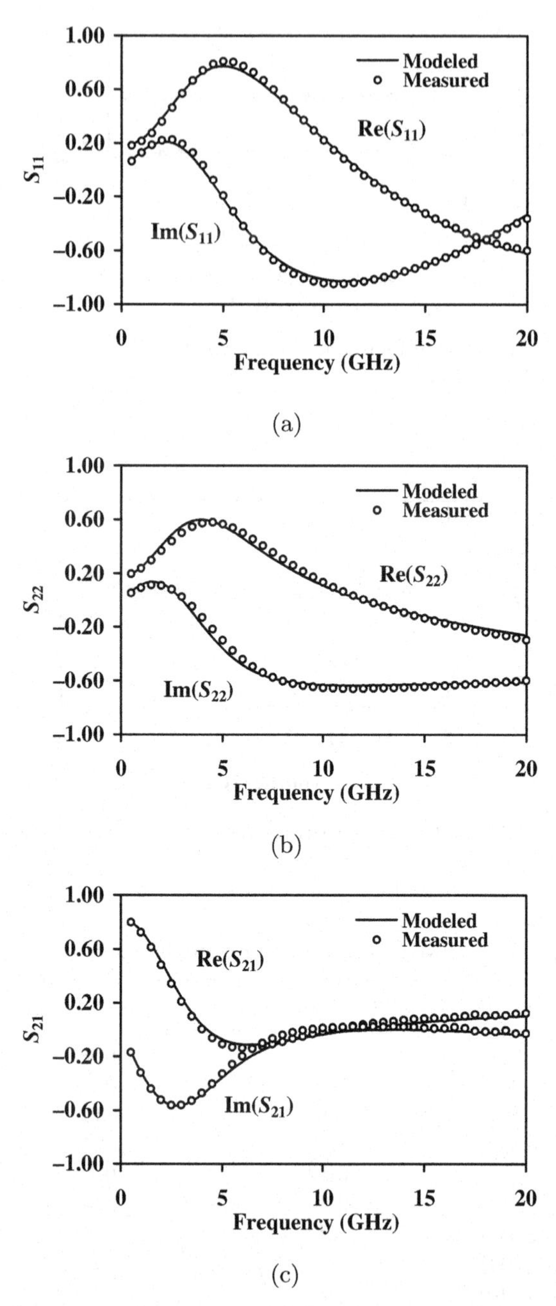

Figure 2.40. Comparison between modeled and measured S parameters for a six metal layers on-chip inductor: (a) S_{11}; (b) S_{22}; (c) S_{21}.

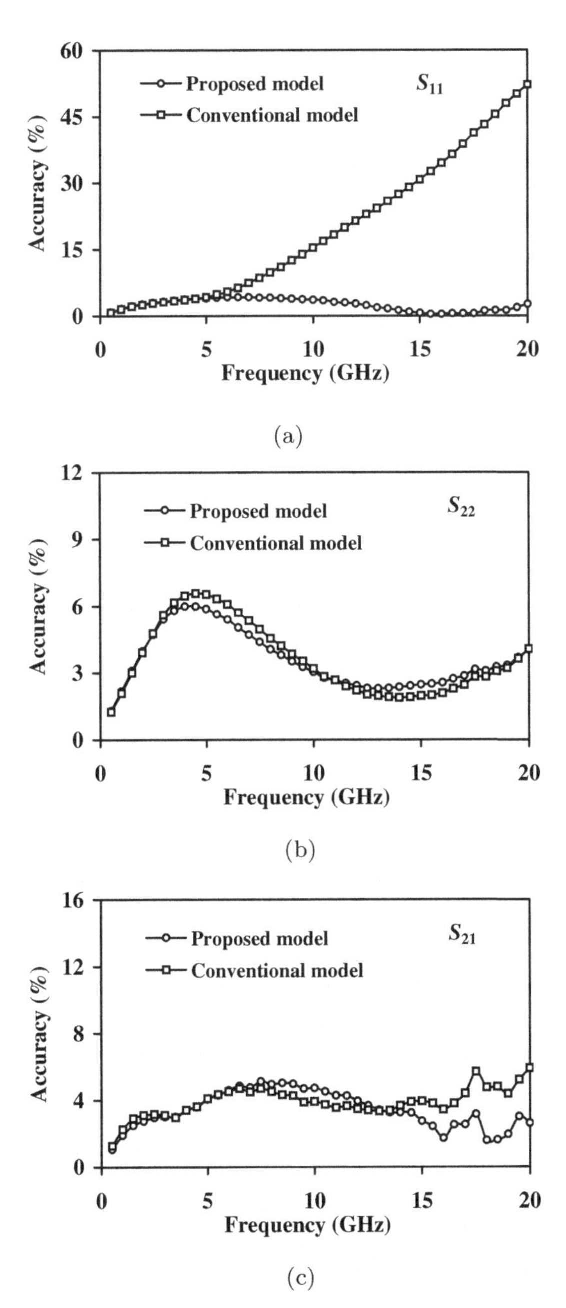

Figure 2.41. Comparison of accuracy between proposed model and conventional model: (a) S_{11}; (b) S_{22}; (c) S_{21}.

2.6 Graphene-Based On-Chip Spiral Inductor

With the emergence of 5G communication system, conventional metal passive devices are facing challenges for the high losses caused by the skin effect in high-frequency circuits. Graphene is a two-dimensional (2D) material with electrical, thermal, and mechanical properties superior to any other semiconductor material. Owing to its extraordinary electrical, thermal, and mechanical properties, graphene is very attractive for densely integrated and flexible RFIC applications. Graphene is expected to replace conventional metal interconnection materials such as aluminum and copper in semiconductor integrated circuits to build all carbon microelectronic devices and circuits.

Graphene has been used as a new material in the design of on-chip spiral inductor. The graphene-based inductors have different performances from the conventional silicon-based inductor. In this section, an improved equivalent circuit model of a graphene-based on-chip spiral inductor is described. Combining the analysis method with the optimization method [18,19], a suitable model parameter extraction method is presented.

2.6.1 *Fabrication and layout design*

To implement the inductor, the prepared multi-layer graphene films are first transferred onto SiO_2 (300 nm)/Si (10 $\Omega\cdot$cm) substrate. Subsequently, graphene films are patterned into ribbon coils and an isolation dielectric layer (Al_2O_3:50 nm) is grown on the graphene multi-turn inductors. In the last step, metal contacts and pads (Ni/Au: 20 nm/80 nm) are deposited and patterned, followed by an annealing process [18]. The fabrication process of graphene-based on-chip spiral inductor is illustrated in Figure 2.42. After that, Figures 2.43(a) and 2.43(b) depict the layout design for one-turn and two-turns of graphene-based inductors, respectively. Moreover, Figures 2.44(a) and 2.44(b) exhibit the scanning electron microscope photographs of two-turns and three-turns graphene inductors, respectively.

The schematic view of the test structure for graphene-based inductor is illustrated in Figure 2.45. At least two metal layers are needed to build the top layer and a via to return the inner terminal

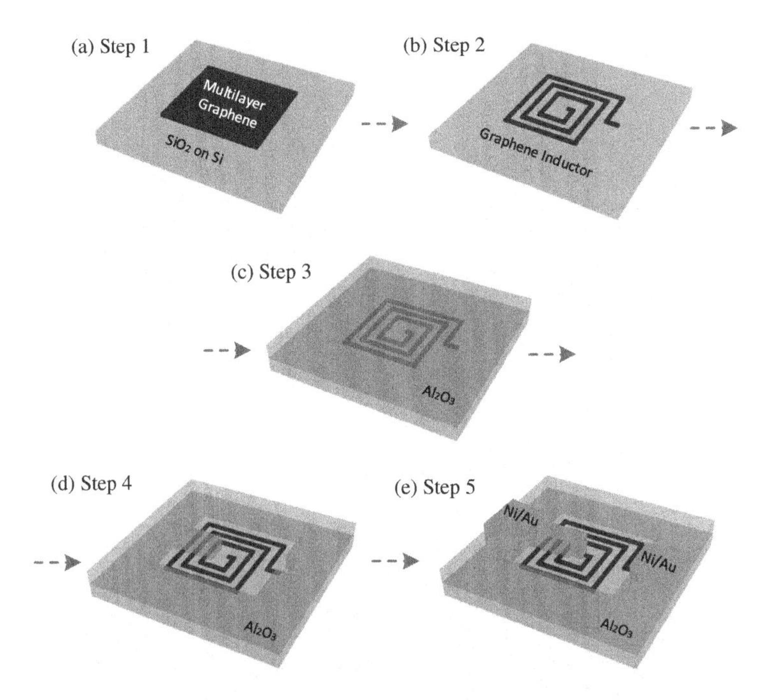

Figure 2.42. Fabrication process of graphene-based on-chip spiral inductor.

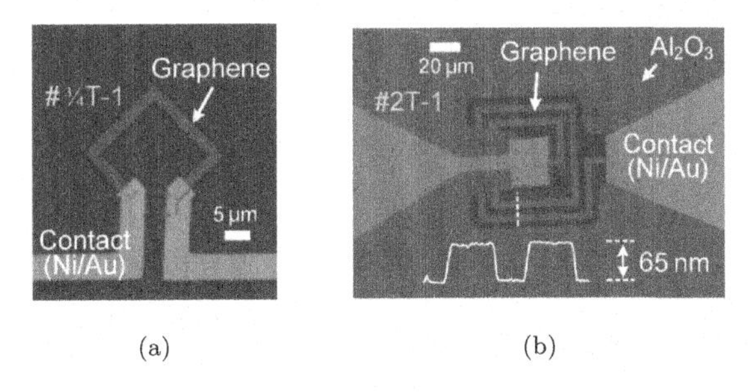

(a) (b)

Figure 2.43. Layout design: (a) one turn; (b) two turns.

of the coil to the outside. In order to measure the S parameters of the square spiral inductor, pads and the feedlines are needed. The verification was made up to 40 GHz by utilizing the N5227A network analyzer. All measurements were carried out on wafer utilizing Cascade Air-Coplanar Probes ACP50-GSG-150.

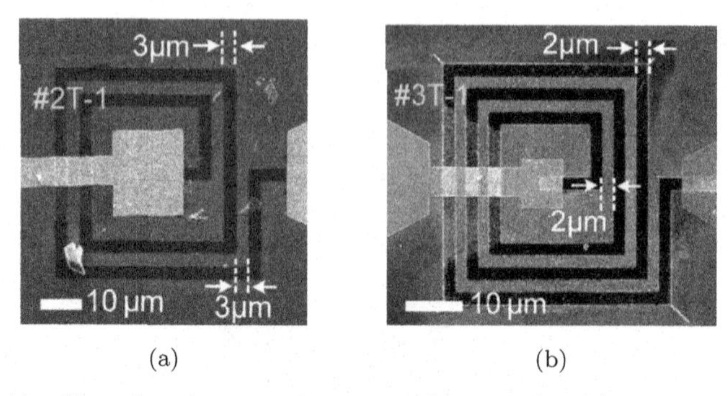

Figure 2.44. Scanning electron microscope photographs: (a) two turns; (b) three turns.

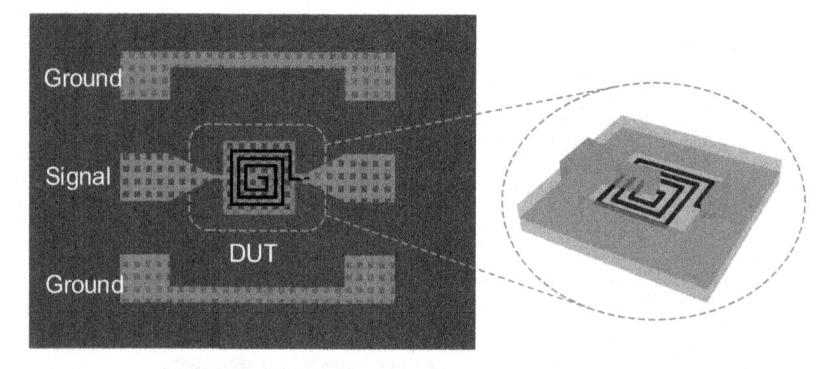

Figure 2.45. Schematic view of the test structure for graphene-based inductor.

2.6.2 *Modeling and parameter extraction*

The improved equivalent circuit model of graphene-based inductor is presented in Figure 2.46. Considering the drop-down of series resistance and the rise of series inductance at high frequency, a capacitance C is added in parallel with R_{s0} to model the high-frequency characteristics [20]. The intrinsic equivalent series resistance $R_{\mathrm{s}}(f)$ and series inductance $L_{\mathrm{s}}(f)$ can be extracted from reverse transmission admittance Y_{12} with input port short-circuited:

$$R_{\mathrm{s}}(f) + j\omega L_{\mathrm{s}}(f) = -\frac{1}{Y_{12}} \tag{2.37}$$

with

$$R_{\mathrm{s}}(\omega) = \frac{R_{\mathrm{S0}}}{1 + \omega^2 C^2 R_{\mathrm{S0}}^2} + \frac{\omega^2 R_{\mathrm{S1}} L_{\mathrm{S1}}^2}{R_{\mathrm{S1}}^2 + \omega^2 L_{\mathrm{S1}}^2}$$

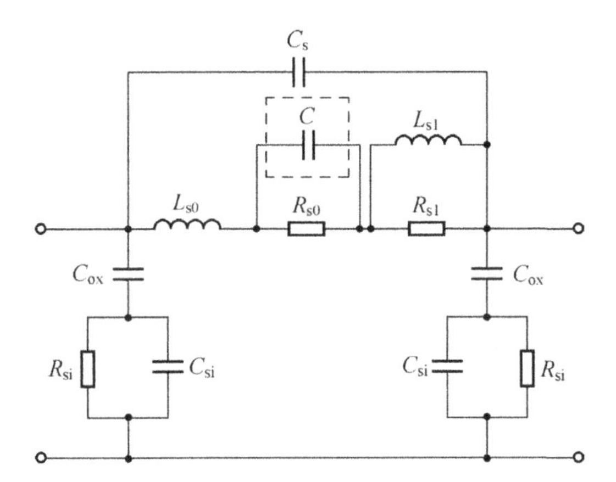

Figure 2.46. Improved equivalent circuit model of graphene-based inductor.

$$L_s(\omega) = L_{S0} - \frac{CR_{S0}^2}{1+\omega^2 C^2 R_{S0}^2} + \frac{R_{S1}^2 L_{S1}}{R_{S1}^2 + \omega^2 L_{S1}^2}$$

For the determination of the given additional capacitance C, first plot the equivalent series resistance versus frequency and then select two different frequency points. Finally, the equivalent series resistance and series inductance can be expressed as follows:

$$R_s(\omega_1) = \frac{\omega_1^2 R_{s1} L_{s1}^2}{R_{s1}^2 + \omega_1^2 L_{s1}^2} + \frac{R_{dc}}{1+\omega_1^2 C^2 R_{dc}^2} \tag{2.38}$$

$$L_s(\omega_2) = L_{dc} - \frac{\omega_2^2 L_{s1}^3}{R_{s1}^2 + \omega_2^2 L_{s1}^2} + \frac{\omega_2^2 C^3 R_{dc}^4}{1+\omega_2^2 C^2 R_{dc}^2} \tag{2.39}$$

When the frequency is close to zero ($\omega \to 0$), the resistance R_{dc} and inductance L_{dc} can be written as

$$R_{dc} = R_{s0} \tag{2.40}$$

$$L_{dc} = L_{s0} + L_{s1} - CR_{s0}^2 \tag{2.41}$$

The flowchart of parameter extraction is illustrated in Figure 2.47. As seen in the flowchart, the first step is to set the initial value of additional capacitance C and then calculate the value of R_{s1}, L_{s1}, R_{s0}, and L_{s0}. The value of C is updated to reduce the error of the simulated and measured S parameters.

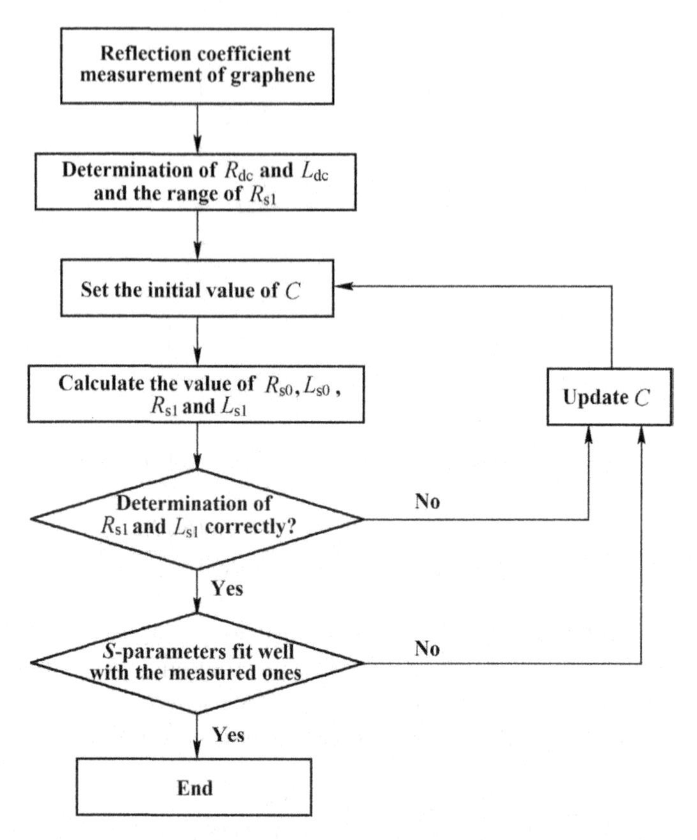

Figure 2.47. Flowchart of parameter extraction.

Note that the definition of optimization error is

$$E_{ij} = \frac{1}{N-1} \sum_{n=0}^{N-1} |S_{ij}^m - S_{ij}^c|^2 \quad (i, j = 1, 2) \tag{2.42}$$

Subscript c represents the simulated S parameters, m represents the measured data, and $n = 0, 1, 2, \ldots, N-1$ is the number of sampling points.

Figures 2.48(a) and 2.48(b) depict the comparison between the measured and modeled data for equivalent series resistance and series inductance. As seen, good agreement between the modeled and measured data is obtained.

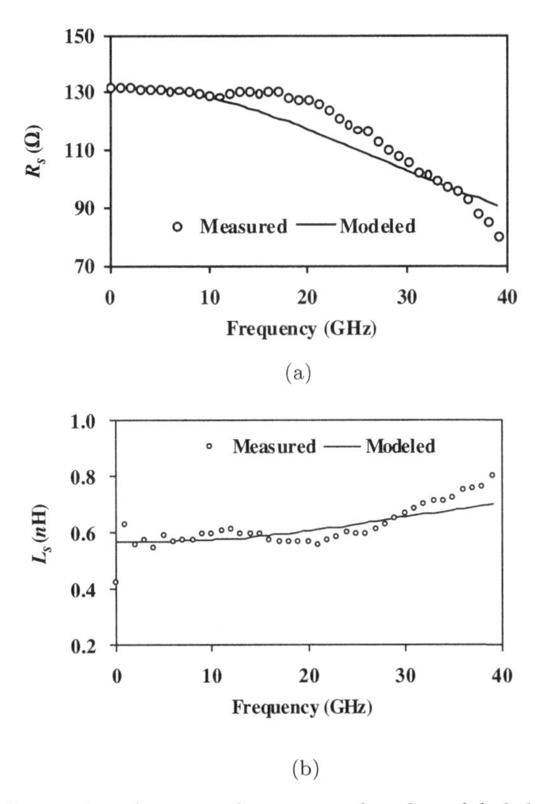

Figure 2.48. Comparison between the measured and modeled data: (a) equivalent series resistance; (b) series inductance.

Table 2.8. Model parameters of graphene-based inductor.

Parameters	C	R_{s0}	R_{s1}	L_{s0}	L_{s1}
Values	30 fF	132Ω	33.9Ω	0.9 nH	0.2 nH

2.6.3 *Experimental results*

To verify the modeling and parameter extraction methodology proposed in this section, an 3/4 turn graphene-based inductor with 24 μm outer diameter, 3 μm trace width, and 62 nm turn pitch was fabricated. The model parameters for the graphene-based inductor are summarized in Table 2.8. Figure 2.49 shows the comparison

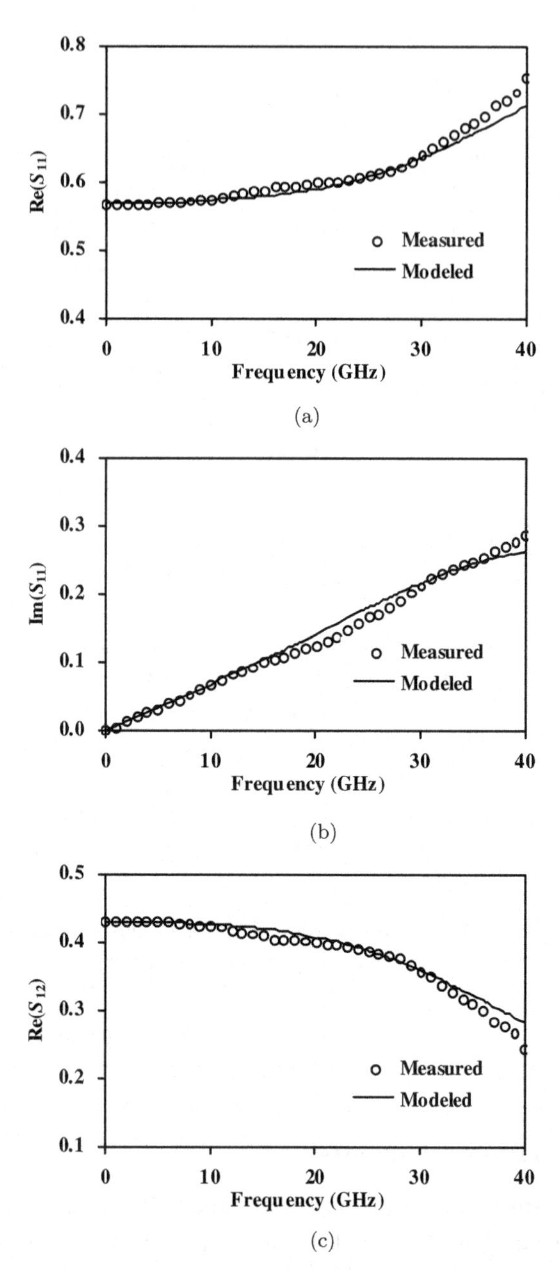

Figure 2.49. Comparison between the modeled and measured S parameters and quality factor: (a) real part of S_{11}; (b) imaginary part of S_{11}; (c) real part of S_{12}; (d) imaginary part of S_{12}; (e) quality factor.

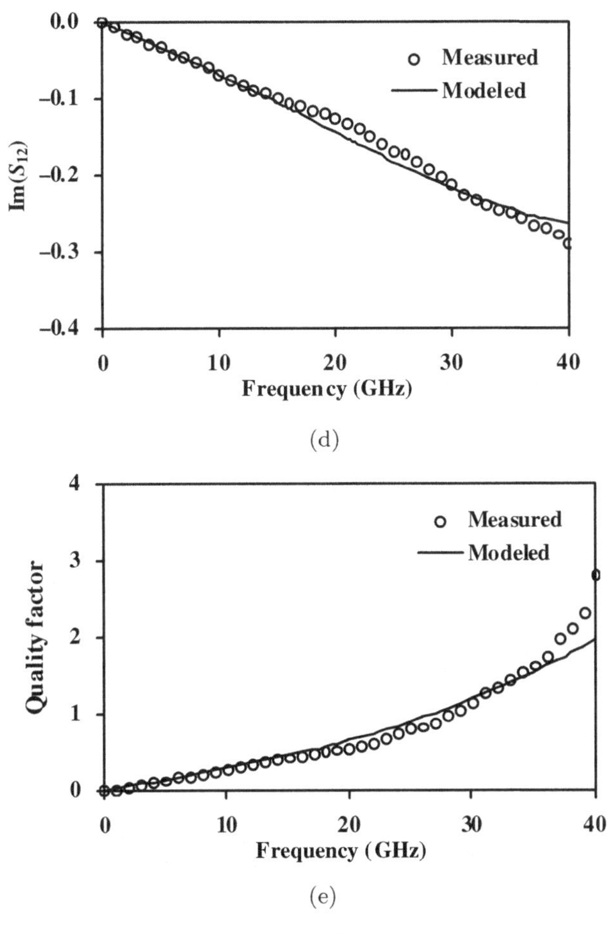

(d)

(e)

Figure 2.49. (*Continued*)

between the modeled and measured S parameters and quality factors up to 40 GHz. As observed, good agreements are obtained over the whole frequency range.

Figure 2.50 plots the comparison of the accuracy (S_{11} and S_{12}) between the conventional model (without additional capacitance) and the proposed model (with additional capacitance). It is obvious that the proposed model is more accurate than the conventional one.

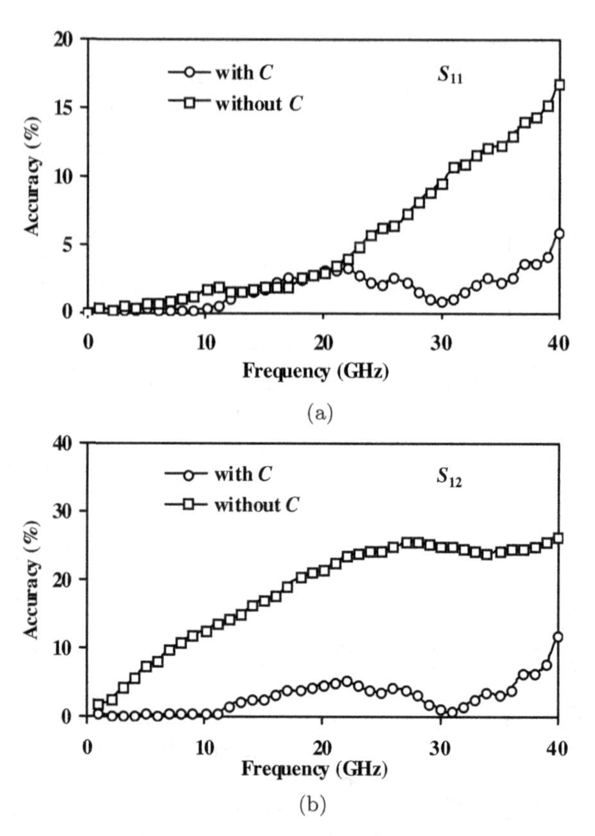

Figure 2.50. Comparison of S-parameters errors between conventional model and proposed equivalent circuit model: (a) S_{11}; (b) S_{12}.

2.7 Effect of Physical Structure Dimensions

Scalability of inductance values is important for optimized layout design. A methodology for scaling the inductance versus layout parameters is needed, with layout parameters including the number of turns, coil conductor width and conductor inter-turn spacing, and inner diameter.

The top view and cross-section of the on-chip spiral inductor are sketched in Figures 2.51(a) and 2.51(b). The coil conductor width and conductor inter-turn spacing are w and s, respectively. The inner diameter is R_{in} and thickness t. The substrate of the inductor is utilized, the silicon substrate with the thickness of t_{sub}, and

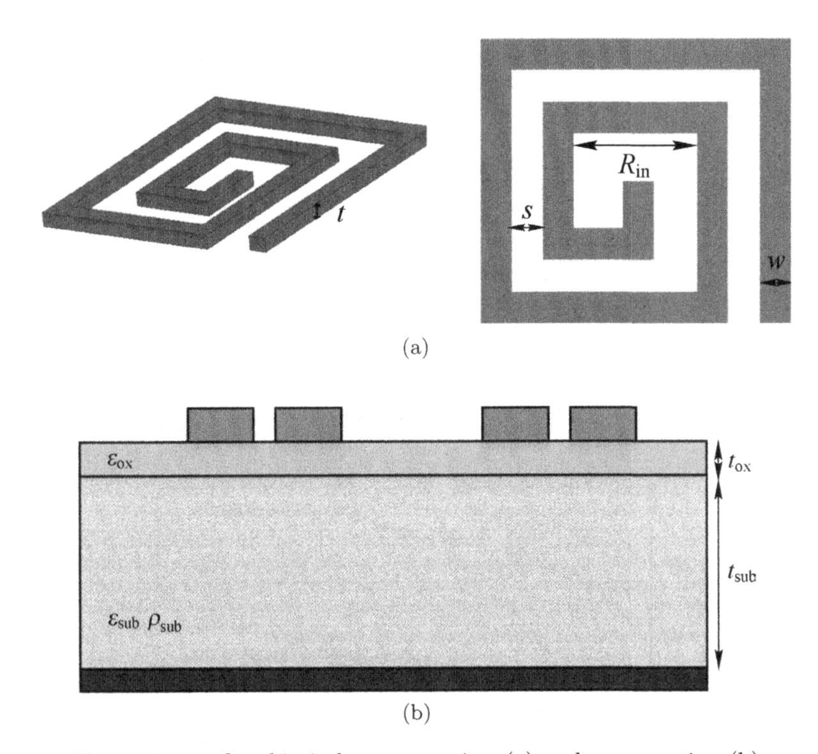

Figure 2.51. On-chip inductor top view (a) and cross-section (b).

the dielectric constant and resistivity are ε_{sub} and ρ_{sub}, respectively. The substrate is covered with a dielectric layer with the thickness of t_{ox}, and its dielectric constant is ε_{ox}.

In order to analyze the influence of geometric parameters on the equivalent series inductance and quality factor, nine on-chip spiral inductors based on $0.18\,\mu$m standard CMOS process and 13 graphene-based on-chip spiral inductors with different geometric parameters are developed [20,21].

2.7.1 *Effect of coil turn*

Three spiral inductors on the silicon substrate with three different numbers of turns are designed. The variation parameters are the number of turns, which are 0.5 turns, 1.5 turns, and 2.5 turns, respectively. The fixed geometric parameters are the coil width ($8\,\mu$m), coil spacing ($2\,\mu$m), and inner diameter ($20\,\mu$m).

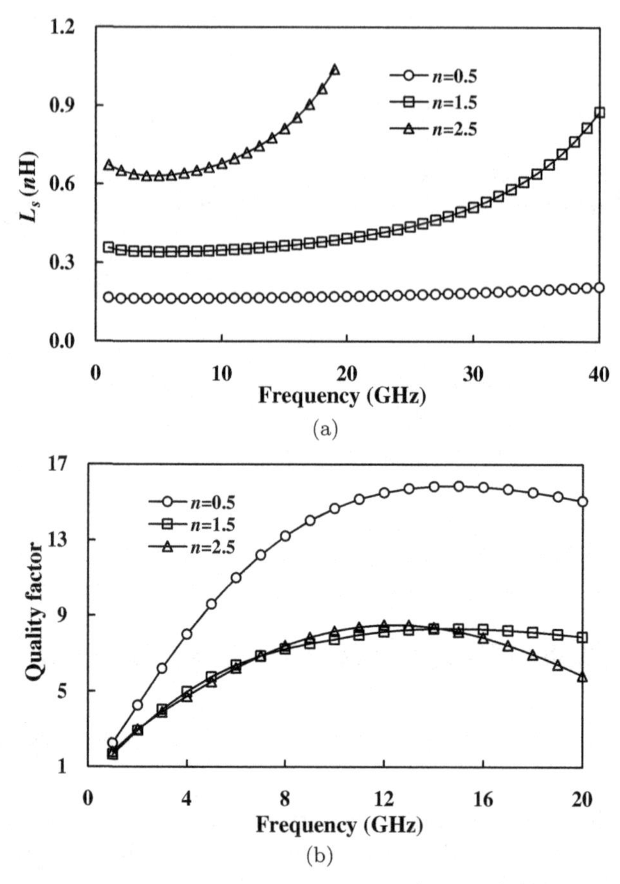

Figure 2.52. Influence of inductance turns on the conventional on-chip spiral inductors: (a) inductance values; (b) quality factor.

The influence of inductance turns on the conventional on-chip spiral inductors is exhibited in Figures 2.52(a) and 2.52(b). It can be clearly seen that the inductance values increase with the increase in the number of turns. The main reason is that the length of the metal wire increases with the increase in the number of turns, thence the number of turns has a great impact on the change of the inductance values. With the increase in the number of turns, the quality factors of the on-chip spiral inductor show a downward trend. The highest values of the quality factors and the operating frequencies corresponding to the highest values decrease significantly.

The reasons for the decrease in quality factor mainly include the following: the increase in the number of turns leads to the increase in the coil resistance, for example, the ohmic loss of the coil is increased. The proximity effect between adjacent conductors and the capacitive coupling between conductors are also more pronounced due to the increased number of turns. At the same time, the effect between the inductor and the substrate results in more substrate losses. Hence, increasing the number of turns for the inductor increases the chip area occupied; the inductance value increases, while the quality factor decreases.

The influence of inductance turns on graphene-based on-chip spiral inductors is plotted in Figures 2.53(a) and 2.53(b). The fixed

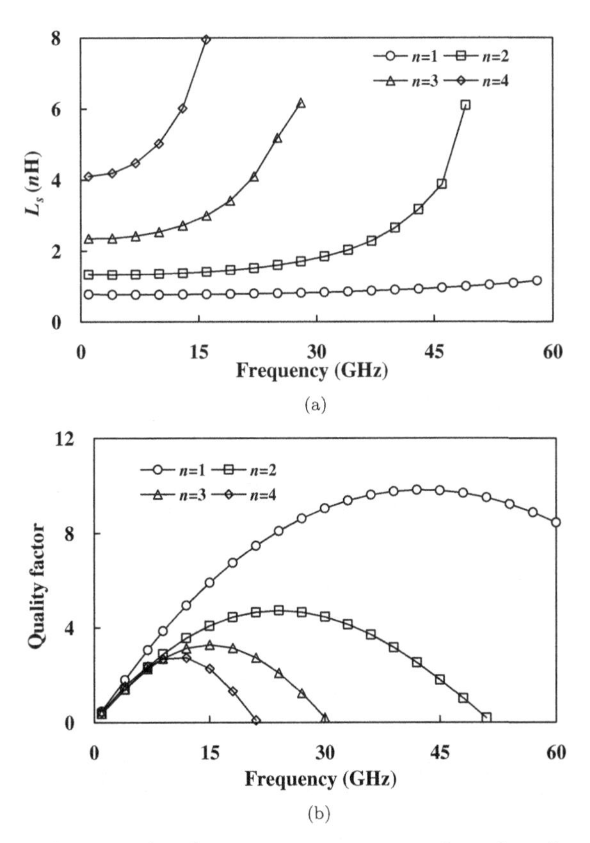

Figure 2.53. Influence of inductance turns on graphene-based on-chip spiral inductors: (a) inductance values; (b) quality factor.

geometric parameters are the coil width (20 μm), coil spacing (8 μm), and inner diameter (90 μm). It can be found that the quality factors of graphene-based on-chip spiral inductors are lower than silicon-based on-chip spiral inductors due to the loss of graphene being larger.

2.7.2 *Effect of coil width*

Three spiral inductors on silicon substrate with three different coil widths are designed. The variation parameter is the coil width, which are 8 μm, 10 μm and 12 μm, respectively; The fixed geometric parameters are the number of turns (1.5 turns), coil spacing (2 μm), and inner diameter (20 μm).

The influence of inductance coil width on conventional on-chip spiral inductors is given in Figures 2.54(a) and 2.54(b). As seen, the inductance increases with slowly increasing coil width and the variations of quality factor are very small for operating frequencies and the maximum values. The conclusion is that the metal line width has little effect on the spiral inductor.

The influence of inductance coil width on graphene-based on-chip spiral inductors is plotted in Figures 2.55(a) and 2.55(b). The variation parameter is the coil width, which is 10 μm, 15 μm, 20 μm and 25 μm, respectively. The fixed geometric parameters are the number of turns (2 turns), coil spacing (10 μm), and inner diameter (90 μm). It can be obviously found that the quality factors of graphene-based on-chip spiral inductors increase with the increase in the coil width.

2.7.3 *Effect of coil spacing*

Three spiral inductors on silicon substrate with three different coil spacing were designed. The variation parameter is the coil spacing, which are 2 μm, 4 μm and 6 μm, respectively. Besides, the fixed geometric parameters are the number of turns (1.5 turns), coil width (8 μm), and inner diameter (20 μm). Figures 2.56(a) and 2.56(b) show the influence of inductance coil spacing on conventional on-chip spiral inductors. As seen, the inductance values decrease with increased coil spacing. The quality factor value of on-chip spiral inductor decreases

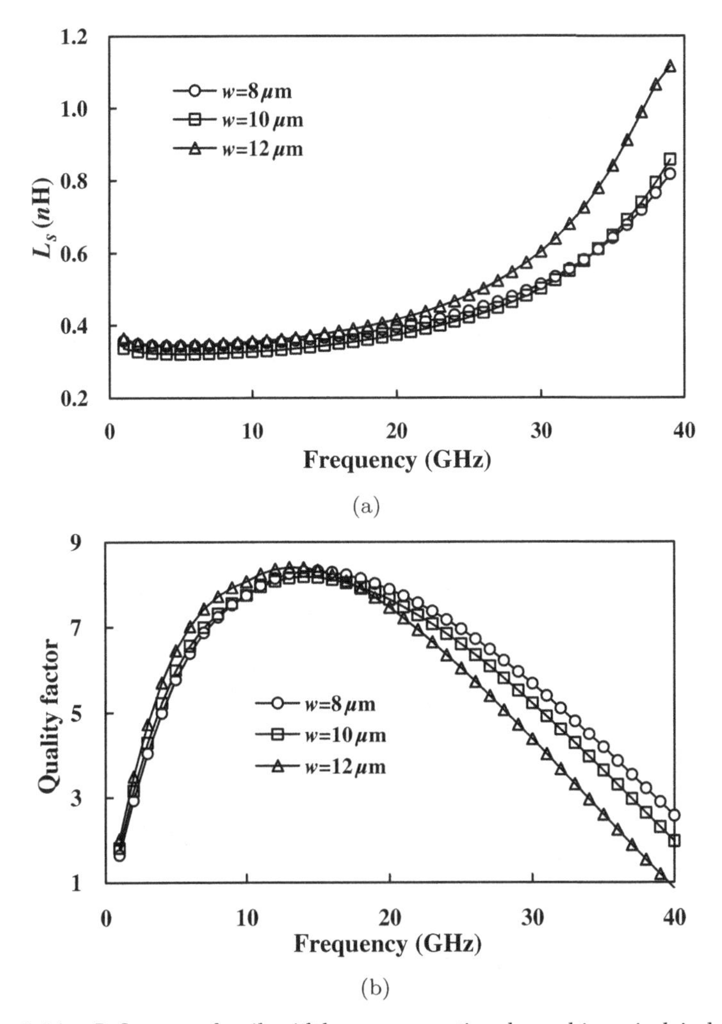

Figure 2.54. Influence of coil width on conventional on-chip spiral inductors: (a) inductance values; (b) quality factor.

with the increase in spacing. The reason is that the proximity effect between adjacent coils increases due to the small spacing, resulting in the increase in losses.

The influence of coil spacing on graphene-based on-chip spiral inductors is illustrated in Figures 2.57(a) and 2.57(b). In this case, the variation parameter is the coil spacing, which are 5 μm, 10 μm,

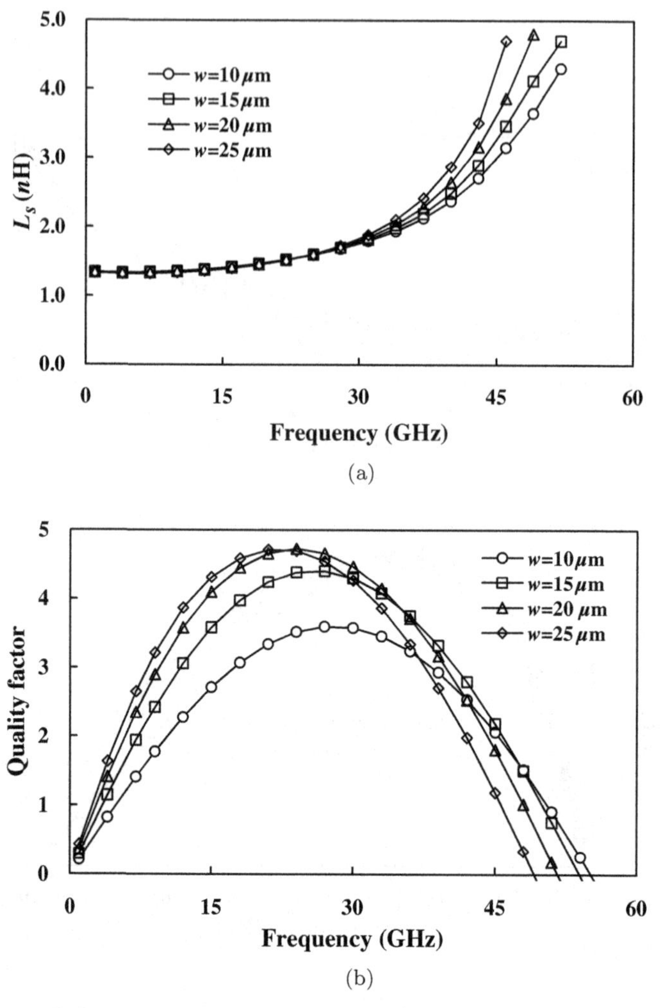

Figure 2.55. Influence of coil width on graphene-based on-chip spiral inductors: (a) inductance values; (b) quality factor.

15 μm, and 2 μm, respectively. Moreover, the fixed geometric parameters are the number of turns (2 turns), coil width (20 μm), and inner diameter (90 μm). It is significant to observe that the quality factors of graphene-based on-chip spiral inductors increase with increased coil width. It can be seen that the change of spacing ratio between coil wires has little effect on the inductance and quality factor of the spiral inductor.

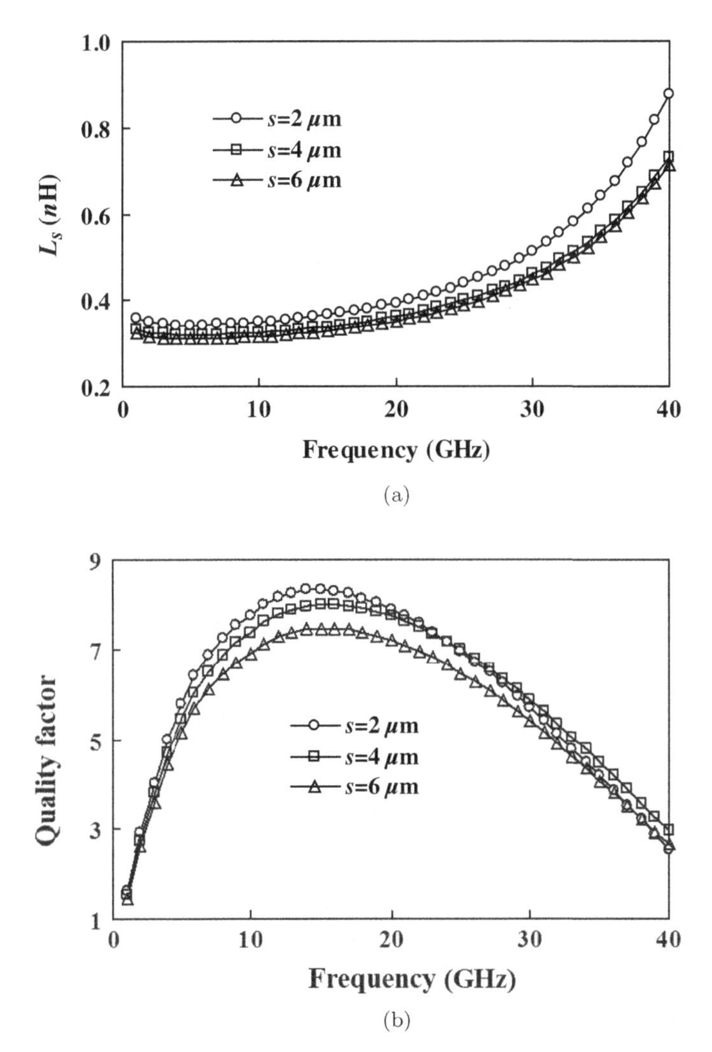

Figure 2.56. Influence of coil spacing on conventional on-chip spiral inductors: (a) inductance values; (b) quality factor.

2.7.4 *Effect of inner diameter*

Three spiral inductors on silicon substrate with three different inner diameters were designed. The variation parameter is the inner diameters, which are 20 μm, 30 μm and 40 μm, respectively. In addition, the fixed geometric parameters are the number of turns (1.5 turns), coil spacing (2 μm), and coil width (8 μm).

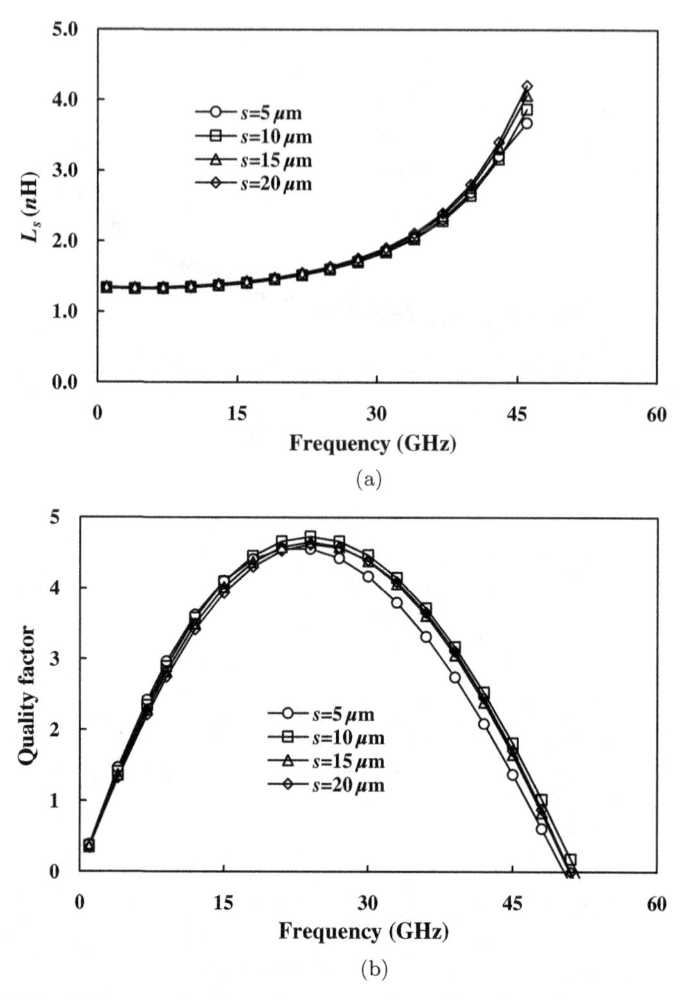

Figure 2.57. Influence of coil spacing on graphene-based on-chip spiral inductors: (a) inductance values; (b) quality factor.

The influence of inner diameter on conventional on-chip spiral inductors is depicted in Figures 2.58(a) and 2.58(b). As seen, the inductance values increase with the increase in the coil spacing due to the inductor area becoming large. By observing the variation of the quality factor with frequency, the operating frequency band of the on-chip spiral inductor can be obtained from the maximum value.

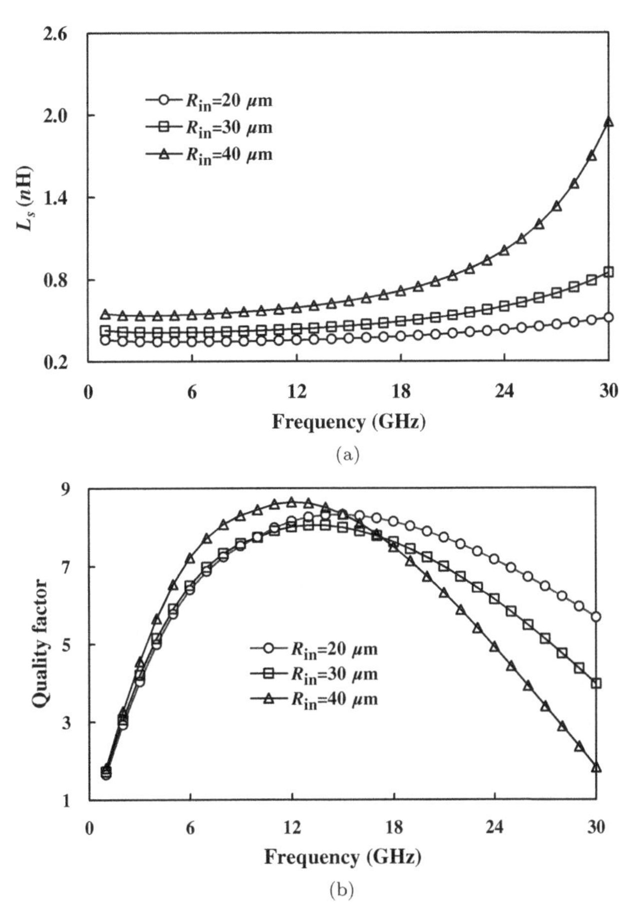

Figure 2.58. Influence of inner diameter on conventional on-chip spiral inductors: (a) inductance values; (b) quality factor.

The influence of inner diameter on graphene-based on-chip spiral inductors is displayed in Figures 2.59(a) and 2.59(b). The variation parameter is the inner diameters, which are 80 μm, 90 μm, 100 μm, and 11 μm, respectively. The fixed geometric parameters are the number of turns (2 turns), coil width (20 μm), and coil spacing (10 μm). As indicated in Figure 2.59, the inductance increases with increased inner diameter, and the quality factors decrease as the inner diameter increases. Larger inductance can be obtained by increasing the inner diameter but will lead to the decline of quality factor value at high frequencies.

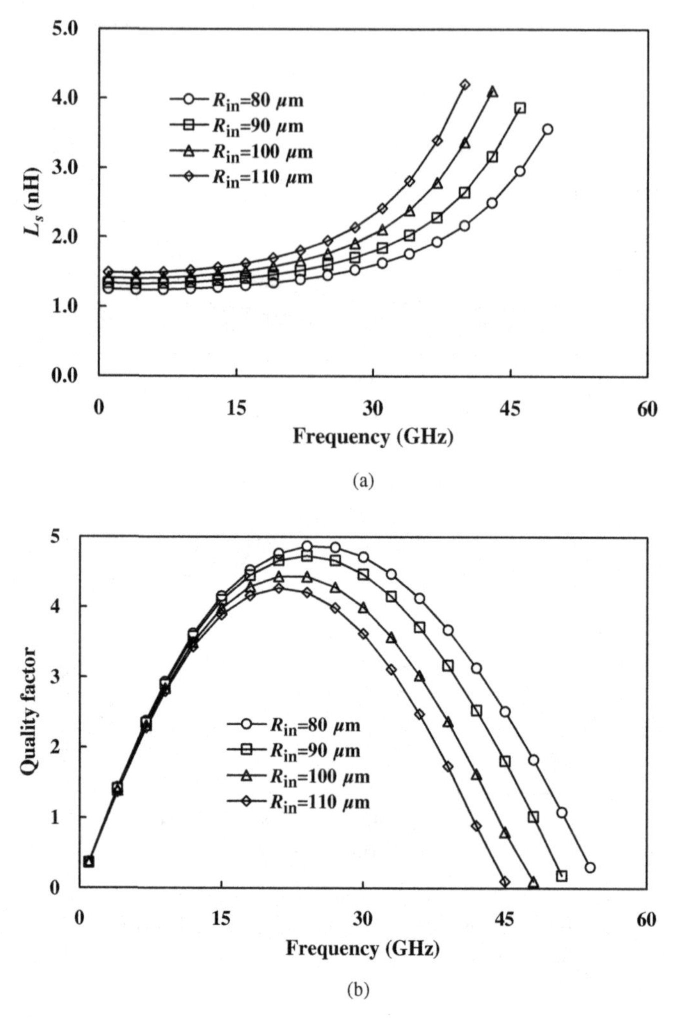

Figure 2.59. Influence of inner diameter on graphene-based on-chip spiral inductors: (a) inductance values; (b) quality factor.

2.8 Summary

This chapter mainly introduces the basic structure and characteristics of three commonly used inductors, including octagonal spiral inductor, three-dimensional inductor, and graphene inductor in the standard CMOS process. The modeling and parameter extraction method of the on-chip spiral inductor are given, and the influence of the main physical structure parameters of the on-chip spiral inductor on the inductance and quality factor is discussed.

References

[1] I. Bahl, *Lumped Elements for RF and Microwave Circuits*. London: Artech House, 2003.

[2] A. Robbins and W. Miller. *Circuit Analysis: Theory and Practice*. USA: Delmar Publishers, 1995.

[3] C. Lin and T. S. Kalkur, "Modeling of current crowding for on-chip multi turn differential -spiral inductors," *IEEE EUROCON 2009*, pp. 178–182, 2009. Doi: 10.1109/EURCON.2009.5167627.

[4] D. Barras, F. Ellinger, H. Jäckel, and W. Hirt, "A low supply voltage SiGe LNA for ultra-wideband frontends," *IEEE Microwave Wireless Components Letters*, 14(10): 469–471, 2004.

[5] Y.-T. Lin, H.-C. Chen, T. Wang, Y.-S. Lin, and S.-S. Lu, "3–10 GHz ultra-wideband low-noise amplifier utilizing miller effect and inductive shunt–shunt feedback technique," *IEEE Trans. Microwave Theory Techniques*, 55(9): 1832–1843, 2007.

[6] S. S. Mohan, M. del Mar Hershenson, S. P. Boyd, and T. H. Lee, "Simple accurate expressions for planar spiral inductances," *IEEE Journal of Solid-State Circuits*, 34(10): 1419–1424, 1999.

[7] W. B. Kuhn and N. M. Ibrahim, "Analysis of current effects in multiturn spiral inductors," *IEEE Trans. Microwave Theory Techniques*, 49(1): 31–38, 2001.

[8] C. P. Yue and S. S. Wong, "Physical modeling of spiral inductors on silicon," *IEEE Trans. Election Devices*, 47(3): 560–568, 2000.

[9] F. Y. Huang and J. X. Lu, "Analytical approach to parameter extraction for on-chip spiral inductors taking into account high-order parasitic effect," *IEEE Journal of Solid-State Circuits*, 49(3): 473–478, 2006.

[10] J. Wei and Z. Wang, "Frequency-independent T equivalent circuit for on-chip spiral inductors," *IEEE Trans. Election Devices*, 31(9): 933–935, 2010.

[11] Y. Cao, R. A. Groves, X. Huang, N. D. Zamdmer, J. O. Plouchart, R. A. Wachnik, T. J. King, and C. Hu, "Frequency-independent equivalent-circuit model for on-chip spiral inductors," *IEEE Journal of Solid-State Circuits*, 38(3): 419–426, March 2003.

[12] N. Yan, C. Yang, and J. Gao, "An approach for determining equivalent circuit model of on-chip inductors," *Microwave and Optical Technology Letters*, 55(10): 2363–2370, 2013.

[13] L. Guo, M. Yu, Z. Chen, H. He, and Y. Zhang, "High Q multilayer spiral inductor on silicon chip for 5-6 GHz," *IEEE Electron Device Letters*, 23(8): 470–472, 2002.

[14] B. Piernas, K. Nishikawa, K. Kamogawa, T. Nakagawa, and K. Araki, "High-Q factor three-dimensional inductors," *IEEE Trans. Microwave Theory Technology*, 50(8): 1942–1949, 2002.

[15] L. Gu and X. Li, "Concave-suspended high-Q solenoid inductors with an RFIC-compatible bulk-micromachining technology," *IEEE Trans. Electron Devices*, 54(4): 882–885, 2007.

[16] Z. Ni, J. Zhan, Q. Fang, X. Wang, Z. Shi, Y. Yang, T.-L. Ren, A. Wang, Y. Cheng, J. Gao, X. Li, and C. Yang, "Design and analysis of vertical nanoparticles-magnetic-cored inductors for RF ICs," *IEEE Trans. Electron Devices*, 60(4): 1427–1435, 2013.

[17] J. Gao and C. Yang, "Microwave modeling and parameter extraction method for multilayer on-chip inductors," *International Journal of RF and Microwave Computer-Aided Engineering*, 23(3): 343–348, 2013.

[18] X. Li, J. Kang, X. Xie, *et al.*, "Graphene inductors for high-frequency applications — design, fabrication, characterization and study of skin effect," *IEEE International Electron Devices Meeting*, pp. 5.4.1–5.4.4, 2014.

[19] Y. Zhang, A. Zhang, B. Wang, and J. Gao, "Radio-frequency modeling and parameter extraction of graphene on-chip spiral inductors," *Journal Infrared Millim. Waves*, 37(4): 393–398, 2018.

[20] Y. Zhang, "Modeling and parameters extraction of graphene on-chip spiral inductors," East China Normal University Master Dissertation, 2019.

[21] L. Yan, "Modeling and parameters extraction of Si-based on-chip spiral inductors for RFICs design," East China Normal University Master Dissertation, 2014.

Chapter 3

On-Chip Spiral Transformer

A transformer is a piece of electrical equipment that changes a voltage (AC) to a higher or lower voltage (AC). Its working principle is based on the electromagnetic coupling mechanism. According to the size of the transformer, it mainly has the following three applications:

(1) Large transformers in the power system. Long-distance power transmission is carried out by utilizing voltage above 10000 V and then reduced to a safe level by using transformers for home and office use.
(2) Small transformer charging device for electronic equipment. The power supply of small electronic equipment such as mobile phones and computers needs to convert the power supply of 110–240 V into the voltage range of 1–5 V to prevent damage to electronic equipment.
(3) The on-chip transformers are utilized in integrated circuit design. Transformers and other basic circuit components are fabricated on the same substrate, and the area occupied is usually less than 1 mm^2. On-chip transformers are widely used to implement functions such as impedance conversion, resonant loads, low-noise feedback, bandwidth enhancement, and differential-to-single conversion. Transformers can be utilized as inter-stage elements in power amplifiers while providing balun functionality, impedance matching, and DC decoupling, eliminating the need for large DC decoupling capacitors.

The RFIC chips in these areas require low power, high yield, and rapid time-to-market. The silicon-based technology is a good candidate to meet these requirements. With the increasing demand for high-performance on-chip transformers in microwave integrated circuits, establishing an accurate equivalent circuit model that can reflect the physical characteristics of on-chip transformers is of great significance to computer-aided circuit design and optimization [1–3]. This chapter mainly introduces the structure of on-chip transformers commonly used in RF and microwave integrated circuits, the basic characteristics of on-chip transformers, as well as the modeling and parameter extraction method of on-chip transformers.

3.1 On-Chip Transformer in RFIC

In the low-frequency analog circuit, the transformer is mainly composed of two coils of metal magnetic core (as shown in Figure 3.1), which can enhance the magnetic coupling between the two coils and reduce the hysteresis loss. Contrary to the single coil, the inductor of the on-chip transformer is composed of two coils to couple or isolate the energy from one coil to the other coil. The main feature of the low-frequency inductors (also called coil inductors) is that it has three-dimension structures and requires a large space volume.

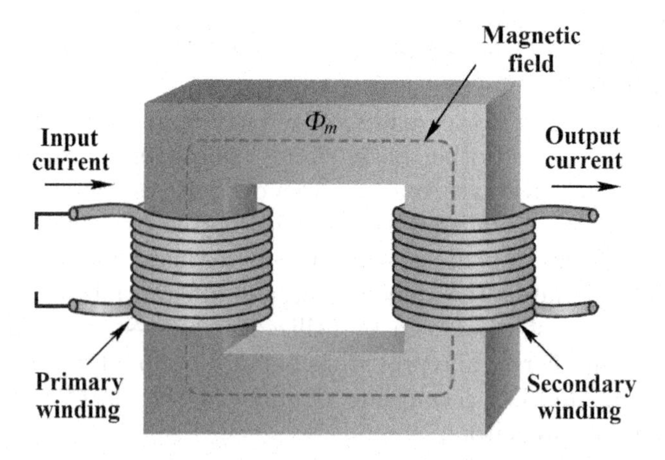

Figure 3.1. Transformer in low-frequency circuit.

It can be soldered directly on the surface of the printed circuit board. Alternating current in one winding establishes a flux that links the other winding and induces a voltage in it. Thence, power flows from one circuit to the other via the medium of the magnetic field, with no electrical connection between the two sides. The input port is called the primary winding, while the output port is called the secondary winding.

In RF microwave integrated circuits, due to the limitation of circuit volume, the transformer is usually composed of two independent planar spiral inductors. As a result, the shape, structure, and implementation of the transformer have actually changed greatly, and the metal magnetic core is no longer used. The integrated spiral transformer based on CMOS technology is an important passive inductance component, which is widely used in various RF front-end circuits. The transformer is two or more inductors coupled to each other. When it is composed of only two coils, it is called the double winding transformer, and when it is composed of multiple coils, it is called the multi-winding transformer. The working principle of the double winding transformer is based on the magnetic field coupling between two windings, converting the AC signal from one end to the other without large energy losses. At the same time, the DC is isolated so that the two coils can be operated at different bias voltages. In the communication circuit, transformers are used for impedance matching and eliminating the DC signal from each part of the system.

Transformers are heavily used in mm-wave circuits mainly for single-ended to differential conversion, vice versa, and power combining. The schematic of the transformer-coupled low-power CMOS power amplifiers is illustrated in Figure 3.2(a) [4]. The matching network of a power amplifier adopts transformers and inductors. The width/length ratio of the first stage and the second stage transistors is $50\,\mu\text{m}/60\,\text{nm}$, and the output stage is $120\,\mu\text{m}/60\,\text{nm}$. The output stage adopts a large width/length ratio to improve the output power of the power amplifier. The die micrograph of the PA prototype is exhibited in Figure 3.2(b). It occupies a total chip area of $0.62\ \text{mm}^2$ (including pads). The amplifier has differential RF input and single-ended RF output.

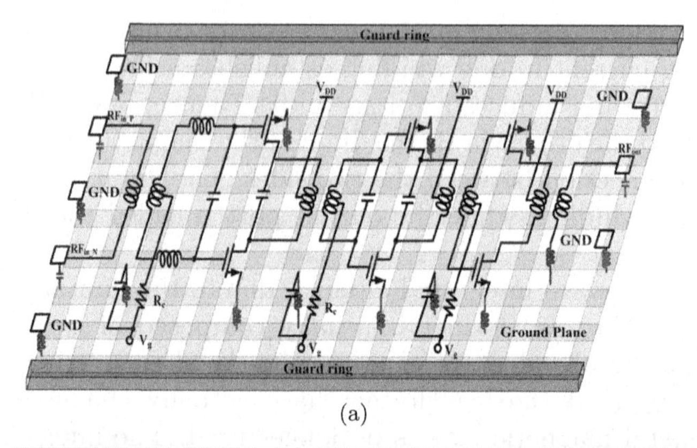

(a)

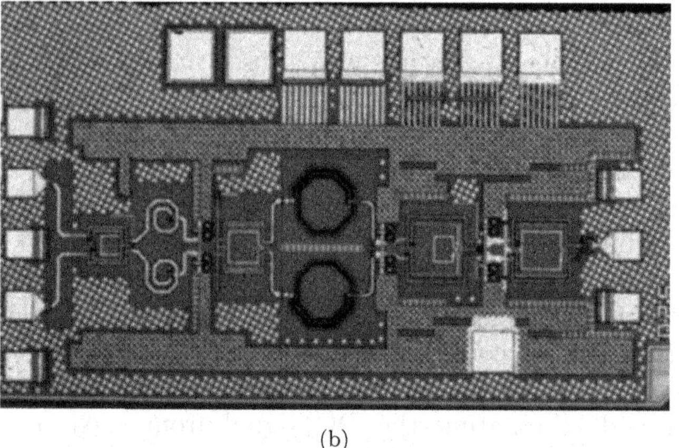

(b)

Figure 3.2. Wide-bandwidth power amplifier fabricated in 65 nm bulk CMOS process. (a) Schematic; (b) die micrograph.

The schematic of the UWB low noise amplifier with dual feedback paths for transformer series feedback and the capacitive shunt feedback is depicted in Figure 3.3. Utilizing shunt and series feedback, broadband matching is achieved with high flat gain [5]. Simplified simulation models of inductive devices are built with the EM simulation results. The optimization utility of EDA tools is used to find the optimal parameters for real devices. With the utilization of a weakly coupled transformer, the chip size is slimmed down dramatically. The 2.4 GHz power amplifier design is illustrated in Figure 3.4. The transformer is used as the output balun in a one-stage push-pull differential power amplifier [6].

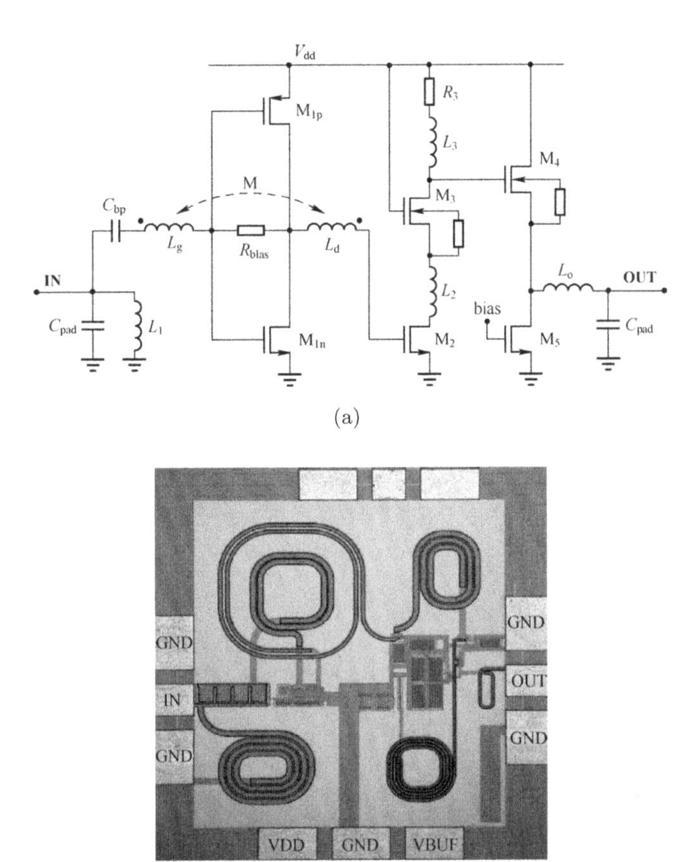

Figure 3.3. UWB low noise amplifier: (a) schematic; (b) die micrograph.

3.2 Basic Structure and Performance

On-chip integrated transformers based on CMOS technology are usually called on-chip spiral transformers, which are mainly based on metal interconnection. The operation of passive transformers is based on the mutual inductance between two or more conductors or windings.

Transformers are designed to couple alternating currents from one winding to the other without significant power loss. Depending on whether the lateral or vertical magnetic coupling is used, the transformer can be categorized into the planar structure and stacked

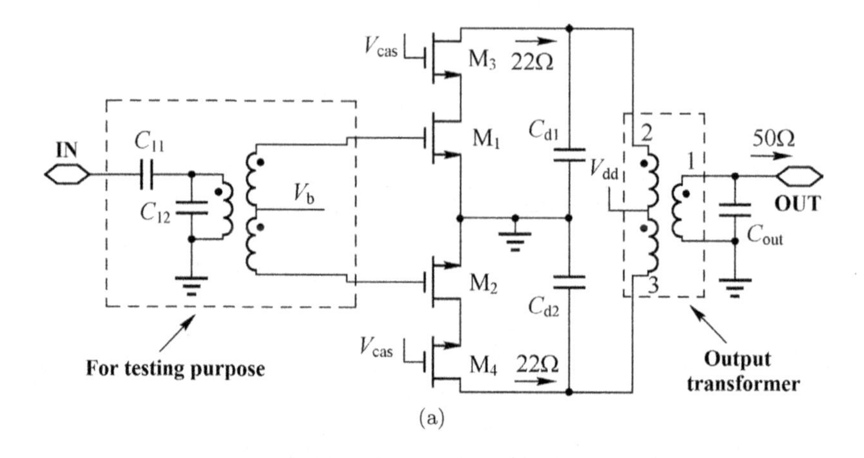

(a)

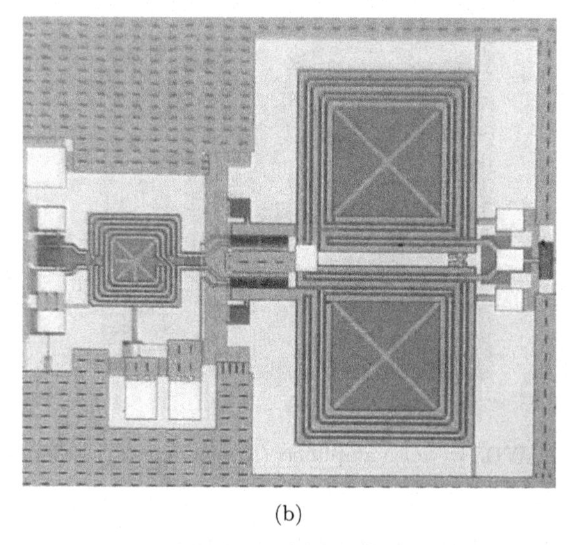

(b)

Figure 3.4. 2.4 GHz power amplifier design: (a) schematic; (b) die micrograph.

structure. The first is the planar spiral transformer, which is wound by two metal wires in parallel. Although the number of turns of the planar spiral transformer is the same, the self-inductances of the primary and secondary coils may be different due to the different sizes of the primary and secondary coils. The second is the stacked transformer, which is wound by two metal wires with different metal layers. Primary and secondary inductance values of the stacked transformer required for mm-wave applications can be readily obtained using single-turn spirals without compromising on the silicon area. Stacked transformers can provide primary and secondary coil ratios greater than one.

The basic structures of a planar spiral transformer and a stacked spiral transformer are described in the following.

3.2.1 *Planar spiral transformer*

There are many shapes of planar transformers, including square transformers, octagonal transformers, hexagonal transformers, and circular transformers. The most common structure is quadrilateral.

The primary and secondary coils of the spiral transformer on the planar structure chip are constructed on the same metal layer, and the lower metal is used to connect the metal segments of the coils. This allows both windings to be perfectly symmetric with regard to inductive and parasitic characteristics. The interleaved topology is traditionally preferred for process technologies that contain only one thick metal layer. In the multi-metal layer process, some of the top metal layers are also connected in parallel to reduce the series resistance. The primary coil and secondary coil of the on-chip spiral transformer can be completely symmetrical, which is very suitable for differential circuits, and the series resistance of the coil is very small. Nevertheless, the disadvantage is that the transformers occupy large chip areas.

According to the winding of primary coil and secondary coils, transformers can be divided into four kinds of structures: parallel winding, inter-twined winding, symmetrical winding, and center tap. Functionally, it can be divided into step-up mode transformer and step-down mode transformer. Eight common planar spiral transformer structures are presented in Figure 3.5 [4], in which (a) is the center tap structure, (b) is the parallel winding structure, (c) is the intertwined structure, (d) is the symmetrical structure, and (e) and (f) are the step-up structures.

(1) The advantage of the parallel winding structure is that the primary and secondary windings can be realized by the same layer of metal simultaneously so that the parasitic capacitance can be minimized and a higher resonant frequency and mutual inductance factor can be obtained. The disadvantage is that the

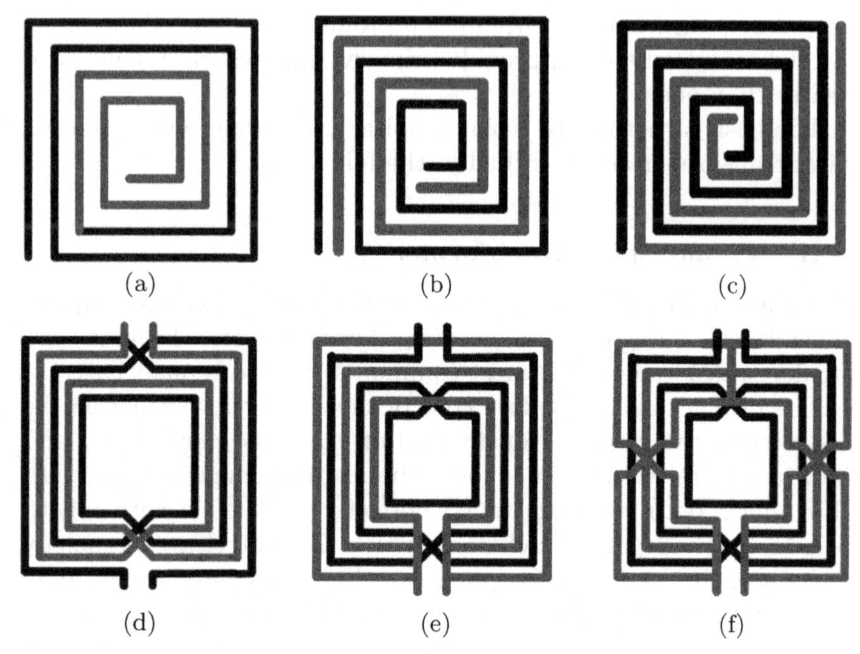

Figure 3.5. Commonly used on-chip planar spiral transformer structures: (a) tapped; (b) parallel; (c) intertwined; (d) symmetric; (e) step-up I; (f) step-up II.

primary and secondary ports are located on the same side, which is inconvenient to connect with other circuit elements. It should be noted that due to the asymmetric structure, the length of the primary coil is different from that of the secondary coil. Thence, although the number of turns is the same, the turn ratio is not one actually.

(2) Intertwined structure transformer adopts two identical coils, which ensures that when the turns of the primary coil and secondary coil are equal, the performance is completely consistent and can provide a perfect 1:1 symmetrical on-chip spiral transformer. The advantage of the intertwined structure is that the primary and secondary coil ports are on different sides, which is easy to connect with other circuit elements. It is very suitable for symmetrical four-port applications and can be realized with the top metal to minimize the capacitance to achieve the maximum resonant frequency. It is worth noting that the mutual inductance factor is not high.

(3) The central tap transformer is also called a self-coupling transformer. Its primary coil is composed of two concentric but independent coils. Tap structure transformer is most suitable for three-port applications. By changing the position of the tap, it can realize various tap ratios. All coils can be realized with top metal, which can minimize the capacitance from the port to the substrate.
(4) The symmetrical structure is most suitable for three-port and four-port applications, and its center tap can be located at the physical midpoint of the winding. In addition, the symmetrical mutual winding structure has multiple cross transitions, and the metal properties of the top layer and the lower layer are quite different.
(5) The symmetrical setup structures can be used to enhance the voltage and current gain and improve the circuit linearity without additional noise. In order to eliminate the proximity effect and make the physical length of the three branches equal, more cross-connections can be used. This structure increases the coupling coefficient but requires more metal layers.

In summary, the planar spiral transformer is often used in differential circuits and as a balun for single-ended to differential conversion in the front-end circuit of RF integrated circuits.

The main advantages are as follows:

- The structure is very simple and easy to optimize the performance.
- The primary and secondary windings can be completely geometrically symmetrical, which is suitable for differential circuits.
- The series resistances of primary and secondary windings are low.

The main disadvantages are as follows:

- A large chip area is occupied.
- The coupling coefficient is small.

3.2.2 *Stacked spiral transformer*

Since the amount of silicon area occupied by transformers can be a limiting factor in most applications, interleaved or tapped structures are often replaced by stacked configurations, which offer higher magnetic coupling and area efficiency, albeit at the expense of increased parasitic capacitances. Patterned ground shields can be profitably exploited in all transformer configurations to reduce losses caused by eddy currents flowing into the substrate. The primary and secondary coils can be magnetically coupled in both lateral and vertical directions simultaneously, so as to obtain a larger coupling coefficient (the coupling coefficient can reach 0.9 at 60 GHz) and reduce the area of the transformer [7].

In order to achieve high self-resonance frequency, the on-chip transformer used in millimeter-wave integrated circuits generally has a small diameter and fewer turns (only 1–2 turns), resulting in small insertion loss (the insertion loss can reach 0.55 dB at 60 GHz frequency).

In the application of low-frequency analog RF, the differential transformer can significantly suppress the noise of the circuit. In the microwave and millimeter-wave band, due to the amplitude mismatch and phase deviation of the differential output signal, the common mode rejection ratio (CMRR) is quite low, resulting in the weakening of the noise suppression function of the differential transformer [8]. Stacked transformers, on the other hand, present the primary and the secondary on different levels, which permits a stronger magnetic coupling and a considerable area reduction. For this structure, though, the parasitics are necessarily asymmetric, since the distances between the windings and the substrate are not the same. The stacked structure is the main form of spiral transformer on the 3D structure chip, and the coil with the 3D space structure is formed by multi-layer metal and through via holes. Indeed, owing to the high magnetic coupling coefficient, the substrate parasitic capacitance associated with the second metal layer is transferred almost entirely to the primary coil that resonates at the same frequency as the secondary one, the turn ratio being approximately equal to unity.

The commonly used stacked spiral transformer structures: asymmetric structure, lateral translation structure, and diagonal translation structure are depicted in Figure 3.6 [9]. It can be seen that the primary coil and secondary coil are of planar structure using different metal layers. The advantage of the asymmetric structure is based on multi-layer metal layers, therefore the high mutual inductance factor can be obtained using edge and lateral coupling. It can save the

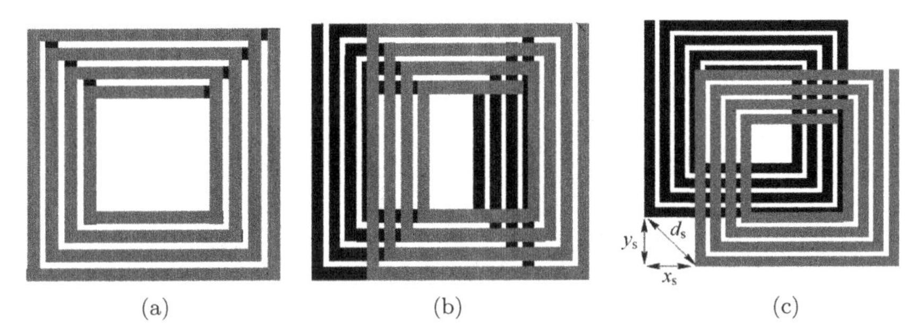

Figure 3.6. Commonly used stacked spiral transformer structures: (a) asymmetric structure; (b) lateral translation structure; (c) diagonal translation structure.

chip area and get high self-feeling simultaneous and is suitable for both three-port and four-port applications. The disadvantage is that the series resistances and coupling parasitic capacitances of primary and secondary coils are obviously different, and the larger parasitic capacitance is caused by the coupling of the input port and output port, resulting in low self-resonance frequency. The parasitic capacitance between the input port and output port can be reduced by utilizing lateral translation and diagonal translation structure. It is worth noting that the mutual inductance coupling will be reduced.

Figures 3.7(a)–3.7(c) illustrate an octagonal flipped transformer with a coil ratio of 1:1, a 1:2 rectangular flipped transformer completed with two-layer metal, and a square flipped transformer completed with three-layer metal coils, respectively.

The main advantages of the stacked transformer are large coupling coefficient and small chip area. The main disadvantages are that the structures of the primary and secondary windings are difficult to symmetrical, and the self-inductance value is inconsistent. The thicknesses of the metal layers of the primary and secondary windings are different, the series resistance of the coil completed with a thinner metal layer is large, and the corresponding quality factor will be low.

Stacked spiral transformers are often used in the inter-stage coupling and feedback network of circuits. These applications require higher coupling and lower symmetry of transformers. The transformer structure can be expanded in 3D structure, which can break through the limitations of planar structure and improve the utilization of chip area.

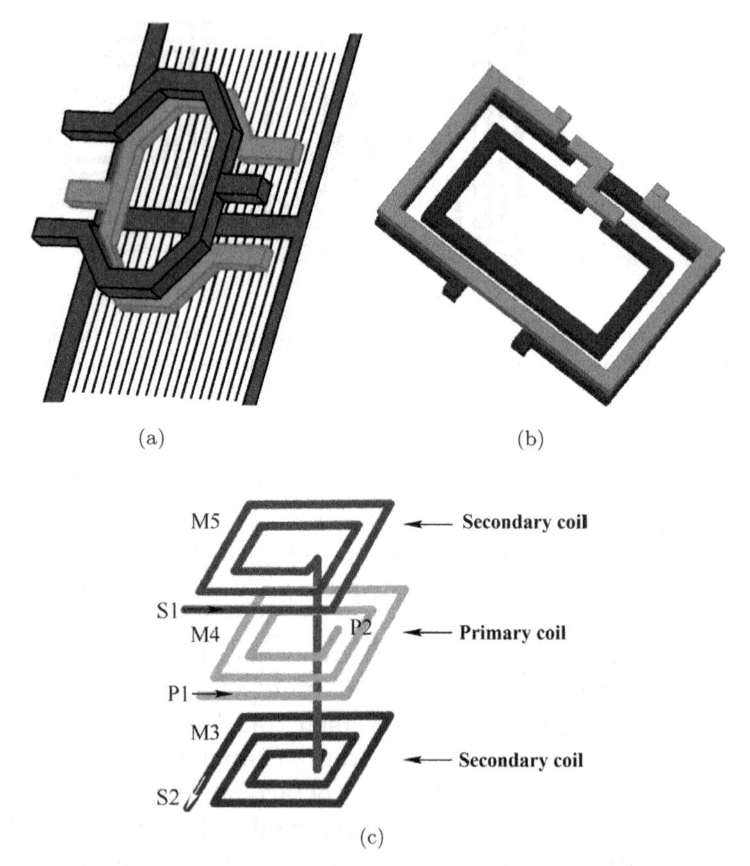

Figure 3.7. 3D stacked spiral transformer: (a) 1:1 octagonal-flipped stacked transformer; (b) 1:2 rectangular-flipped stacked transformer completed with two metal layers; (c) 1:2 square-flipped stacked transformer completed with two metal layers.

3.3 Equivalent Circuit Model of Transformer

In the computer-aided design of RF circuits, the equivalent circuit models of on-chip integrated transformers are very useful to understand the device physical mechanism. The accuracy of the models will directly affect the simulation accuracy and performance of the circuit. At the same time, the model also needs to have strong scalability and be able to cover the transformer design with changing geometric dimensions. Therefore, high-precision on-chip transformer model library will play a very important role in RF CMOS circuit

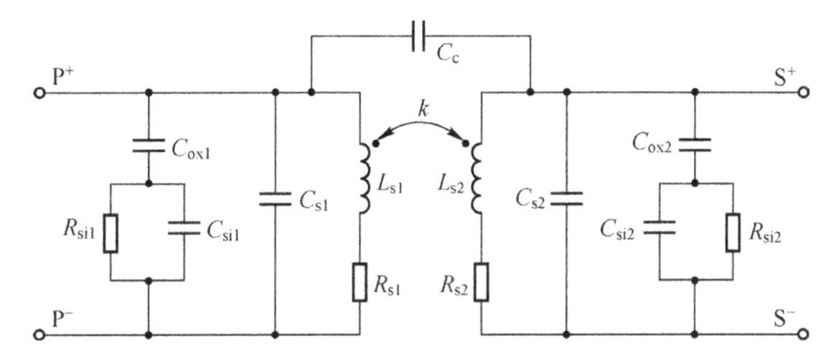

Figure 3.8. Single π equivalent circuit model.

design. The commonly used equivalent circuit models of on-chip integrated transformers include the single π model, double π model, and T model, which are introduced, respectively, in the following.

The single π equivalent circuit model is depicted in Figure 3.8 [10], where L_{s1} and L_{s2} are the intrinsic inductances (in the low-frequency ranges) of the primary and secondary windings, respectively. R_{s1} and R_{s2} represent the series metal resistance at low frequencies of the primary and the secondary coil, respectively. C_c models the parasitic capacitance between the primary and the secondary coils. C_{s1} and C_{s2} model the parasitic capacitance between metal wires of the primary and the secondary coils, respectively. Obviously, the single π model structure of the transformer is based on the inductance model. The C-R-C network is used to characterize the substrate coupling effect. There are inductive coupling and capacitive coupling between the primary and secondary coils of the on-chip transformer. The strength of the magnetic coupling between windings is indicated by the k-factor ($k = M/\sqrt{L_{s1}L_{s2}}$), where M is the mutual inductance between the primary and secondary windings.

At low frequencies, the series inductance and resistance of the primary and the secondary coils are smaller than the impedance of the coupling capacitance, hence a single capacitance can be used to characterize the capacitive coupling between coils. Nevertheless, in the millimeter frequency ranges, the distribution effect becomes more and more obvious, and the impedance of the coupling capacitance becomes smaller, while the impedances of the series inductances of the coils can be compatible or greater than that of the coupling capacitance. Therefore, in order to accurately characterize the high-frequency coupling effect between coils, it is necessary to

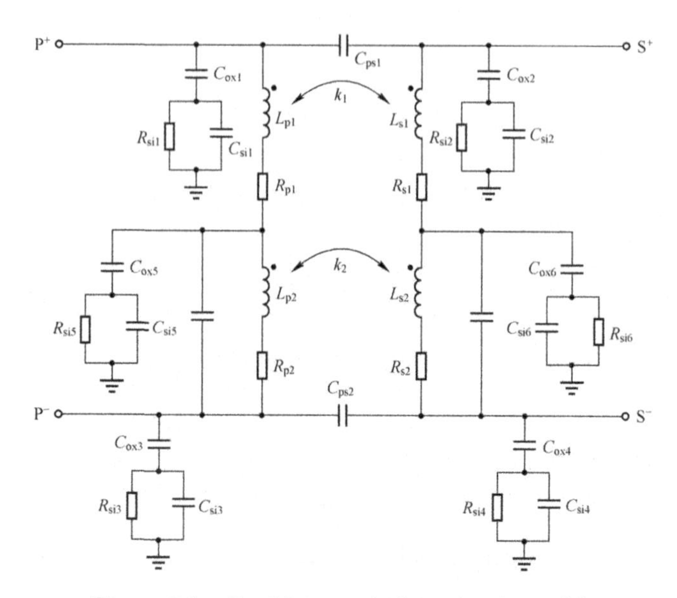

Figure 3.9. Double π equivalent circuit model.

add inductance and resistance to the coupling branch in the single π equivalent circuit model.

The single π model has the following advantages:

- simple circuit structure,
- easy model parameters extraction,
- less number of components.

But its shortcomings are also obvious, the accuracy between the modeled and measured Y parameters has been found to be unsatisfactory.

The double π equivalent circuit model is drawn in Figure 3.9 [11]. The equivalent circuit for two-coil transformers consists of a pair of double π subcircuits, accounting for each inductor coil (the primary and the secondary coils, respectively). In addition, these two coils are coupled by both mutual inductances and parasitic capacitances. Besides, the skin and proximity effects and the lossy capacitive substrate couplings are also well modeled. Since the primary and secondary coils of the transformer can be regarded as two coupled planar spiral inductors, respectively, so the inductance double

π model can consider the parasitics of individual coils and the coupling of magnetic field and electric field. The two input ports of the primary coil and the two input ports of the secondary coil of the model are symmetrical, respectively. From the transformer model, it is easy to find the center of the primary coil and the secondary coil, therefore this model also can be used to represent a transformer with the central tap.

Both the single π model and the double π model have physical meanings and are easy to understand. However, due to the magnetic coupling coefficient in the model, this model is not suitable for the computer-aided design; a decoupled circuit model is needed. Figures 3.10(a) and 3.10(b) show the basic transformer models: coupled circuit model and decoupled circuit model. The model is completed with a shunt magnetizing branch which is associated with the primary and secondary windings. The mutual inductance L_m in Figure 3.10(b) is used to replace the coupling coefficient k in Figure 3.10(a) to characterize the magnetic field coupling between coils [3].

Figure 3.11 depicts the equivalent circuit of a two-port transformer that includes the self-inductances of primary and secondary coils, mutual components, parasitic components, and substrate RC components [12]. This model considers the parasitic capacitance between the primary coil and the secondary coil and the influence of the substrate parasitic effect. L_m and R_m series branches are used to represent the coupling between the primary coil and secondary coil instead of the coupling coefficient, and they can be adjusted to fit the network parameters.

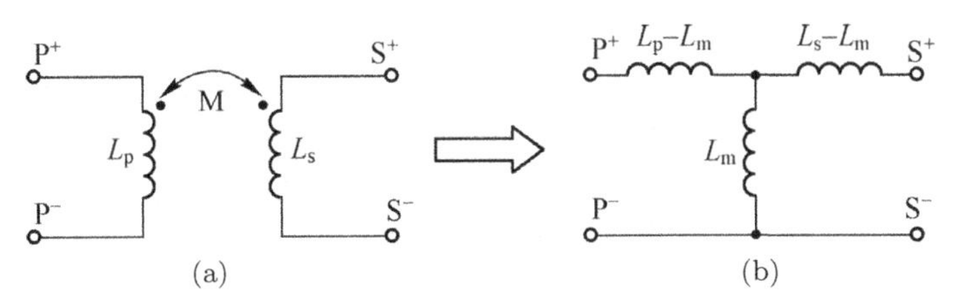

Figure 3.10. Basic transformer models: (a) coupled circuit model; (b) decoupled circuit model.

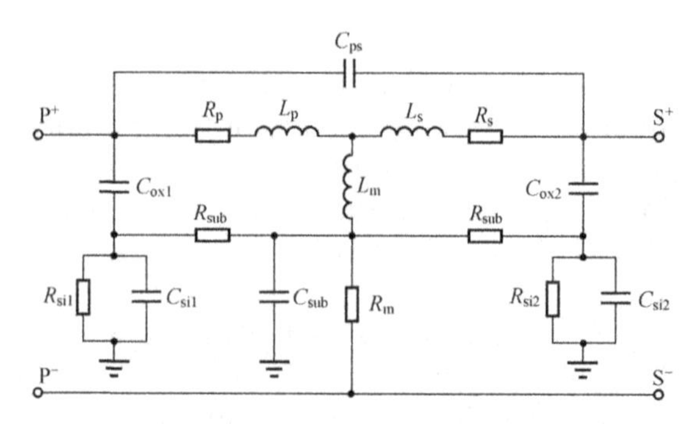

Figure 3.11. *T*-type model of transformer.

3.4 Transformer Design and Model Parameter Extraction

High-performance transformers mainly include the following features:

- low parasitic resistances in the main coil and secondary coils,
- high magnetic field coupling coefficient between the main coil and the secondary coils,
- small capacitive coupling between the main coil and the secondary coil,
- small capacitive coupling between the coils and the substrate.

Integrated transformers can present several different topologies. The most conventional distinction is between planar and stacked transformers. Compared with the planar spiral structure, the transformer with a stacked spiral structure has a larger coupling coefficient and smaller chip area, which has become one of the main forms of microwave and RF integrated circuit design. Hence, this section mainly discusses the structural design and model parameter extraction method of a stacked spiral structure transformer.

3.4.1 *Transformer design*

Transformers have the advantage of being very compact, impedance matching, and differential to single-ended conversion. In a multi-stage amplifier design, they can provide easy DC biasing through center-tap, without using DC block capacitance. Nevertheless, as the loss of the bulk silicon substrate, the quality factor is always a low quality

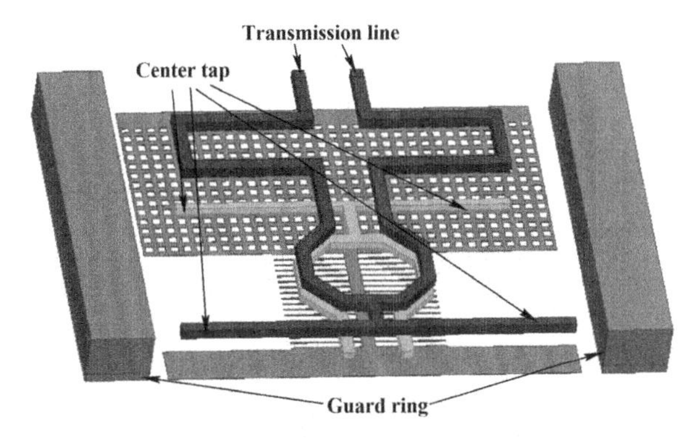

Figure 3.12. Stacked on-chip transformer with transmission line and guard ring.

factor. In order to reduce losses in the silicon substrate due to the time-varying electric field at a high frequency, one practical way is to employ an appropriate patterned ground shield (PGS). PGS provides a short terminal to the electric field leaking into the substrate.

The top view of stacked on-chip transformer with transmission line and guard ring is sketched in Figure 3.12 [13]. In order to maintain the symmetry of the transformer, the tap is led out from the center of the transformer and transmits signals to both sides of the transformer simultaneously. The added guard ring will decrease the coupling between the transformer and other devices. Transmission lines are connected to the transformer top metal layout directly without via holes to minimize connection loss [14–17].

The primary coil of the transformer is directly connected with the transmission line, which not only reduces the connection loss between the active device and the transformer but also completes the circuit matching. The shield layer is adopted under the transformer to reduce the influence of the substrate, therefore the corresponding quality factor of the transformer can be increased. The cubes on both sides of the transformer are the protection rings of the voltage transformer, which are used to reduce the interference of transformer signals to other modules through magnetic field coupling. If the output power of the power amplifier is too large, the high power will be coupled to other circuits (such as voltage-controlled oscillator and low noise amplifier) through the electromagnetic field, which will cause frequency drift and noise increase. In order to reduce the influence of the power amplifier on other circuits, the rectangular protection ring in Figure 3.12 is used to reduce the coupling of the magnetic field and electric field to other circuit modules.

Figure 3.13 depicts the three different structures of the stacked transformer: conventional transformer, transformer with guard ring structure not connected to PGS, and transformer with guard ring structure connected to PGS.

The electromagnetic simulation curve of the isolation between the transformer and feedline is illustrated in Figure 3.14. As seen, the transformer with no connection between PGS and guard ring has the best isolation. The isolation is 14 dB better than the conventional transformer structure, and 16.6 dB better than the transformer with PGS and guard ring connected structure. The reason is that the current generated by the electromagnetic field coupled to the PGS will flow directly to the guard ring, and the isolation of the transformer is reduced.

3.4.2 *Model parameter determination method*

The parameter extraction method for the on-chip stacked transformer is described here, which combines the advantages of the direct extraction method and optimization method. The chip layout of the stacked transformer is exhibited in Figure 3.15. The stacked transformer is fabricated by RF 130 nm CMOS process, including two layers of metal and pads that need to be connected with the measurement system.

The T-type equivalent circuit model of the on-chip stacked transformer is sketched in Figure 3.16. The equivalent circuit model can be divided into two parts: the intrinsic elements in the dashed box and the parasitic elements outside the dashed box.

The capacitances $C_{\text{ox}i}$ and $C_{\text{ox}o}$ represent the oxide capacitances between the metal segments and Si substrate. The capacitance C_{io} represents the capacitive coupling between the input and output ports. The substrate resistance $R_{\text{sub}i}$ and capacitance $C_{\text{sub}i}$ model the ohmic loss in the conductive silicon substrate at the input port. The substrate resistance $R_{\text{sub}o}$ and capacitance $C_{\text{sub}o}$ model the ohmic loss in the conductive silicon substrate at the output port.

For the intrinsic part, three independent inductors are used to model the coupling between the primary coil and the secondary coil, where L_p is the primary coil self-inductance, L_s is the secondary

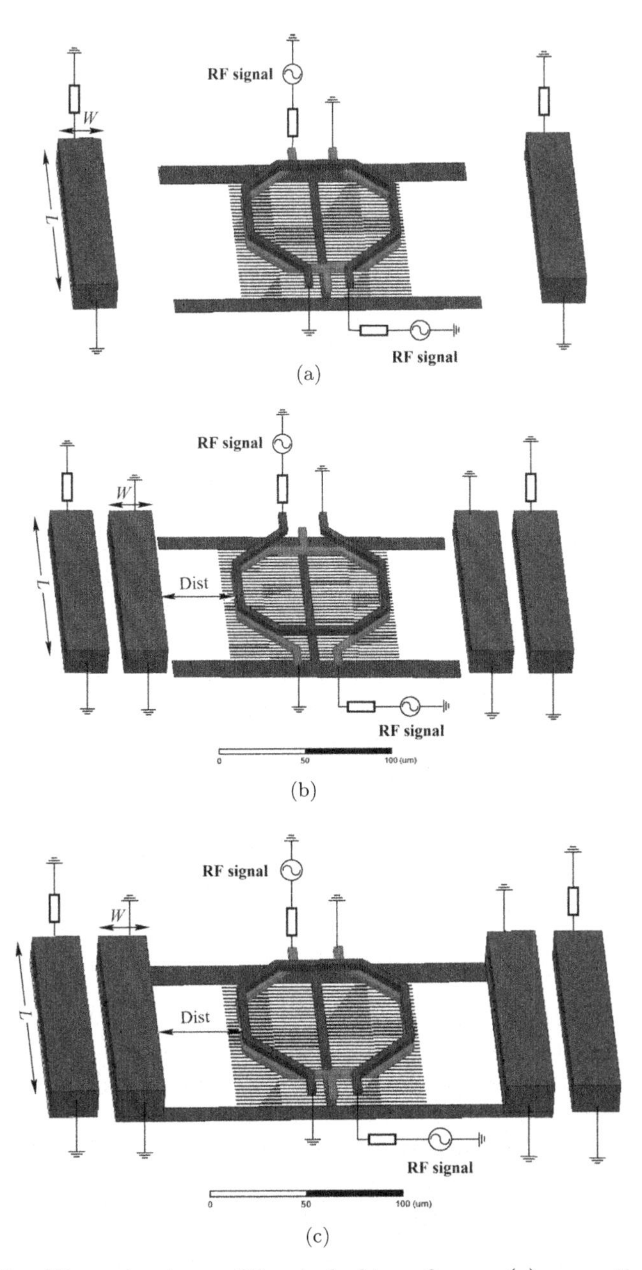

Figure 3.13. Three structures of the stacked transformer: (a) conventional transformer; (b) transformer with guard ring structure not connected with PGS; (c) transformer with guard ring structure connected to PGS.

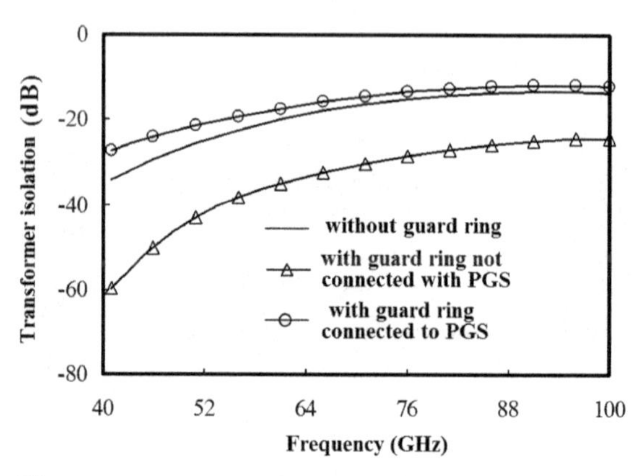

Figure 3.14. Electromagnetic simulation of the isolation between transformer and feedline.

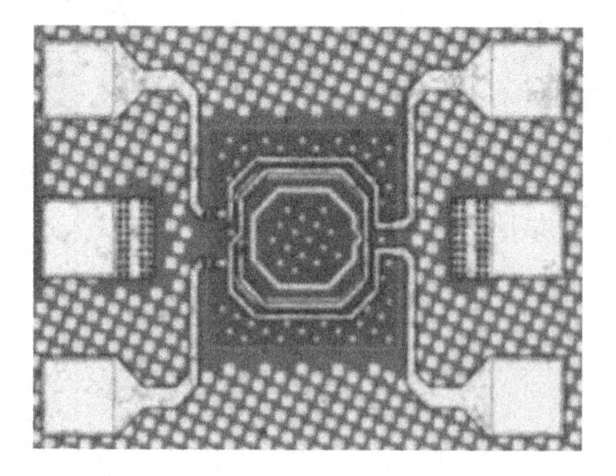

Figure 3.15. Die micrograph of stacked transformer.

coil self-inductance, and L_M is used to represent the mutual inductance between the two coils, which can represent the coupling degree between the primary and secondary coils. R_p and R_s represent the losses (including skin effect loss and proximity effect loss) of primary and secondary coils, respectively.

The on-chip transformer equivalent circuit parameter extraction methods mainly include direct model parameter extraction method and numerical analysis optimization model parameter extraction method. The direct extraction method is based on the measured S parameters. First, the influence of parasitic elements is removed, and then the elements versus frequency can be directly calculated

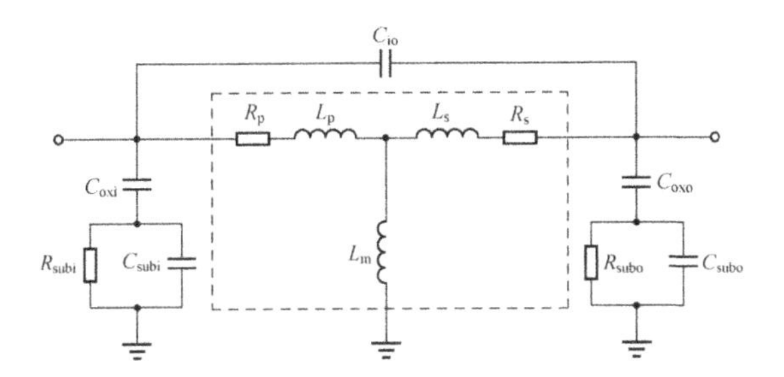

Figure 3.16. T-type model of stacked transformer.

by using the analytical expressions. The numerical optimization method is mainly to fit the measured S parameters of the transformer with the simulation data to obtain the best values of the elements. Although the method of directly extracting model parameters is very fast, it is easy to be affected by the measurement accuracy, further optimization is necessary to improve the accuracy. The numerical optimization method has higher requirements for the initial optimization value. The setting of the initial optimization value directly determines whether the parameter value of the device model has physical significance.

The flowchart of model parameter extraction is depicted in Figure 3.17. This is a mixed parameter extraction method combining the direct extraction method and optimization method. The parameter values obtained by the direct extraction method are used as the initial value of the optimization procedure leading to the final model parameters. This method may retain the advantages of the direct extraction method and optimization method and overcome their shortcomings, so as to accurately extract the equivalent circuit model parameters of the on-chip transformer. The extraction process is as follows:

(1) measurement of the S parameters of the open circuit test structure, determination of pad capacitance,
(2) de-embedding the parasitics, determination of the intrinsic parameters using analytical expressions,
(3) comparison of the measured S parameters and simulation results, calculation of calculating the accuracy,

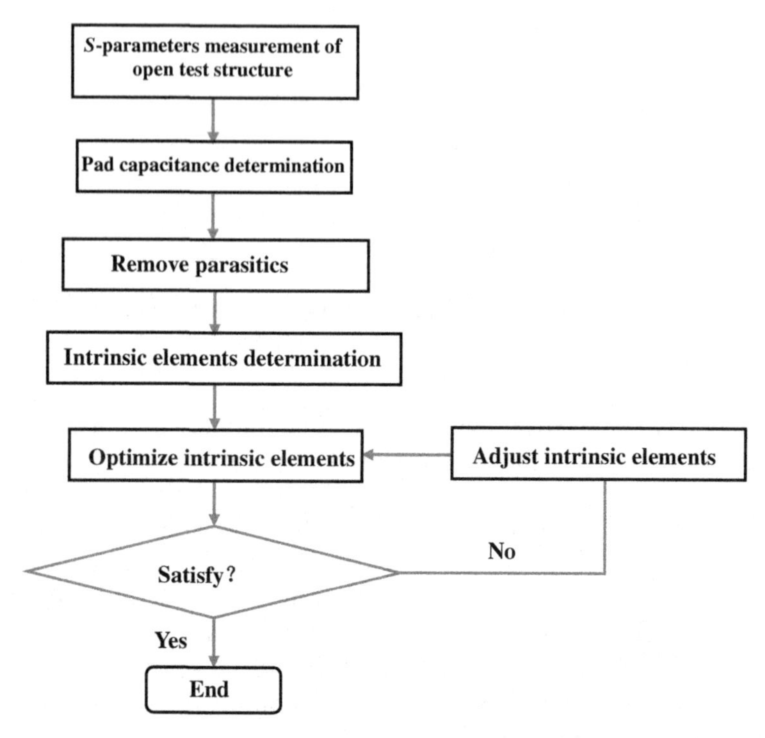

Figure 3.17. Flowchart of model parameter extraction.

(4) if the accuracy meets the requirements, the extraction process is completed; otherwise, adjustment of the intrinsic parameters is needed.

3.4.3 *Extrinsic elements determination*

De-embedding is one of the important technologies in microwave measurement. Its purpose is to eliminate the influence of parasitic components on the devices to be measured. The measurement system cannot be directly connected to the device under test (DUT); pads and metal feedlines are required to place between DUT and measurement. In order to test DUT with the microwave RF measurement instrument, a coplanar waveguide structure connected with the coaxial waveguide needs to be designed on the chip, which is composed of the input signal pad, output signal pad, and ground.

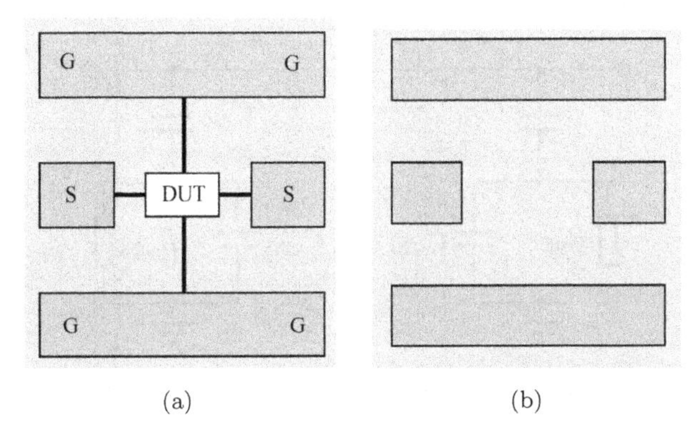

Figure 3.18. (a) Transformer test structure; (b) open test structure.

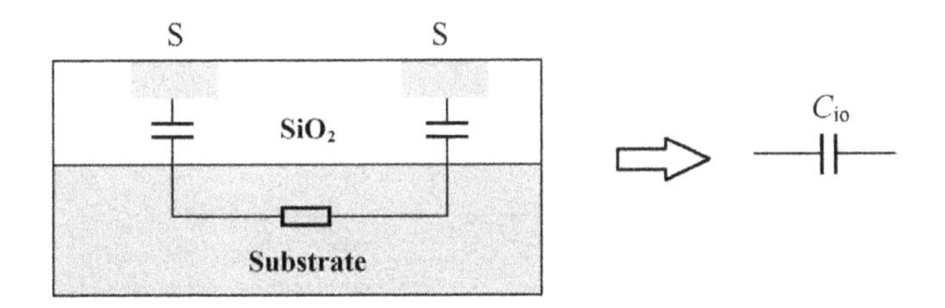

Figure 3.19. Coupled capacitance model between input and output ports.

The transformer test structure and corresponding open test structure are presented in Figures 3.18(a) and 3.18(b). It is clear that in order to obtain the performance of the transformer, the influence of pads and metal feedings on the test structure must be eliminated.

The coupled capacitance model between input and output ports is depicted in Figure 3.19. The corresponding equivalent circuit model of open test structure is illustrated in Figure 3.20.

Y-parameters of the open test structure can be expressed as follows:

$$Y_{11}^{\mathrm{o}} = Y_i + j\omega C_{io} \tag{3.1}$$

$$Y_{22}^{\mathrm{o}} = Y_o + j\omega C_{io} \tag{3.2}$$

$$Y_{12}^{\mathrm{o}} = Y_{21}^{\mathrm{o}} = -j\omega C_{io} \tag{3.3}$$

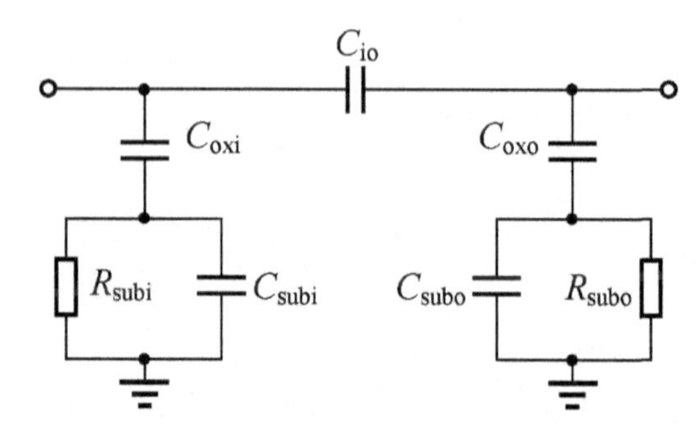

Figure 3.20. Equivalent circuit model of the open test structure.

with

$$Y_i = \frac{1}{\frac{1}{j\omega C_{\mathrm{ox}i}} + \frac{R_{\mathrm{sub}i}}{1+j\omega R_{\mathrm{sub}i} C_{\mathrm{sub}i}}}$$

$$Y_o = \frac{1}{\frac{1}{j\omega C_{\mathrm{ox}o}} + \frac{R_{\mathrm{sub}o}}{1+j\omega R_{\mathrm{sub}o} C_{\mathrm{sub}o}}}$$

where Y^{o}_{ij} $(i = 1, 2,\ j = 1, 2)$ are the Y-parameters; the relationship between S-parameters and Y-parameters are

$$Y^{\mathrm{o}}_{11} = Y_o \frac{(1 - S_{11})(1 + S_{22}) + S_{12}S_{21}}{(1 + S_{11})(1 + S_{22}) - S_{12}S_{21}} \tag{3.4}$$

$$Y^{\mathrm{o}}_{12} = Y_o \frac{-2S_{12}}{(1 + S_{11})(1 + S_{22}) - S_{12}S_{21}} \tag{3.5}$$

$$Y^{\mathrm{o}}_{21} = Y_o \frac{-2S_{21}}{(1 + S_{11})(1 + S_{22}) - S_{12}S_{21}} \tag{3.6}$$

$$Y^{\mathrm{o}}_{22} = Y_o \frac{(1 + S_{11})(1 - S_{22})S_{12}S_{21}}{(1 + S_{11})(1 + S_{22}) - S_{12}S_{21}} \tag{3.7}$$

where Y_{o} is the characteristic admittance.

The coupling capacitance can be determined from (3.3):

$$C_{\mathrm{io}} = -\frac{\mathrm{Im}(Y^{\mathrm{o}}_{12})}{\omega} \tag{3.8}$$

In the low-frequency ranges, the equations can be obtained from (3.1)–(3.3):

$$R_{\mathrm{sub}i} + \frac{1}{j\omega C_{\mathrm{ox}i}} = \frac{1}{Y_{11}^{\mathrm{o}} + Y_{12}^{\mathrm{o}}} \tag{3.9}$$

$$R_{\mathrm{sub}o} + \frac{1}{j\omega C_{\mathrm{ox}o}} = \frac{1}{Y_{22}^{\mathrm{o}} + Y_{12}^{\mathrm{o}}} \tag{3.10}$$

The oxide capacitances and substrate resistances can be determined as follows:

$$C_{\mathrm{ox}i} = -\frac{1}{\omega\,\mathrm{Im}\left(\frac{1}{Y_{11}^{\mathrm{o}}+Y_{12}^{\mathrm{o}}}\right)} \tag{3.11}$$

$$R_{\mathrm{sub}i} = \mathrm{Re}\left(\frac{1}{Y_{11}^{\mathrm{o}} + Y_{12}^{\mathrm{o}}}\right) \tag{3.12}$$

$$C_{\mathrm{ox}o} = -\frac{1}{\omega\,\mathrm{Im}\left(\frac{1}{Y_{22}^{\mathrm{o}}+Y_{12}^{\mathrm{o}}}\right)} \tag{3.13}$$

$$R_{\mathrm{sub}o} = \mathrm{Re}\left(\frac{1}{Y_{22}^{\mathrm{o}} + Y_{12}^{\mathrm{o}}}\right) \tag{3.14}$$

In the high-frequency ranges, the substrate capacitances ($C_{\mathrm{sub}i}$ and $C_{\mathrm{sub}o}$) can be determined from the imaginary parts of $1/(Y_{11}^{\mathrm{o}} + Y_{12}^{\mathrm{o}})$ and $1/(Y_{22}^{\mathrm{o}} + Y_{12}^{\mathrm{o}})$.

To illustrate the above model and parameter extraction method, a stacked on-chip transformer using 130 nm CMOS process has been used. The transformer has been designed with one turn as the primary coil on the first metal layer and two turns as the secondary on another layer. The metal width is 3 μm, the thickness of the primary coil is 1.325 μm, the thickness of the secondary coil is 3.3 μm, and the gap between the two metal layers is 1.45 μm. Figure 3.21 illustrates the variation of $C_{\mathrm{ox}i}$ and $C_{\mathrm{ox}o}$ with frequency calculated by analytical expression in the low-frequency ranges. As seen, the curve of capacitance $C_{\mathrm{ox}o}$ is very close to $C_{\mathrm{ox}i}$, which means the input port and output port are symmetrical. In the frequency range from 0.5 GHz to 8 GHz, the capacitance value of $C_{\mathrm{ox}i}$ varies from 12 fF to 15 fF, and the capacitance $C_{\mathrm{ox}o}$ varies from 10 fF to 15 fF, and their fluctuations are very small.

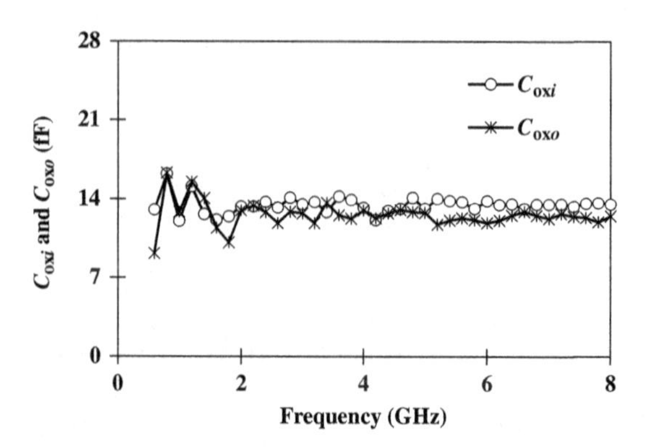

Figure 3.21. Capacitances $C_{\text{ox}i}$ and $C_{\text{ox}o}$ versus frequency.

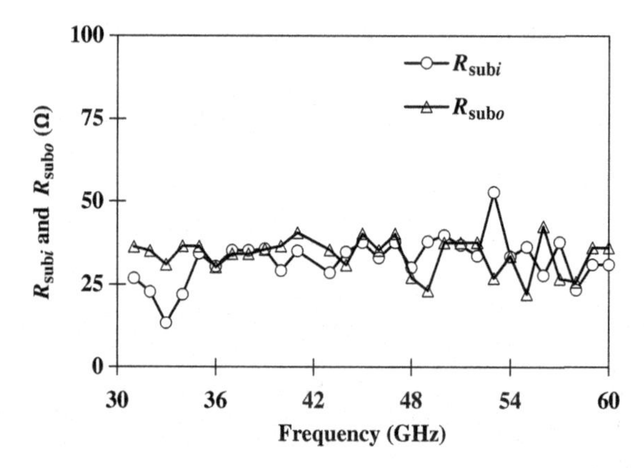

Figure 3.22. Resistances $R_{\text{sub}i}$ and $R_{\text{sub}o}$ versus frequency.

The variation of parasitic resistances $R_{\text{sub}i}$ and $R_{\text{sub}o}$ calculated from analytical expression at high frequencies is plotted in Figure 3.22. It can be seen that the resistances can be extracted only in the millimeter wave band, and their variation range can be used as the initial values for further optimization.

After extracting the initial value of parasitic parameters of the equivalent circuit model elements, in order to further improve the accuracy of the equivalent circuit model, the subsequent optimization of each parameter value is carried out in a small range on the initial value. Table 3.1 tabulates the comparison between the extracted values and the optimized values of the open circuit test model parameters.

Table 3.1. Comparison between extraction values and optimized values.

Parameters	Extraction values	Optimized values	Units
C_{io}	0.06	0.04	fF
$C_{\text{ox}i}$	13.4	13.5	fF
$C_{\text{ox}o}$	13.2	13.0	fF
$R_{\text{sub}i}$	36	35	Ω
$R_{\text{sub}o}$	35.5	35	Ω
$C_{\text{sub}i}$	44	43	fF
$C_{\text{sub}o}$	49	50.1	fF

Figure 3.23. (a) illustrates the comparison of the absolute errors before and after the optimization of S_{11} parameter of the equivalent circuit model for open test structure, and the comparison of the absolute errors for S_{22} is exhibited in Figure 3.23 (b). As seen, the error of S_{11} is less than 0.2% after optimization. For S_{22}, after optimization, the error is reduced to less than 0.3%.

In the frequency ranges of 0.1–60 GHz, the comparison between the measured and modeled S-parameters is depicted in Figure 3.24. Figures 3.24(a) and 3.24(b) are the S_{11} amplitude and phase, and (c) and (d) are the comparison curves of S_{22} amplitude and phase. Good agreement is obtained to prove the accuracy of the equivalent circuit model of open test structure.

3.4.4 *Intrinsic elements determination*

Once the circuit model parameters of the open test structure are obtained, the intrinsic model parameters can be directly obtained by using the analytical expression [18]. The pad network and the DUT network are in parallel, so de-embedding is needed:

$$Y^{\text{DUT}} = Y^{\text{M}} - Y^{\text{O}} \tag{3.15}$$

where Y^{DUT} is the Y parameter of DUT (on-chip transformer), Y^{M} is the measured Y parameter, and Y^{O} is the Y parameter of the open test structure.

A convenient de-embedding method is to use the negative element method to eliminate the influence of parasitic elements on device

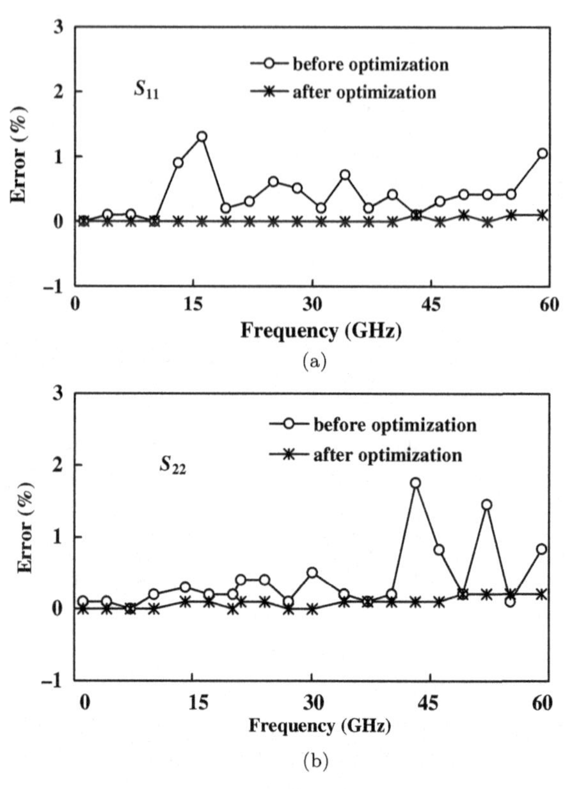

Figure 3.23. Comparison of the absolute errors between before and after the optimization in the frequency ranges of 0.1–60 GHz: (a) S_{11}; (b) S_{22}.

characteristics. The corresponding circuit topology using negative elements is presented in Figure 3.25. The DUT is the intrinsic structure of the device, the Y parameter of the input network is $-Y_i$, and the Y parameter of the output network is $-Y_o$. The intrinsic part of the equivalent circuit model for the transformer is illustrated in Figure 3.26, which is a typical T-type structure.

According to the equivalent circuit model in Figure 3.26, the open circuit Z parameter can be expressed as

$$Z_{11} = R_p + j\omega(L_p + L_m) \tag{3.16}$$

$$Z_{22} = R_s + j\omega(L_s + L_m) \tag{3.17}$$

$$Z_{12} = Z_{21} = j\omega L_m \tag{3.18}$$

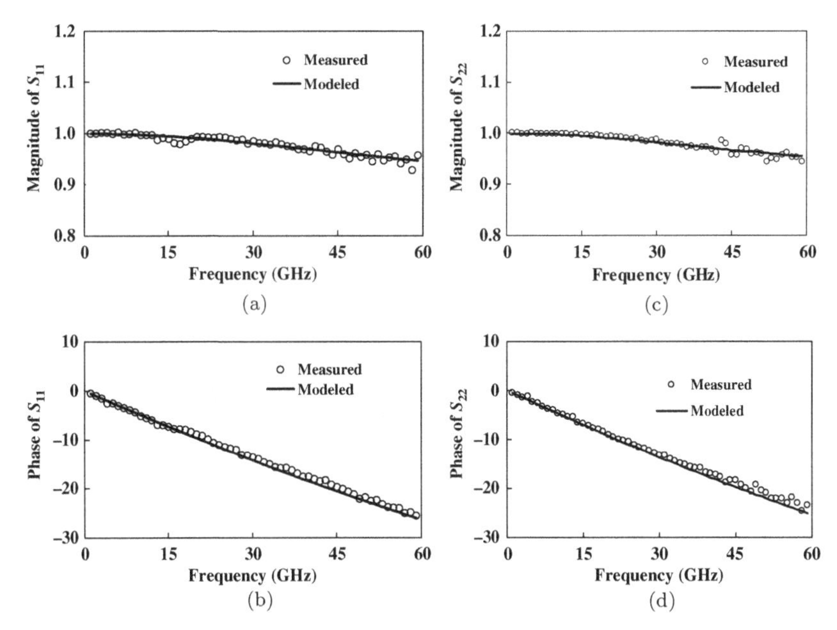

Figure 3.24. Comparison between the measured and modeled S-parameters in the frequency ranges of 0.1–60 GHz: (a) magnitude of S_{11}; (b) phase of S_{11}; (c) magnitude of S_{22}; (d) phase of S_{22}.

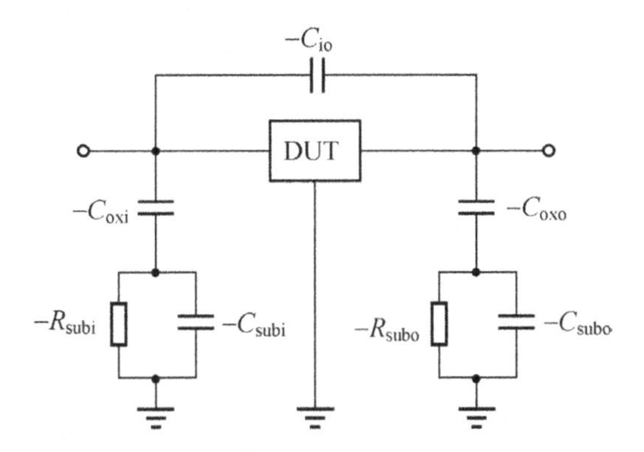

Figure 3.25. De-embedding procedure using negative element.

Three intrinsic inductances can be calculated directly from the imaginary parts of Z parameters:

$$L_p = \frac{\text{Im}(Z_{11} - Z_{12})}{\omega} \tag{3.19}$$

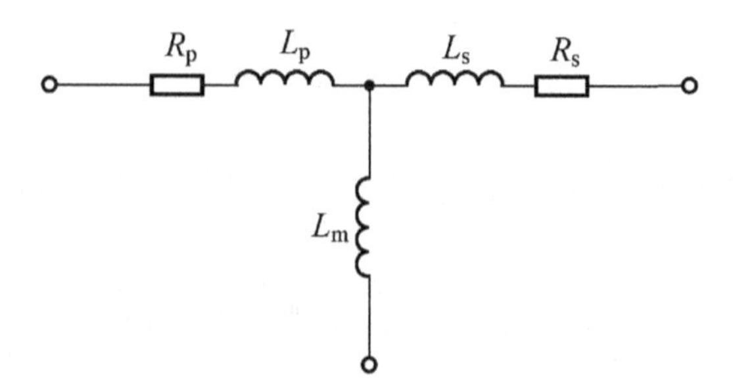

Figure 3.26. Intrinsic part of equivalent circuit model for transformer.

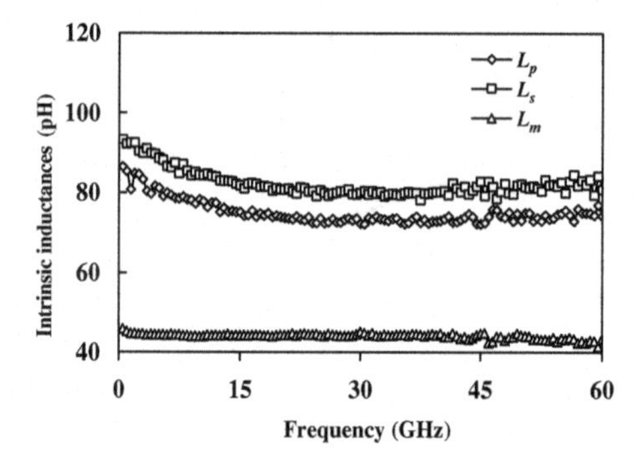

Figure 3.27. Extracted inductances L_p, L_s, and L_m versus frequency.

$$L_s = \frac{\text{Im}(Z_{22} - Z_{12})}{\omega} \tag{3.20}$$

$$L_m = \frac{\text{Im}(Z_{12})}{\omega} \tag{3.21}$$

Two intrinsic resistances can be calculated directly from the real parts of Z parameters:

$$R_p = \frac{\text{Re}(Z_{11})}{\omega} \tag{3.22}$$

$$R_s = \frac{\text{Re}(Z_{22})}{\omega} \tag{3.23}$$

The extracted inductances L_p, L_s, and L_m versus frequency are plotted in Figure 3.27. It can be observed that the inductance values are the most constants. L_p is about 82 pH, L_s is about 77 pH, and L_m is 44 pH roughly. Thence, the self-inductance of the primary coil is the sum of L_p and L_m, i.e., 126 pH. The self-inductance of the secondary coil is the sum of L_s and L_m, i.e., 121 pH.

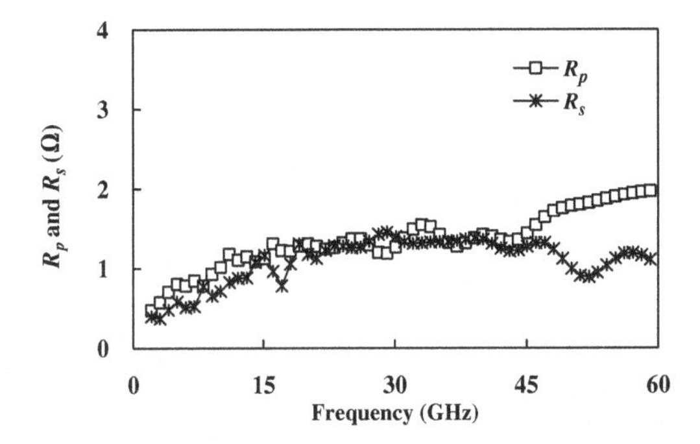

Figure 3.28. Extracted resistances R_p and R_s versus frequency.

Table 3.2. Intrinsic model parameters.

Parameters	Values	Units
L_p	75	pH
L_s	80	pH
L_m	44.1	pH
R_p	1.44	Ω
R_s	1.32	Ω

The extracted resistances R_p and R_s versus frequency are exhibited in Figure 3.28. As seen, the variation of the resistances R_p and R_s is between 1–2 ohms. After extracting the intrinsic parameters of the on-chip spiral transformer, in order to further improve the accuracy of the equivalent circuit model, the extracted parameters' values are used as the initial values and subsequent optimization is needed. The final intrinsic circuit model parameters are tabulated in Table 3.2. In the frequency ranges from 0.1 GHz to 60 GHz, the comparison between the modeled and measured S parameters of the stacked transformer is demonstrated in Figure 3.29. Figures 3.29(a) and 3.29(b) are the S_{11} amplitude and phase, (c) and (d) are the comparison curves of S_{22} amplitude and phase, and (e) and (f) are the comparison curves of S_{12} amplitude and phase. Good agreement is obtained to prove the model accuracy of the transformer.

3.5 Comparison of Commonly Used Transformers

According to the winding style, the structure of an on-chip spiral transformer can be divided into three categories: center

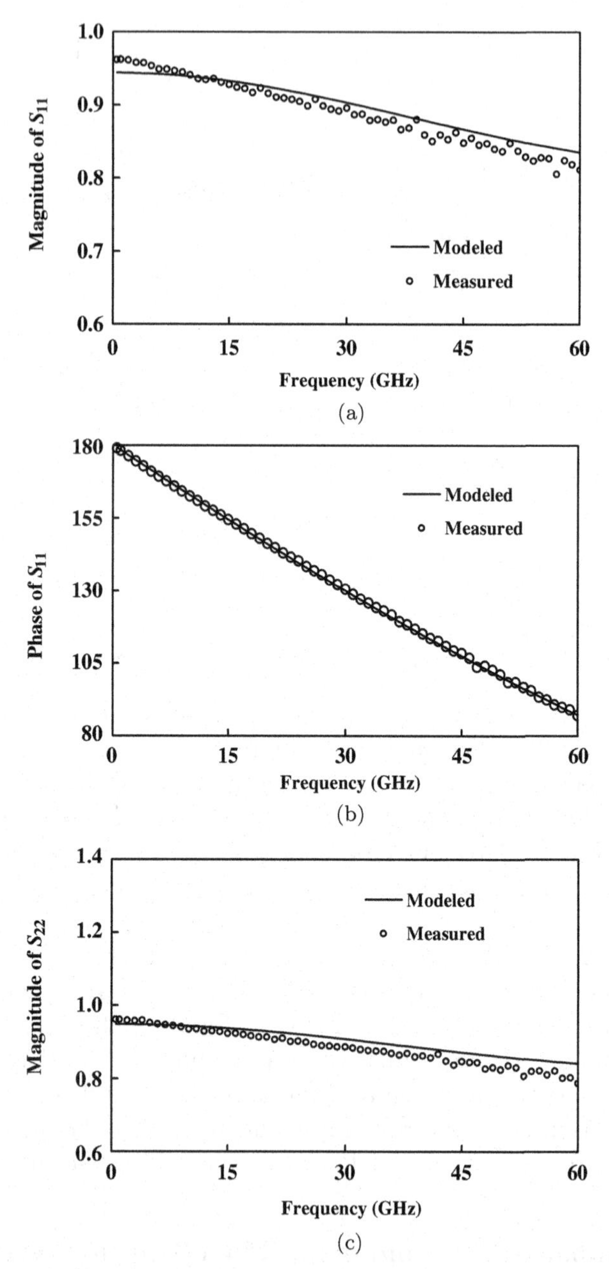

Figure 3.29. Comparison between the modeled and measured S parameters of stacked transformer: (a) magnitude of S_{11}; (b) phase of S_{11}; (c) magnitude of S_{22}; (d) phase of S_{22}; (e) magnitude of S_{12}; (f) phase of S_{12}.

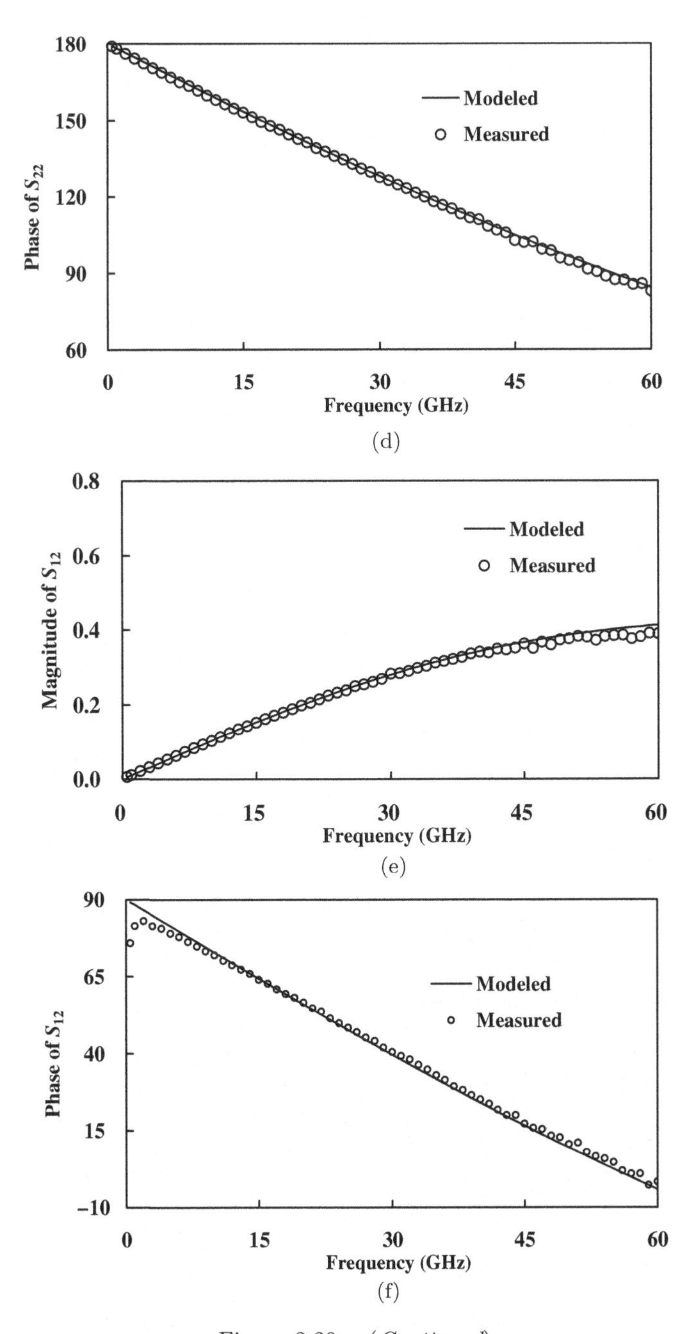

Figure 3.29. (*Continued*)

tap transformer, interleaved transformer, and stacked transformer. Obviously, different structures have their own unique advantages in self-inductance, coupling coefficient, series resistances of primary and secondary coils, resonant frequencies, and chip area occupied. In order to design on-chip spiral transformers suitable for different circuits, it is particularly important to understand the influence of spiral structure on various performance parameters for transformers. In this section, the simulation results of three typical spiral structures are given and discussed.

The transformer with a central tap structure designed based on electromagnetic simulation software is depicted in Figure 3.30. The primary coil is designed on the top metal (aluminum) and 9.585 μm away from the substrate. The metal width of the primary coil is 3 μm, the spacing is 2 μm, the thickness is 1.325 μm, and the inner diameter is 30 μm. The number of turns of the primary coil is 3, and the ratio of turns is 1:1.

The interleaving transformer structure is sketched in Figure 3.31. The geometric parameters are similar to the structure of the center tap. The top metal is used for the primary coil, which is made of aluminum and 9.585 μm away from the substrate. The metal width of the primary coil is 3 μm, the spacing is 2 μm, the inner diameter is 30 μm, and the thickness is 1.325 μm. The number of turns of the primary coil is 3 μm, and the ratio of turns is 1:1.

The stacked structure transformer is sketched in Figure 3.32, which is slightly different from the central tap and interleaved transformers. The primary coil is designed on the top metal and the secondary coil on the subtop metal, respectively. The primary coil uses

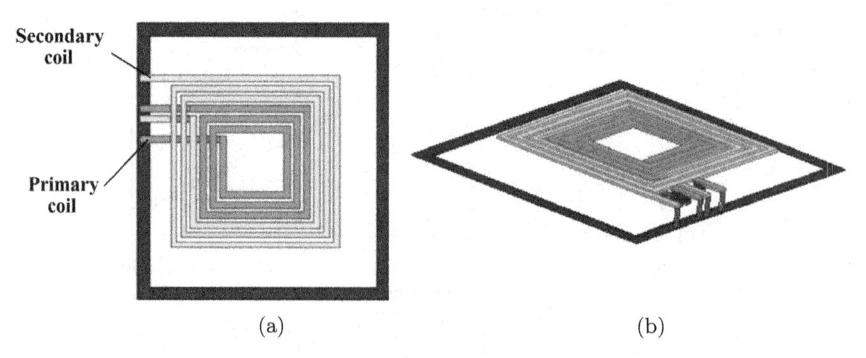

Figure 3.30. Center tap transformer: (a) top view; (b) cubic view.

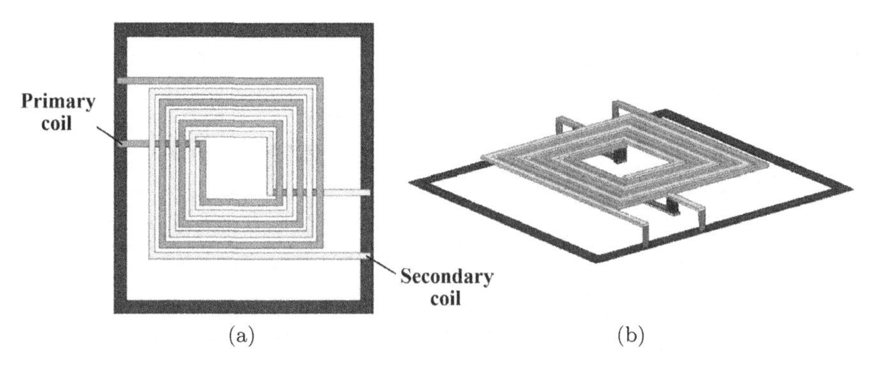

Figure 3.31. Interleaved transformer: (a) top view; (b) cubic view.

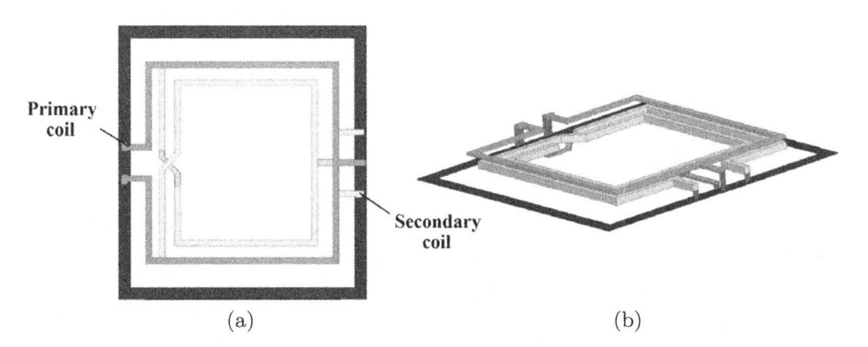

Figure 3.32. Stacked transformer: (a) top view; (b) cubic view.

the top metal, which is fabricated by aluminum with a thickness of 1.325 μm, the metal width is 3 μm, the spacing is 2 μm, the inner diameter is 96 μm, and the number of turns is 1. The secondary coil uses the subtop metal, which is fabricated by copper with a thickness of 3.3 μm, the metal width is 3 μm, the metal spacing is 2 μm, the inner diameter is 78 μm, and the number of turns is 2.

The magnetic coupling k-factors versus the frequency of three transformers are plotted in Figure 3.33. As seen, the variation of coupling coefficient of interleaved transformer is the smallest and k-factor is the largest, the stacked transformer in the middle and center tap structure is the worst. The main reason is that the primary and secondary coils of the interleaved transformer are close to each other, while the primary and secondary coils of the center tap structure are far away.

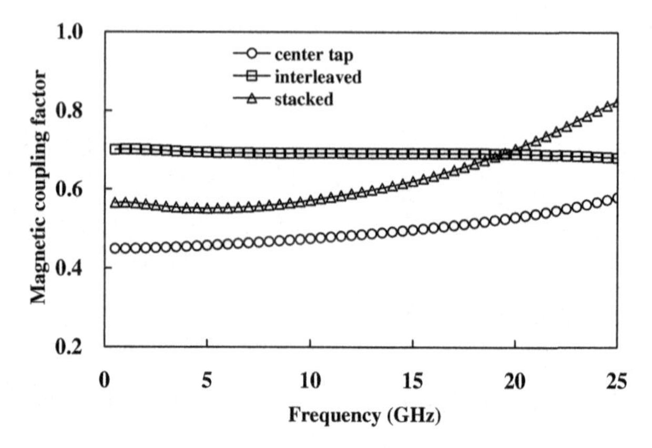

Figure 3.33. Magnetic coupling factors versus frequency of three transformers.

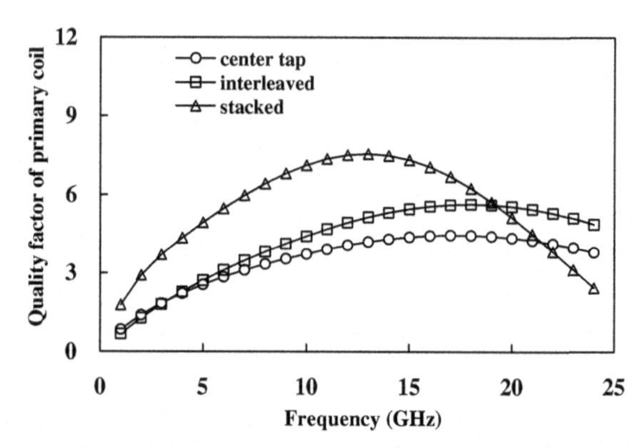

Figure 3.34. Quality factors of the primary coil versus frequency for three transformers.

The quality factors Q_p of the primary coil versus the frequency for three transformers is depicted in Figure 3.34, and it can be seen that the stacked structure transformer works best. The transmission coefficients versus the frequency of three on-chip spiral transformers are exhibited in Figure 3.35. The interleaved transformer works best. The comparison of three on-chip spiral transformers is tabulated in Table 3.3. L_p and L_s are the low-frequency self-inductance of primary and secondary coils, respectively, and $Q_{p\max}$ and $Q_{s\max}$ are the maximum quality factors of primary and secondary coils, respectively.

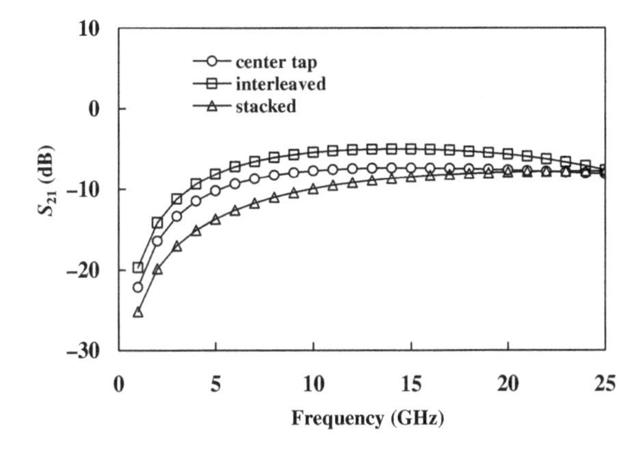

Figure 3.35. Transmission coefficients versus frequency of three transformers.

Table 3.3. Comparison of three on-chip spiral transformers.

Structures	k	$L_p(n\mathrm{H})$	$L_s(n\mathrm{H})$	$Q_{p\max}$	$Q_{s\max}$
Center tap	0.45	1.11	0.57	4.45	3.47
Interleaved	0.70	0.82	0.81	5.64	5.37
Stacked	0.56	0.34	0.86	7.56	4.60

3.6 Effect of Physical Structure Dimensions

This section mainly discusses the influence of the geometric structure parameters of the transformer on the coupling coefficient and transmission characteristics of the transformer, including the two most commonly used structures of interleaved transformer and stacked transformer [18,19].

3.6.1 *Interleaved transformer*

The geometric structure parameters of an interleaved transformer are illustrated in Figure 3.36. The coil conductor width and conductor spacing are w and s, respectively. ID is the inner diameter, and n is the number of turns. Tables 3.4– 3.7 summarize the variation parameters and fixed geometric parameters of interleaved transformers with different metal widths, metal spacings, inner diameters, and number of turns, respectively.

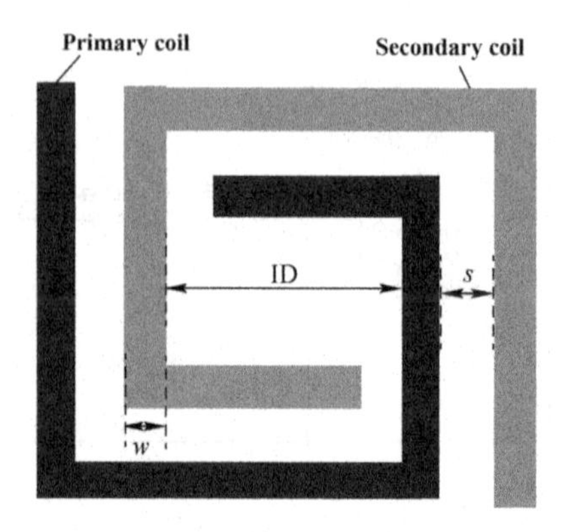

Figure 3.36. Geometric structure parameters of interleaved transformer.

Table 3.4. Interleaved transformer with different metal widths.

Variation parameter			Fixed parameters		
Parameter	Range	Step	n	s (μm)	ID (μm)
w (μm)	2.5–5.5	1	2	2	30

Table 3.5. Interleaved transformer with different metal spacings.

Variation parameter			Fixed parameters		
Parameter	Range	Step	n	w (μm)	ID (μm)
s (μm)	1–4	1	2	3	30

Table 3.6. Interleaved transformer with different inner diameters.

Variation parameter			Fixed parameters		
Parameter	Range	Step	n	w (μm)	s (μm)
ID (μm)	20–50	10	2	3	2

The magnetic coupling factor and quality factor of primary coil versus coil width are plotted in Figure 3.37. It can be clearly seen that the coupling coefficient decreases slightly with increased metal width in the low-frequency range. Nevertheless, in the high-frequency

Table 3.7. Interleaved transformer with different number of turns.

Variation parameter			Fixed parameters		
Parameter	Range	Step	ID (μm)	w (μm)	s (μm)
n	1–4	1	20	3	2

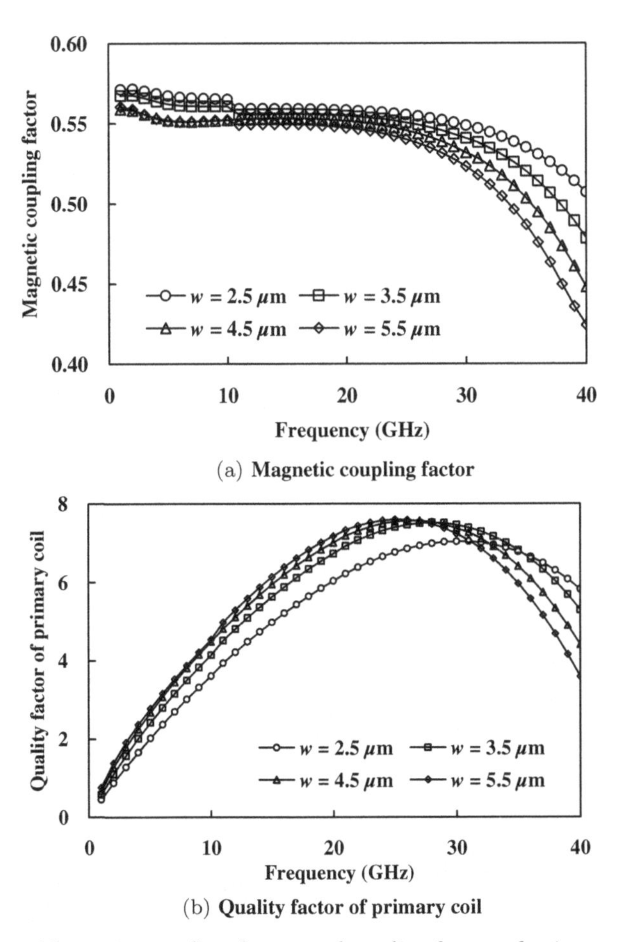

Figure 3.37. Magnetic coupling factor and quality factor of primary coil versus coil width.

ranges, the coupling coefficient decreases rapidly with the increase in metal width.

Table 3.8. Effect of coil width for interleaved transformer.

$w\,(\mu\text{m})$	k	L_p (nH)	L_s (nH)	Q_{pmax}	Q_{smax}
2.5	0.56	0.38	0.37	7.04	7.27
3.5	0.55	0.39	0.38	7.52	7.73
4.5	0.54	0.39	0.38	7.57	7.79
5.5	0.54	0.40	0.39	7.60	7.86

Table 3.8 summarizes the effect of metal width on the magnetic coupling factor and quality factor of the primary coil. As seen, the metal width has little effect on the performance of the interleaved transformer, including the self-inductance of the primary and secondary coils.

The magnetic coupling factor and quality factor of primary coil versus conductor spacing are illustrated in Figure 3.38. It is significant to observe that the magnetic coupling factors increase with decreased conductor spacing. The magnetic coupling factors decrease from 0.60 (spacing is $1\,\mu$m) to 0.47 (spacing is $4\,\mu$m). It is worth noting that the magnetic coupling factor of the on-chip transformer is limited by the process.

Table 3.9 summarizes the effect of conductor spacing on the magnetic coupling factor and quality factor of the primary coil. The quality factors of transformer primary coils can be improved by increasing conductor spacing, and the quality factors tend to saturation with the increase in spacing.

The magnetic coupling factor and quality factor of the primary coil versus inner diameter are depicted in Figure 3.39. It can be seen that the coupling coefficients increase with the increase in inner diameter. The coupling coefficient increases from 0.5 (inner diameter is $20\,\mu$m) to 0.65 (spacing is $50\,\mu$m). It is worth noting that the area occupied by the transformer increases when the inner diameter increases.

Table 3.10 summarizes the effect of conductor spacing on the magnetic coupling factor and quality factor of the primary coil. With the increase in coil inner diameter, the self-inductance of primary and secondary coils increases, and the quality factors decrease simultaneously.

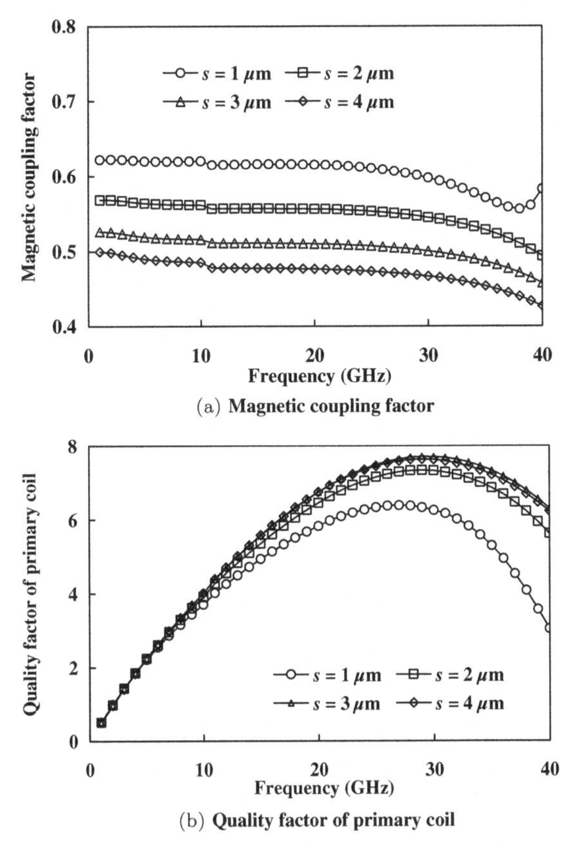

(a) Magnetic coupling factor

(b) Quality factor of primary coil

Figure 3.38. Magnetic coupling factor and quality factor of primary coil versus conductor spacing for interleaved transformer.

Table 3.9. Effect of conductor spacing for interleaved transformer.

s (μm)	k	L_p (nH)	L_s (nH)	Q_{pmax}	Q_{smax}
1	0.62	0.37	0.36	6.39	6.57
2	0.57	0.39	0.37	7.34	7.55
3	0.52	0.41	0.39	7.68	7.85
4	0.49	0.43	0.41	7.64	7.93

The magnetic coupling factor and quality factor of the primary coil versus the number of turns are exhibited in Figure 3.40. Since the coil ratio of the transformers is 1:1, which means the number

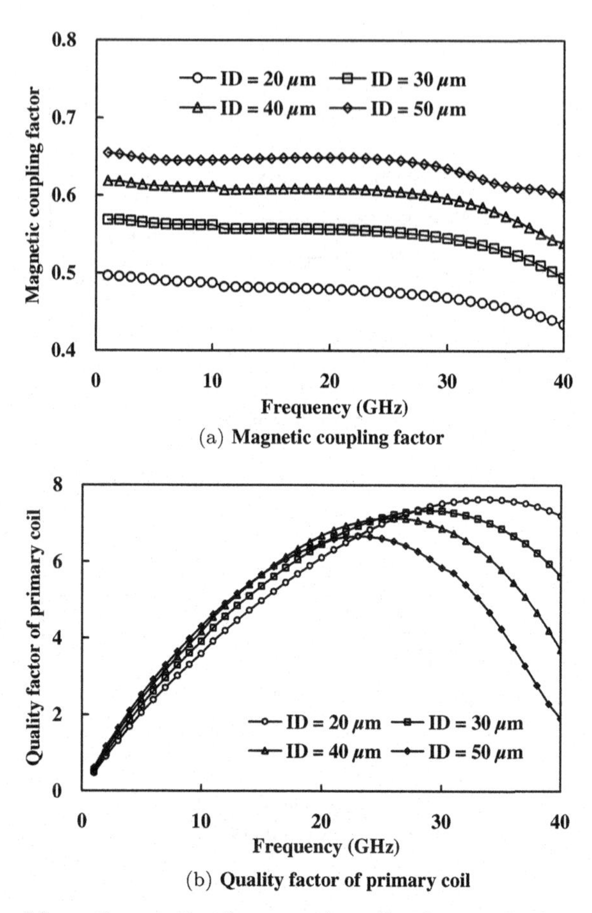

(a) **Magnetic coupling factor**

(b) **Quality factor of primary coil**

Figure 3.39. Magnetic coupling factor and quality factor of primary coil versus inner diameter for interleaved transformer.

of turns of the primary coil is equal to the number of turns of the secondary coil (one turn for both primary and secondary coils). The number of turns determines the magnetic flux passing through the primary and secondary coils. Too few coils will result in magnetic coupling factors that are too small, and too many turns will result in an oversized transformer, increasing loss and reducing performance.

The effect of the number of turns on magnetic coupling factors and quality factors of the primary coil is summarized in Table 3.11. Obviously, with the increase in the number of coils, the coupling coefficient increases almost linearly, and the self-inductances of the

Table 3.10. Effect of inner diameter for interleaved transformer.

ID (μm)	k	L_p (nH)	L_s (nH)	$Q_{p\max}$	$Q_{s\max}$
20	0.50	0.29	0.28	7.63	7.84
30	0.55	0.39	0.37	7.34	7.55
40	0.61	0.50	0.49	7.14	7.36
50	0.65	0.64	0.62	6.67	6.87

Table 3.11. Effect of the number of turns for interleaved transformer.

n	k	L_p (nH)	L_s (nH)	$Q_{p\max}$	$Q_{s\max}$
1	0.21	0.11	0.10	5.30	5.30
2	0.48	0.28	0.27	6.37	6.42
3	0.64	0.60	0.59	5.82	5.90
4	0.74	1.22	1.20	4.82	4.94

primary and secondary coils increase rapidly, while the maximum value of the quality factor fluctuates slightly.

3.6.2 *Stacked transformer*

In order to analyze the influence of the geometric structure parameters on the performance of the stacked transformer, a group of stacked transformers with a fixed number of turn ratio of 1:1 (primary and secondary coils are 1 or 2 turns) is designed, and the corresponding structures are shown in Figure 3.41. The coil width, inner diameter ID, and the number of turns n are mainly deliberated here.

For laminated on-chip transformers, the variation of magnetic coupling factors and quality factors of primary coils with coil width (double turns) are given in Figure 3.42. It can be seen that the magnetic coupling factors rise with the increase in frequency and slightly increase with the increase in coil width. The optimum operating frequency is the frequency corresponding to the maximum value of the quality factor for the primary coil. Tables 3.12 and 3.13 tabulate the effect of coil width on the performance of stacked transformer (single turn and double turns). The inductances of primary and secondary

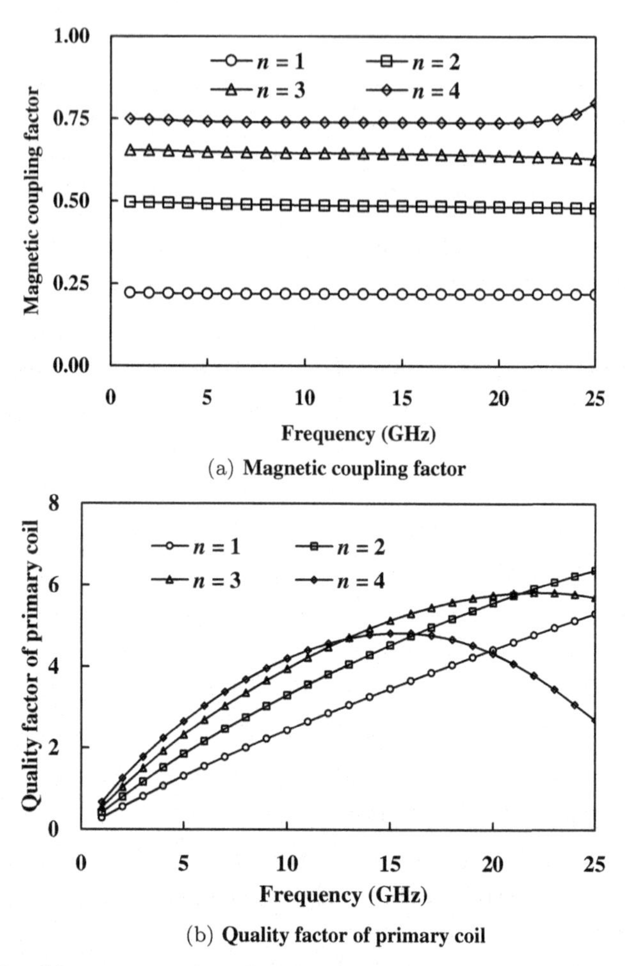

Figure 3.40. Magnetic coupling factor and quality factor of primary coil versus the number of turns for interleaved transformer.

coils decrease slightly with the increase in coil width and the variations of maximum quality factors of primary and secondary coils are very small.

The magnetic coupling factor and quality factor of primary coil versus inner diameter for stacked transformer are given in Figure 3.43, and Table 3.14 summarizes the effect of inner diameter on the performance of stacked transformer (double turns). It is significant to observe that the magnetic coupling factors rise with the

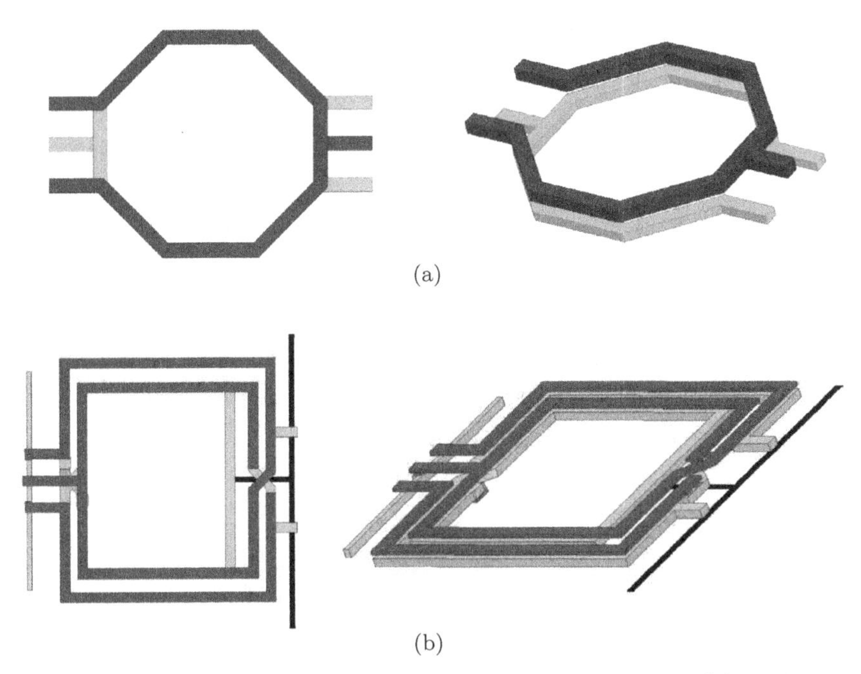

Figure 3.41. Top view and cubic view of stacked transformers: (a) single turn structure; (b) double turns structure.

Table 3.12. Effect of coil width of stacked transformer (single turn).

w (μm)	k	L_p (nH)	L_s (nH)	Q_{pmax}	$Q_{s\text{max}}$
2.5	0.56	0.12	0.13	22.6	14.4
3.5	0.58	0.11	0.11	23.5	14.7
4.5	0.60	0.11	0.11	24.3	15.6
5.5	0.61	0.10	0.10	24.8	16.6

increase in frequency, and increase with the increase in inner diameter, and the variations of maximum quality factors of primary and secondary coils are very small.

The magnetic coupling factor and quality factor of primary coil versus the number of turns for stacked transformer are shown in

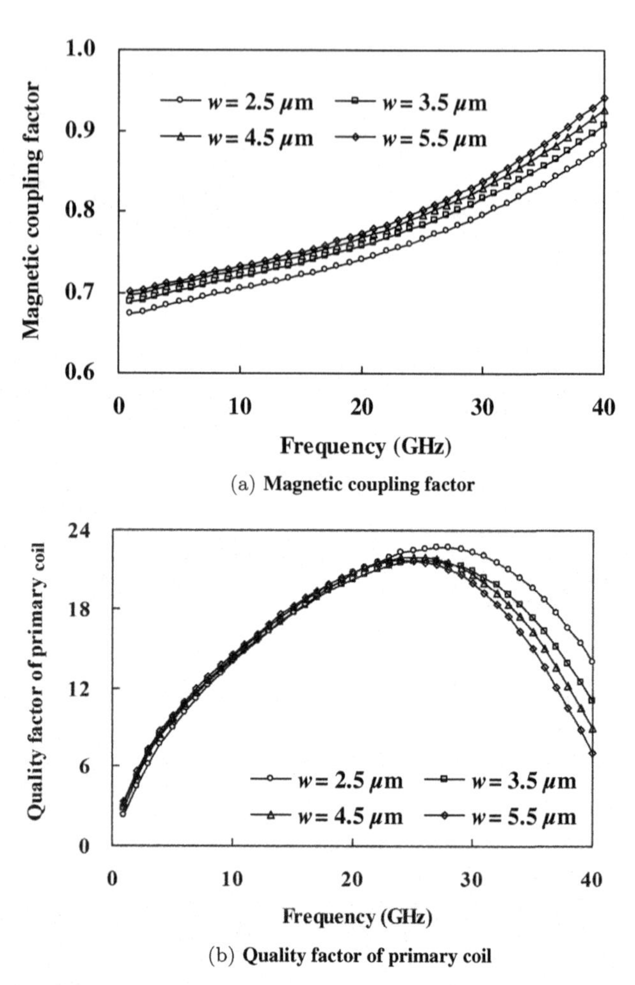

(a) **Magnetic coupling factor**

(b) **Quality factor of primary coil**

Figure 3.42. Magnetic coupling factor and quality factor of primary coil versus coil width (double turns).

Figure 3.44. Moreover, the effect of the number of turns on the performance of the stacked transformer is tabulated in Table 3.15. As seen, the magnetic coupling factors increase with increased frequency and the number of turns. The inductances of primary and secondary coils increase with the increase in the number of turns, and the quality factors of primary and secondary coils decrease with the increase in the number of turns rapidly.

Table 3.13. Effect of coil width of stacked transformer (double turns).

w (μm)	k	L_p (nH)	L_s (nH)	$Q_{p\max}$	$Q_{s\max}$
2.5	0.67	0.36	0.41	22.6	10.2
3.5	0.68	0.35	0.38	21.7	11.6
4.5	0.69	0.32	0.36	21.9	11.5
5.5	0.70	0.30	0.32	21.7	11.7

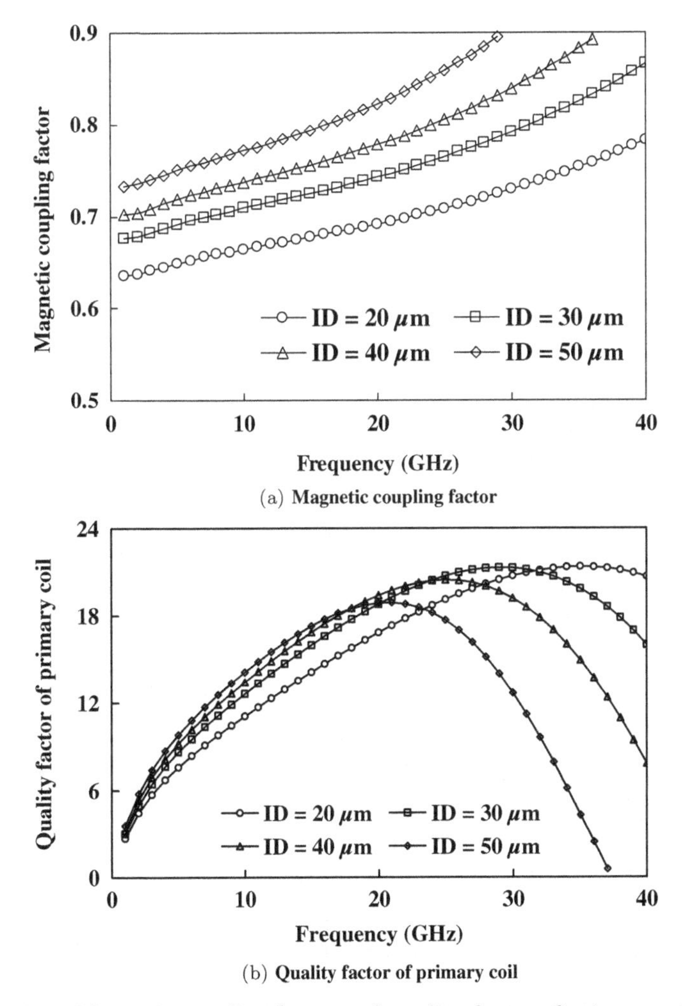

(a) **Magnetic coupling factor**

(b) **Quality factor of primary coil**

Figure 3.43. Magnetic coupling factor and quality factor of primary coil versus inner diameter for the stacked transformer (double turns).

Table 3.14. Effect of inner diameter for stacked transformer (double turns).

ID (μm)	k	L_p (nH)	L_s (nH)	$Q_{\rm pmax}$	$Q_{\rm smax}$
20	0.70	0.22	0.25	18.0	8.9
30	0.71	0.29	0.31	19.8	9.8
40	0.79	0.36	0.39	20.1	10.1
50	0.82	0.45	0.48	19.5	10.1

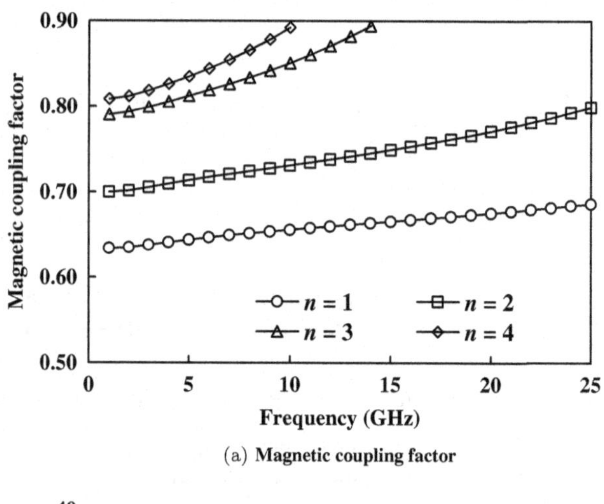

(a) Magnetic coupling factor

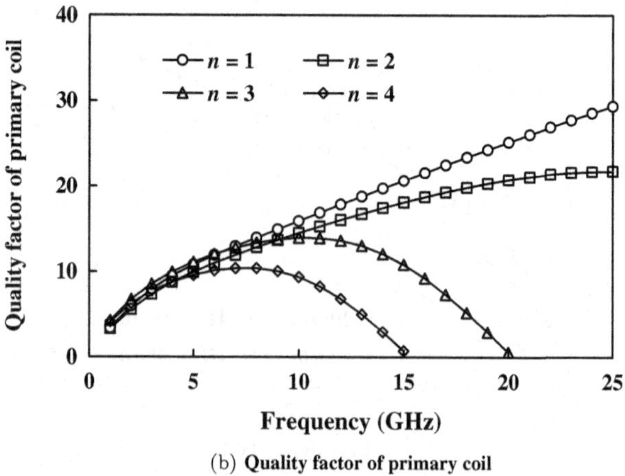

(b) Quality factor of primary coil

Figure 3.44. Magnetic coupling factor and quality factor of primary coil versus the number of turns for stacked transformer.

Table 3.15. Effect of the number of turns for stacked transformer.

n	k	L_p (nH)	L_s (nH)	Q_{pmax}	Q_{smax}
1	0.65	0.18	0.20	38.0	19.5
2	0.70	0.48	0.52	21.8	11.6
3	0.79	0.75	0.81	14.0	7.0
4	0.81	1.04	1.10	10.3	5.4

3.7 Summary

In the design of RF integrated circuit, the on-chip transformer can complete impedance matching, feedback, single-end to double-end conversion, and AC coupling between inter-stages of the circuit. Owing to these important functions, the on-chip transformers have become a research hotspot. This chapter mainly introduces the typical structure of on-chip transformer, the equivalent circuit model of on-chip transformer, and the corresponding parameter extraction method. Moreover, the influence of physical and geometric parameters on the characteristics of interleaved transformer and stacked transformer is analyzed.

References

[1] T. O. Dickson, M.-A. L. Croix, S. Boret, D. Gloria, R. Beerkens, and S. P. Voinigescu, "30–100-GHz inductors and transformers for millimeter-wave (Bi)CMOS integrated circuits," *IEEE Transactions on Microwave Theory and Techniques*, 53(1): 123–133, January 2005.
[2] J. R. Long, "Monolithic transformers for silicon RF IC design," *IEEE Journal Solid-State Circuits*, 35(9): 1368–1382, September 2000.
[3] O. El-Gharniti, E. Kerherve, and J.-B. Begueret, "Modeling and characterization of on-chip transformers for silicon RFIC," *IEEE Transactions Mcrow. Theory Tech*, 55(4): 607–615, April 2007.
[4] B. Chen, L. Lou, K. Tang, Y. Wang, and J. Gao, "A 13.5–19 GHz 20.6-dB Gain CMOS power amplifier for FMCW radar application," *IEEE Microwave and Wireless Components Letters*, 27(4): 377–379, 2017.
[5] C. T. Fu and C. N. Kuo, "3–11 GHz CMOS UWB LNA using dual feedback for broadband matching," in *IEEE Radio Frequency Integrated Circuits Symposium*, June 2006, pp. 53–56.

[6] H. Gan, "On-chip transformer modeling, characterization, and applications in power and low noise amplifiers," Ph.D Dissertation, Stanford University, Stanford, 2006.

[7] B. Leite, E. Kerhervé, J.-B. Bégueret, and D. Belot, "Transformer topologies for mmW integrated circuits," *European Microwave Conference*, pp. 181–184, October 2009.

[8] T. O. Dickson, M.-A. LaCroix, S. Boret, D. Gloria, R. Beerkens, and S. P. Voinigescu, "30–100-GHz inductors and transformers for millimeter-wave (Bi) CMOS integrated circuits," *IEEE Transactions on Microwave Theory and Techniques*, 53(1): 123–133, January 2005.

[9] H. Wang, "A study on the modeling techniques of millimeter-wave (mmW) on-chip passive components based on transfer function analysis," Ph.D Dissertation, East China Normal University, 2012.

[10] Y.-S. Lin, C.-Z. Chen, H.-B. Liang, and C.-C. Chen, "High-performance on-chip transformers with partial polysilicon patterned ground shields (PGS)," *IEEE Transactions on Electron Devices*, 54(1): 157–160, January 2007.

[11] W. Gao, C. Jiao, T. Liu, and Z. Yu, "Scalable compact circuit model for differential spiral transformers in CMOS RFICs," *IEEE Transactions on Electron Devices*, 53: 2187–2193, September 2006.

[12] Y. J. Lee and C. S. Kim, "Q-enhanced 5 GHz CMOS VCO using 4-port transformer," *Topical Meeting on Silicon Monolithic Integrated Circuits in RF Systems*, pp. 119–122, January 2007.

[13] B. Chen, "CMOS millimeter wave transmitter chip design," Ph.D Dissertation, East China Normal University, 2016.

[14] W. Simburger, H. Wohlmuth, and P. Weger, "A monolithic transformer coupled 5-W silicon power amplifier with 59% PAE at 0.9 GHz," *IEEE Journal of Solid-State Circuits*, 34(12): 1881–1892, 1999.

[15] J. G. McRory, *et al.* "Transformer coupled stacked FET power amplifiers," *IEEE Journal of Solid-State Circuits*, 34(2): 157–161, 1999.

[16] T. LaRocca and F. Chang, "60 GHz CMOS differential and transformer-coupled power amplifier for compact design," *IEEE RFIC Symposium*, pp. 65–86, 2008.

[17] D. Huang, R. Wong, Q. Gu, N. Wang, T. Ku, C. Chien, and F. Chang, "A 60 GHz CMOS differential receiver front-end using on-chip transformer for 1.2 volt operation with enhanced gain and linearity," *VLSI Circuits Symposium*, pp. 144–145, 2006.

[18] R. Chen, B. Chen, D. Luo, and J. Gao, "Direct extraction method of equivalent circuit parameters for stacked transformer," *Journal of Infrared Millimeter Waves*, 35(2): 172–176, 2016.

[19] R. Cheng, "Modeling and parameters extraction of on-chip spiral transformers for RFICs design," Master Dissertation, East China Normal University, 2016.

Chapter 4

MOSFET Small-Signal Model

The model parameters of the small-signal equivalent circuit are directly related to the physical structure and performance of the MOSFET device. Moreover, the small-signal equivalent circuit of the device can help researchers readily understand the physical mechanism of the device and deeply analyze the radio frequency (RF) behavior of the device. Meanwhile, the small-signal equivalent circuit of the device is also the basis for the large-signal and noise modeling. The part of parameters in the large-signal and noise model needs to be extracted from the small-signal model. Hence, an accurate small-signal model is particularly significant for the subsequent large-signal and noise modeling [1–3]. When the transistor operates at the microwave and radio frequencies (RF), the influence of extrinsic elements such as the gate, source, and drain extrinsic network and the loss of the silicon substrate on the transistor cannot be ignored. Therefore, compared with modeling at low frequencies, the MOSFET device modeling is still challenging at the microwave and radio frequencies.

4.1 Small-Signal Model

When the amplitude of an alternating current (AC) signal applied to the quiescent operation points of the device is less than the thermal voltage (kT/q), the device can be considered to operate in the small-signal state, and the small-signal behaviors of the device can be

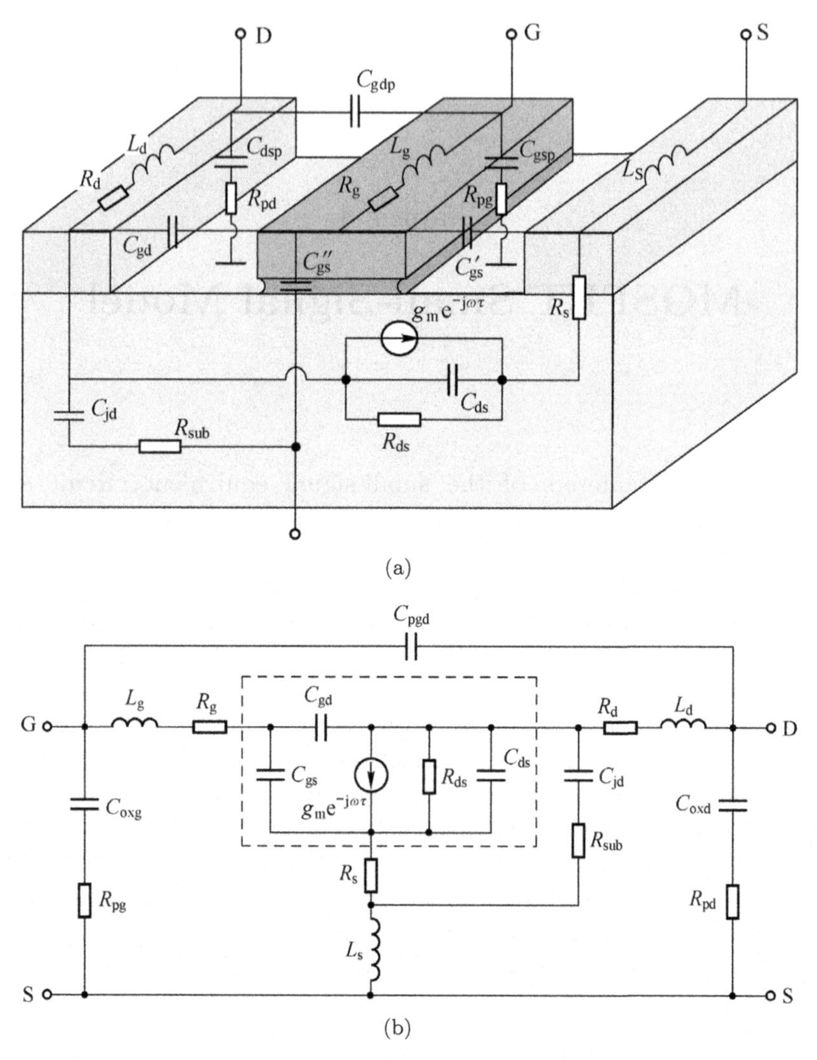

Figure 4.1. Small-signal equivalent circuit of MOSFET device: (a) three-dimensional structure; (b) plane structure.

approximately analyzed by linear methods. For a MOSFET device, the three-dimensional and plane structure of the small-signal equivalent circuit is illustrated in Figure 4.1 [1,2]. As seen, the small-signal equivalent circuit consists of two parts, the extrinsic elements out of the dashed box and the intrinsic elements in the dashed box, including the following:

(1) Pad extrinsic capacitances.
$C_{\mathrm{ox}g}$ represents the capacitance of the input signal pad to the ground, $C_{\mathrm{ox}d}$ represents the capacitance of the output signal pad to the ground, and $C_{\mathrm{pg}d}$ is the coupling capacitance between the input signal pad and the output signal pad.
(2) Pad extrinsic resistances.
R_{pg} and R_{pd} are the resistances that characterize the substrate loss of the input signal pad and the output signal pad, respectively.
(3) Feedline extrinsic inductances.
L_g, L_d, and L_s represent the gate, drain, and source feedline extrinsic inductance, respectively.
(4) MOSFET device contact resistances.
R_g represents the gate resistance, which is mainly composed of the polysilicon resistance on the channel gate oxide, the contact resistance between the polysilicon and the silicide, the resistance of the polysilicon extension around the active area, and the contact hole resistance caused by the contact between the polysilicon and the metal. R_d and R_s represent the drain and source series resistance, which are mainly composed of contact hole resistance, silicide resistance, silicide and drain-source junction contact resistance, and diffusion resistance in the drain-source port.
(5) Extrinsic coupling element of the drain to substrate network.
C_{jd} and R_{sub} represent the drain-substrate coupling capacitance and substrate loss, respectively.
(6) Intrinsic capacitances.
The gate-source capacitance C_{gs} is composed of the gate-channel capacitance and the gate-source overlapping capacitance, while the gate-drain capacitance C_{gd} is composed of the gate-drain overlapping capacitance. C_{ds} represents the drain-source capacitance.
(7) Intrinsic transconductance and output conductance.
g_m and g_{ds} (equal to $1/R_{\mathrm{ds}}$) represent the transconductance and output conductance of the device, respectively. τ represents the time delay associated with the transconductance.

Considering the influence of the lossy silicon substrate region under the drain-substrate junction of the MOSFET device, the coupling network consisting of C_{jd} and R_{sub} in series is utilized to characterize. Notably, the gate substrate body capacitance is included in the intrinsic gate-source capacitance.

The open Z parameter of a small-signal equivalent circuit for a MOSFET device can be expressed as follows [4]:

$$Z_{11} = \frac{Z_{11}^{\text{INT}} + R_{\text{s}} + Y_{\text{jd}}N}{1 + (Z_{22}^{\text{INT}} + R_{\text{s}})Y_{\text{jd}}} + j\omega(L_{\text{g}} + L_{\text{s}}) + R_{\text{g}} \tag{4.1}$$

$$Z_{12} = \frac{Z_{12}^{\text{INT}} + R_{\text{s}}}{1 + (Z_{22}^{\text{INT}} + R_{\text{s}})Y_{\text{jd}}} + j\omega L_{\text{s}} \tag{4.2}$$

$$Z_{21} = \frac{Z_{21}^{\text{INT}} + R_{\text{s}}}{1 + (Z_{22}^{\text{INT}} + R_{\text{s}})Y_{jd}} + j\omega L_{\text{s}} \tag{4.3}$$

$$Z_{22} = \frac{Z_{22}^{\text{INT}} + R_{\text{s}}}{1 + (Z_{22}^{\text{INT}} + R_{\text{s}})Y_{jd}} + j\omega(L_{\text{d}} + L_{\text{s}}) + R_{\text{d}} \tag{4.4}$$

wherein

$$Y_{\text{jd}} = \frac{j\omega C_{\text{jd}}}{1 + j\omega R_{\text{sub}} C_{\text{jd}}}$$

$$N = Z_{11}^{\text{INT}} Z_{22}^{\text{INT}} - Z_{12}^{\text{INT}} Z_{21}^{\text{INT}} + R_s(Z_{11}^{\text{INT}} + Z_{22}^{\text{INT}} - Z_{12}^{\text{INT}} - Z_{21}^{\text{INT}})$$

Note that $Z_{ij}^{\text{INT}}(i, j = 1, 2)$ represents the Z parameter of the intrinsic network in the above equations and can be written as follows:

$$Z_{11}^{\text{INT}} = \frac{g_{\text{ds}} + j\omega(C_{\text{gd}} + C_{\text{ds}})}{M} \tag{4.5}$$

$$Z_{12}^{\text{INT}} = \frac{j\omega C_{\text{gd}}}{M} \tag{4.6}$$

$$Z_{21}^{\text{INT}} = \frac{-g_{\text{m}} e^{-j\omega\tau} + j\omega C_{\text{gd}}}{M} \tag{4.7}$$

$$Z_{22}^{\text{INT}} = \frac{j\omega(C_{\text{gs}} + C_{\text{gd}})}{M} \tag{4.8}$$

wherein

$$\begin{aligned} M = -\omega^2(&C_{\text{gs}}C_{\text{ds}} + C_{\text{gs}}C_{\text{gd}} \\ &+ C_{\text{gd}}C_{\text{ds}}) + j\omega[g_m e^{-j\omega\tau} C_{\text{gd}} + g_{\text{ds}}(C_{\text{gs}} + C_{\text{gd}})] \end{aligned}$$

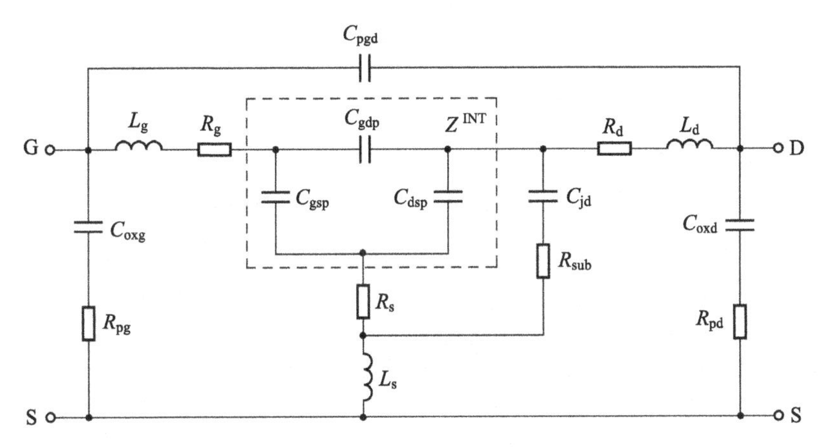

Figure 4.2. Small-signal equivalent circuit of MOSFET device under cut-off condition.

When the device operates under the none bias condition or the channel is not conducting, the current source controlled by the drain voltage disappears, and the gate, source, and drain are capacitive. Figure 4.2 illustrates the small-signal equivalent circuit of the MOSFET device under cut-off condition and the Z-matrix of the network can be expressed as follows:

$$Z_{11}^{\mathrm{c}} = \frac{j\omega(C_{\mathrm{gdp}} + C_{\mathrm{dsp}}) + M_c R_s + Y_{\mathrm{jd}}[1 + j\omega R_s(C_{\mathrm{gsp}} + C_{\mathrm{dsp}})]}{M_c + Y_{\mathrm{jd}}[j\omega(C_{\mathrm{gdp}} + C_{\mathrm{gsp}}) + M_c R_s]} + j\omega(L_g + L_s) + R_g \tag{4.9}$$

$$Z_{12}^{\mathrm{c}} = Z_{21}^{\mathrm{c}} = \frac{j\omega C_{\mathrm{gdp}} + M_c R_s}{M_c + Y_{\mathrm{jd}}[j\omega(C_{\mathrm{gdp}} + C_{\mathrm{gsp}}) + M_c R_s]} + j\omega L_s \tag{4.10}$$

$$Z_{22}^{\mathrm{c}} = \frac{j\omega(C_{\mathrm{gdp}} + C_{\mathrm{gsp}}) + M_c R_s}{M_c + Y_{\mathrm{jd}}[j\omega(C_{\mathrm{gdp}} + C_{\mathrm{gsp}}) + M_c R_s]} + j\omega(L_d + L_s) + R_d \tag{4.11}$$

wherein

$$M_c = -\omega^2(C_{\mathrm{gsp}} C_{\mathrm{dsp}} + C_{\mathrm{gsp}} C_{\mathrm{gdp}} + C_{\mathrm{gdp}} C_{\mathrm{dsp}})$$

At low frequencies, the extrinsic elements of the pad and feedline can be neglected. A simplified small-signal equivalent circuit under

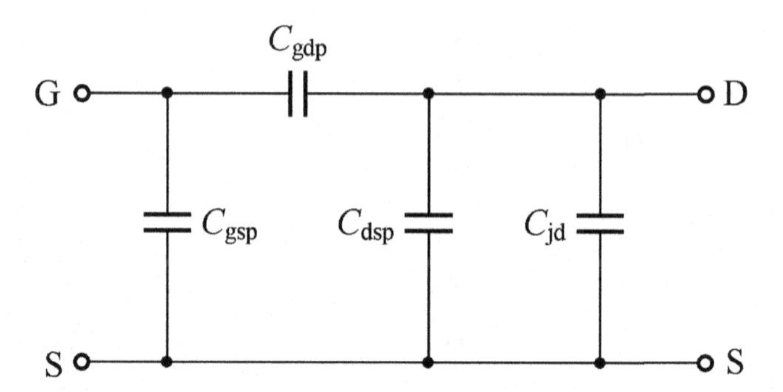

Figure 4.3. Simplified small-signal equivalent circuit under cut-off condition at low frequencies.

cut-off condition at low frequencies is illustrated in Figure 4.3, and the corresponding Y-matrix can be written as follows:

$$Y_{11}^{cl} = j\omega(C_{gsp} + C_{gdp}) \tag{4.12}$$

$$Y_{22}^{cl} = j\omega(C_{gdp} + C_{dsp}) \tag{4.13}$$

$$Y_{12}^{cl} = Y_{21}^{cl} = -j\omega C_{gdp} \tag{4.14}$$

wherein the superscript cl denotes the cut-off condition at low frequencies.

4.2 De-embedding Method

Demonstrating good convergence on the simulation platform based on device physical characteristics is the first step to meeting the designer's needs. The complexity of parameter extraction will directly depend on the choice of the model but also on the accuracy of the measurement equipment, the reliability of the de-embedding technology, and the effectiveness of the model parameter extraction method. With the device features dimensions continuous down and the continuous improvement of integrated circuit characteristic frequency, the influence of pad and feedline parasitics on the performance of on-wafer test devices has become increasingly essential. It is crucial to accurately de-embed the influence of parasitic elements and obtain the characteristics of the transistor itself. This process is called de-embedding technology.

4.2.1 *De-embedding procedure*

Currently, the commonly de-embedding methods are given as follows [5–8]:

(1) *Open De-embedding Method*: In the open de-embedding method, the parasitic effect of the test structure is mainly caused by the parasitic elements between the pad and the substrate and between the pads, ignoring series parasitic elements such as interconnects and only considering the influence of the pad. Hence, this method is suitable for the low frequencies (below 10 GHz) and is less accurate at high frequencies.
(2) *Open-Short De-embedding Method*: With increased frequency, especially in the millimeter-wave frequency ranges, so as to improve the accuracy of the de-embedding method, it is necessary to use the open-short de-embedding method. Compared with the open de-embedding method, the independent short test structure is added, and more accurate results can be obtained at high frequencies (above 50 GHz). The open-short de-embedding method assumes that all parasitics in parallel are placed at the pads, and all parasitics in series are equivalent to the interconnecting lines.
(3) *Thru De-embedding Method*: Thru de-embedding method only requires the device under test and the thru test structure. This method can effectively save the area of chips, however the premise is that the two-port network of thru structure must be reciprocal and mirror symmetry, and as known, the general test structure is not absolutely symmetry, hence this de-embedding method has great limitations.

In general, the open-short de-embedding method is the most commonly used method. The measured data obtained by the vector network analyzer (VNA) are all expressed as S-parameters, while the impedance parameters (Z-parameters) and admittance parameters (Y-parameters) are often utilized in parameter extraction of MOSFET devices. For example, the π-type network can be characterized by Y parameter matrix, while the T-type network can choose the Z-parameter matrix. Tables 4.1–4.3 summarize the relationship between Z parameters, Y parameters, and S parameters. As seen, three parameters can be directly converted to each other.

Table 4.1. Relationship between Z parameter and Y parameter.

Z parameter	Y parameter
$Z_{11} = \frac{Y_{22}}{Y_{11}Y_{22} - Y_{12}Y_{21}}$	$Y_{11} = \frac{Z_{22}}{Z_{11}Z_{22} - Z_{12}Z_{21}}$
$Z_{12} = -\frac{Y_{12}}{Y_{11}Y_{22} - Y_{12}Y_{21}}$	$Y_{12} = -\frac{Z_{12}}{Z_{11}Z_{22} - Z_{12}Z_{21}}$
$Z_{21} = -\frac{Y_{21}}{Y_{11}Y_{22} - Y_{12}Y_{21}}$	$Y_{21} = -\frac{Z_{21}}{Z_{11}Z_{22} - Z_{12}Z_{21}}$
$Z_{22} = \frac{Y_{11}}{Y_{11}Y_{22} - Y_{12}Y_{21}}$	$Y_{22} = \frac{Z_{11}}{Z_{11}Z_{22} - Z_{12}Z_{21}}$

Table 4.2. Relationship between Z parameter and S parameter.

Z parameter	S parameter
$Z_{11} = Z_o\frac{(1+S_{11})(1-S_{22})+S_{12}S_{21}}{(1-S_{11})(1-S_{22})-S_{12}S_{21}}$	$S_{11} = \frac{(Z_{11}-Z_o)(Z_{22}+Z_o)-Z_{12}Z_{21}}{(Z_{11}+Z_o)(Z_{22}+Z_o)-Z_{12}Z_{21}}$
$Z_{12} = Z_o\frac{2S_{12}}{(1-S_{11})(1-S_{22})-S_{12}S_{21}}$	$S_{12} = \frac{2Z_{12}Z_o}{(Z_{11}+Z_o)(Z_{22}+Z_o)-Z_{12}Z_{21}}$
$Z_{21} = Z_o\frac{2S_{21}}{(1-S_{11})(1-S_{22})-S_{12}S_{21}}$	$S_{21} = \frac{2Z_{21}Z_o}{(Z_{11}+Z_o)(Z_{22}+Z_o)-Z_{12}Z_{21}}$
$Z_{22} = Z_o\frac{(1-S_{11})(1+S_{22})+S_{12}S_{21}}{(1-S_{11})(1-S_{22})-S_{12}S_{21}}$	$S_{22} = \frac{(Z_{11}+Z_o)(Z_{22}-Z_o)-Z_{12}Z_{21}}{(Z_{11}+Z_o)(Z_{22}+Z_o)-Z_{12}Z_{21}}$

Table 4.3. Relationship between Y parameter and S parameter.

Y parameter	S parameter
$Y_{11} = Y_o\frac{(1-S_{11})(1+S_{22})+S_{12}S_{21}}{(1+S_{11})(1+S_{22})-S_{12}S_{21}}$	$S_{11} = \frac{(Y_o-Y_{11})(Y_o+Y_{22})+Y_{12}Y_{21}}{(Y_{11}+Y_o)(Y_{22}+Y_o)-Y_{12}Y_{21}}$
$Y_{12} = Y_o\frac{-2S_{12}}{(1+S_{11})(1+S_{22})-S_{12}S_{21}}$	$S_{12} = \frac{-2Y_oY_{12}}{(Y_{11}+Y_o)(Y_{22}+Y_o)-Y_{12}Y_{21}}$
$Y_{21} = Y_o\frac{-2S_{21}}{(1+S_{11})(1+S_{22})-S_{12}S_{21}}$	$S_{21} = \frac{-2Y_oY_{21}}{(Y_{11}+Y_o)(Y_{22}+Y_o)-Y_{12}Y_{21}}$
$Y_{22} = Y_o\frac{(1+S_{11})(1-S_{22})+S_{12}S_{21}}{(1+S_{11})(1+S_{22})-S_{12}S_{21}}$	$S_{22} = \frac{(Y_o+Y_{11})(Y_o-Y_{22})+Y_{12}Y_{21}}{(Y_{11}+Y_o)(Y_{22}+Y_o)-Y_{12}Y_{21}}$

The test structure of the open and short de-embedding method is illustrated in Figure 4.4. $S1$ and $S2$ represent the signal input and output pads, respectively, and the gate and drain of the MOSFET

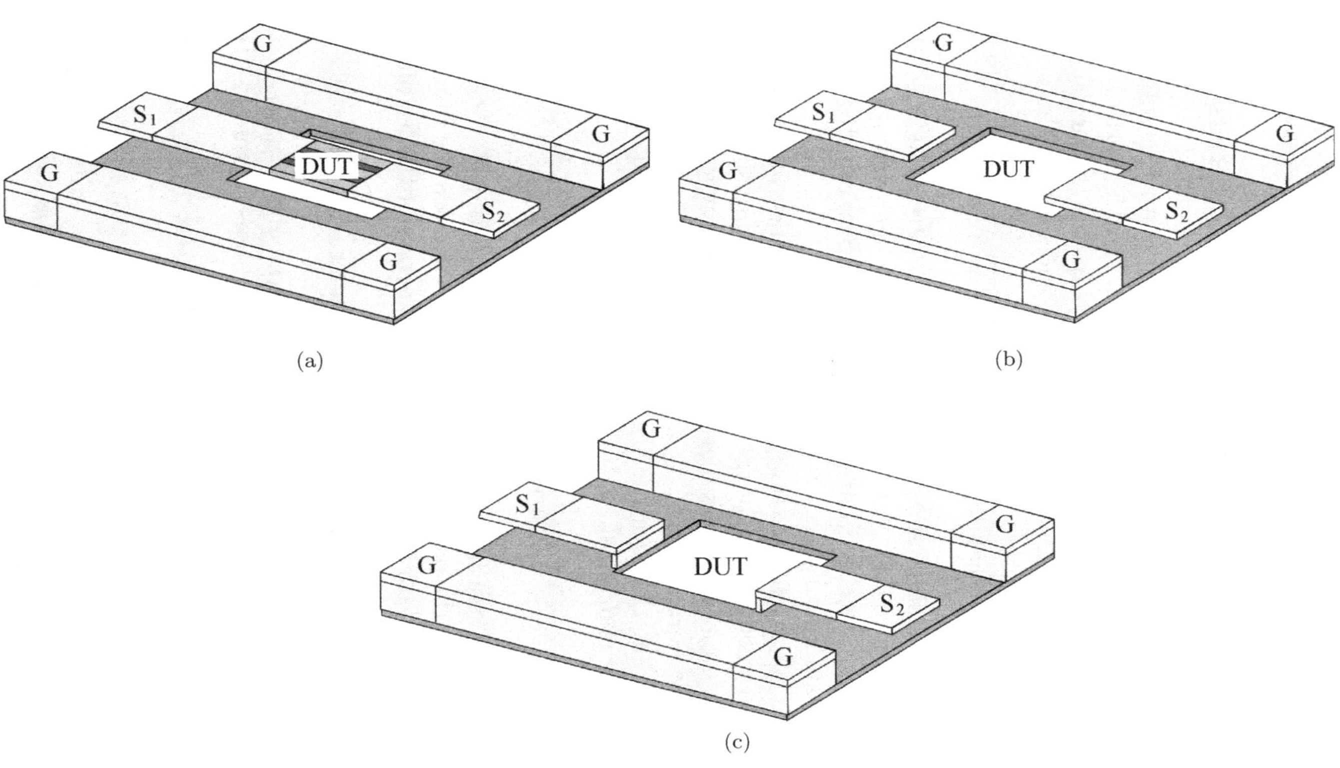

Figure 4.4. 3D view of the open and short test structure: (a) Test structure including DUT; (b) Open test structure; (c) Short test structure.

device are connected by metal interconnects. G represents the ground pad; through interconnects connect the source and substrate of DUT. The test structure including the DUT is illustrated in Figure 4.4(a). Figure 4.4(b) depicts the open test structure excluding the DUT and interconnecting lines, and Figure 4.4(c) shows the short test structure in which the input and output ports are shorted to the substrate layer, and the test structure excludes the DUT. It is worth noting that when fabricating two structures on the chip, the two device test structures should with the same dimension and exclude the active devices.

The specific de-embedding steps are listed as follows:

(1) First, the S parameters S_{meas} of the test structure including DUT shown in Figure 4.4(a) are measured by the microwave on-wafer test system and then convert the measured S parameters S_{meas} to Y parameters Y_{meas}.
(2) Measure the S-parameter S_{open} of the open test structure and the S-parameter S_{short} of the short test structure shown in Figures 4.4(b) and 4.4(c). Similarly, convert S parameters S_{open} and S_{short} to Y parameters Y_{open} and Y_{short}, respectively.
(3) De-embedding the parasitic elements in parallel are used to characterize the pads on the DUT

$$Y_{\text{DUT1}} = Y_{\text{meas}} - Y_{\text{open}}$$

and convert Y parameters Y_{DUT1} to Z parameters Z_{DUT1}.
(4) De-embedding the parasitic elements of the pads on the short test structure to obtain the parasitic Y parameters of the feedline

$$Y_{\text{short1}} = Y_{\text{short}} - Y_{\text{open}}$$

and convert Y parameters Y_{short1} to Z parameters Z_{short1}.
(5) According to the Z parameters obtained in steps (3) and (4), the Z parameters of DUT can be achieved:

$$Z_{\text{DUT}} = Z_{\text{DUT1}} - Z_{\text{short1}}$$

(6) Convert the obtained Z parameters to S parameters, and the S parameters S_{DUT} of the device can be obtained; the de-embedding flowchart is shown in Figure 4.5.

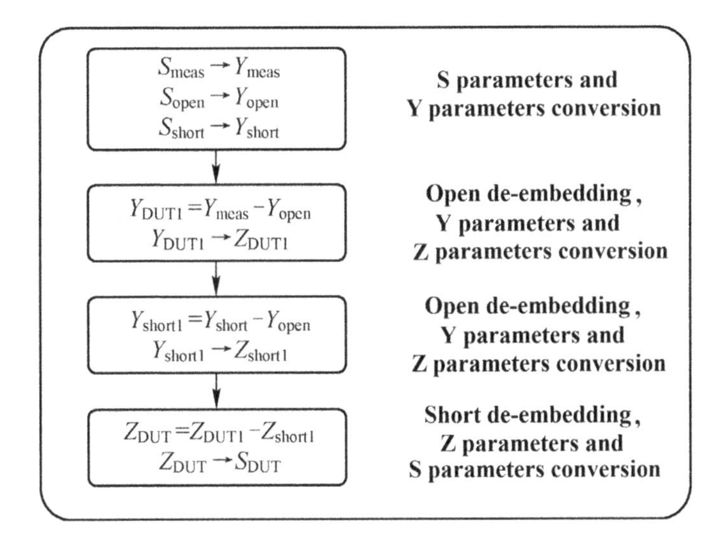

Figure 4.5. Flowchart of de-embedding method.

In a bid to compare the difference between the open de-embedding method and the open-short de-embedding method, the comparison of Y_{11} and Y_{22} embedding and de-embedding utilizing the two different de-embedding methods is illustrated in Figure 4.6 [9–11]. The MOSFET device with 0.13 μm gate length, 5 μm gate width, and 16 gate fingers was investigated to demonstrate the de-embedding method. As seen in Figure 4.6, in the low-frequency ranges (below 10 GHz), the Y parameters obtained by utilizing the open de-embedding method and the open-short de-embedding method are close. The reason is that the influence of the extrinsic resistance and inductance of the interconnecting lines can be neglected at low frequencies.

Nevertheless, with increased frequency, the Y parameters obtained using the two de-embedding methods are quite different. Due to increased frequency, the influence of the extrinsic inductance for interconnecting lines increases so that the extrinsic inductance cannot be ignored. In order to obtain the accurate model parameters, the open-short de-embedding method is essential to use, as the frequency is above 10 GHz. Thence, before extracting the model parameters of the RF microwave MOSFET device, the data must be de-embedded to eliminate the influence of the test structure.

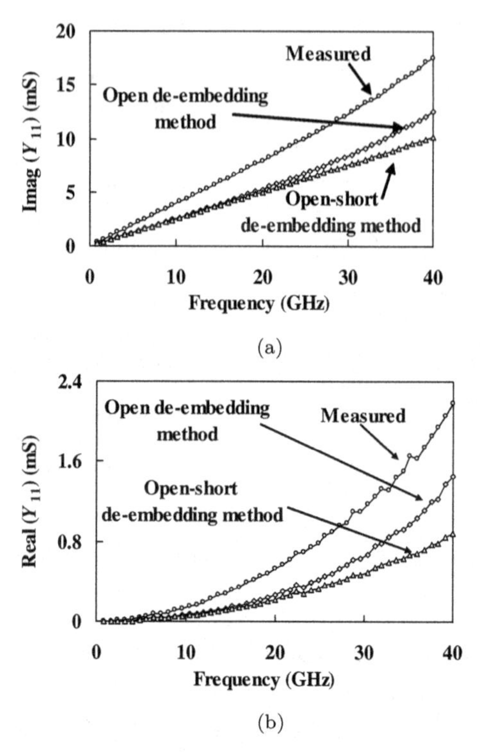

Figure 4.6. Y parameters versus frequency: (a) real part of Y_{11}; (b) imaginary part of Y_{11}; (c) real part of Y_{22}; (d) imaginary part of Y_{22}.

4.2.2 *Open test structure*

The layout of test structure for the MOSFET device is sketched in Figure 4.7. As seen, the layout of the pads and interconnecting lines is identical for the test structure and the de-embedding structures. The length of the test structure is 235 μm and the width is 230 μm. Starting on the left-hand side, there are input pads in GSG arrangement with 100 μm pitch and output pads with the same configuration as input pads. The open test structure excluding the DUT and interconnecting lines is depicted in Figure 4.7(a), wherein G and D, connected to the gate and drain of the MOSFET device through interconnecting lines, represent the signal input and output pads, respectively. Besides, S connected to the source and substrate of DUT through interconnecting lines represents the pad to ground. In addition, the open test structure including interconnecting lines

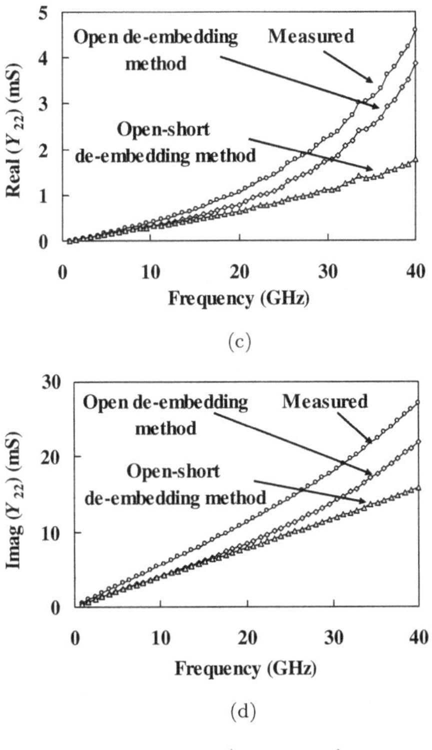

Figure 4.6. *(Continued)*

is illustrated in Figure 4.7(b). As seen, the gate, source, and drain interconnecting lines are open circuit, hence it is called open test structure, and d represents the width of the interconnecting lines. The corresponding equivalent circuit model of open test structure is depicted in Figure 4.8.

The typical 0.13 μm RF CMOS technology with eight levels of metal interconnects is used to fabricate the test structure. The cross-section of the test structure is illustrated in Figure 4.9.

It can be obviously seen that the signal pad in contact with the probe is fabricated with the top metal layer M8. When fabricating the ground pad, the top metal layer M8 is connected to the bottom metal layer M1 through the vias and the middle metal layer. Moreover, the bottom metal layer M1 extends below the signal pad connecting the source and drain. This structure can cut down the coupling of the signal pads to the substrate for the test structure [12].

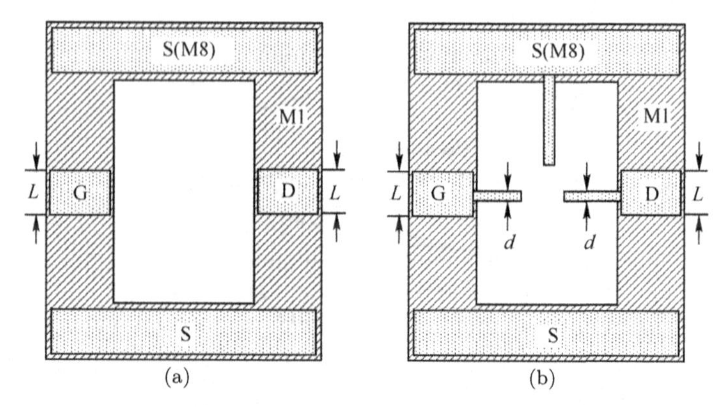

Figure 4.7. Layout of the open test structure: (a) exclude the device under test and interconnecting lines; (b) include the device under test and interconnecting lines.

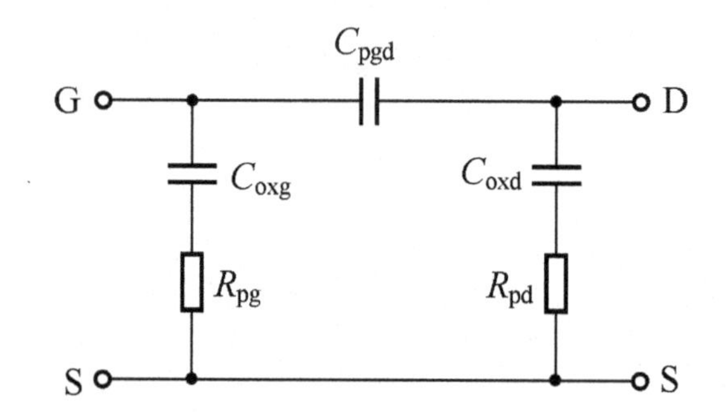

Figure 4.8. Equivalent circuit of open test structure.

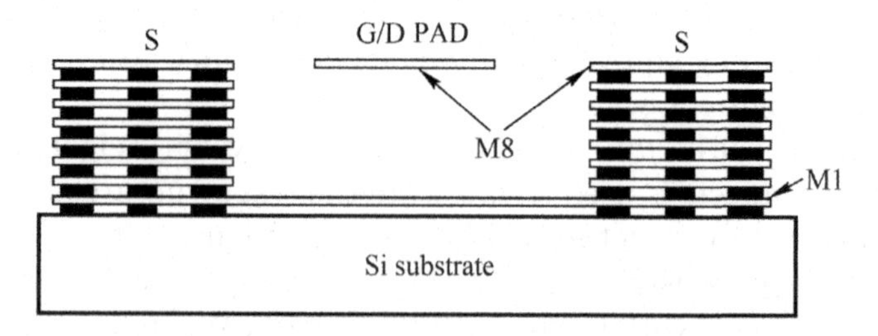

Figure 4.9. Cross-section of the test structure.

Figure 4.10 sketches the detailed structure of each metal, contact, via, and dielectric layer of the test structure. As can be obviously seen, there are eight layers of metals with different thicknesses

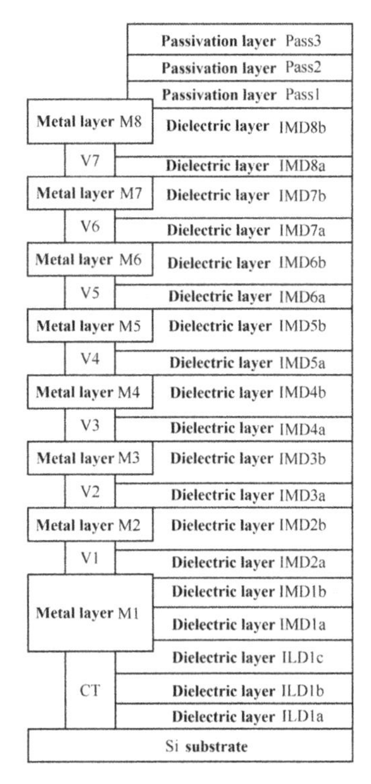

Figure 4.10. Schematic diagram of the metal layer, through hole, and dielectric layer of the test structure.

and give numbers from M1 to M8. Note that the M1 and M8 are the bottom and the top metal, respectively. Besides, CT represents the contact hole, and V1–V7 represents the vias between the metal layers. Pass1, Pass2, and Pass3 layers with different thicknesses and dielectric constant are the passivation layers. Additionally, this test structure has 19 dielectric layers in total, which are ILD1a, ILD1b, ILD1c, IMD1a, IMD1b–IMD8a, and IMD8b layers.

Analyzing the test structure, it can be found that the extrinsic capacitances ($C_{\text{ox}g}$ and $C_{\text{ox}d}$) of the oxide layer between the signal pad and the ground metal layer can be approximated using the expressions for parallel plate capacitance.

The dielectric of extrinsic capacitance consists of the insulation layers with different thicknesses d and different dielectric constants ε. The insulation layers can be equivalent to a dielectric layer with a thickness of d_{total} and a dielectric constant of $\varepsilon_{r,\text{total}}$. There are

14 dielectric layers between the metal layer M8 of the signal pad and the ground shield layer M1, which are IMD2a, IMD2b, IMD3a, IMD3b,..., IMD8a, and IMD8b layers. The thickness d_{total} of the equivalent dielectric layer is equal to the sum of the thickness of the vias V1–V7 and the thicknesses of the metal layers M2–M7, which can be written as

$$d_{\text{total}} = \sum_{i=1}^{7} d_{\text{V}i} + \sum_{j=2}^{7} d_{\text{M}j} \tag{4.15}$$

wherein $d_{\text{V}i}$ and $d_{\text{M}j}$ are the thicknesses of the via and the metal layer, respectively.

When calculating the equivalent dielectric constant, by assuming that the extrinsic capacitance between the metal layer M8 and the grounding shield layer M1 can be regarded as the subcapacitances in series, and the dielectric of each subcapacitance is a dielectric layer with the fixed dielectric constant. As shown in Figure 4.10, since there are 14 layers dielectric from M8 to M1 layers, the extrinsic capacitance between M8 and M1 layers can be regarded as 14 extrinsic capacitances in series. According to the definition of parallel plate capacitance, we have

$$C_{\text{total}} = \varepsilon_o \varepsilon_{r,\text{total}} \frac{A}{d} \tag{4.16}$$

wherein ε_o represents the dielectric constant in vacuum, $\varepsilon_{r,\text{total}}$ represents the relative dielectric constant, A represents the area of the parallel plate, and d represents the distance between two parallel plates.

Figure 4.11 depicts the calculation method of equivalent dielectric constant. It can be found from Figure 4.11 that the reciprocal of the total capacitance (on the left side of the figure) is equal to the sum of the inverses of subcapacitances (on the right side of the figure). Hence, we have

$$\frac{1}{C_{\text{total}}} = \sum_{i=2}^{8} \frac{1}{C_{ia}} + \sum_{j=2}^{8} \frac{1}{C_{ib}} \tag{4.17}$$

Accordingly,

$$\frac{d_{\text{total}}}{\varepsilon_o \varepsilon_{r,\text{total}} A} = \sum_{i=2}^{8} \frac{d_{ia}}{\varepsilon_o \varepsilon_{r,ia} A} + \sum_{j=2}^{8} \frac{d_{jb}}{\varepsilon_o \varepsilon_{r,jb} A} \tag{4.18}$$

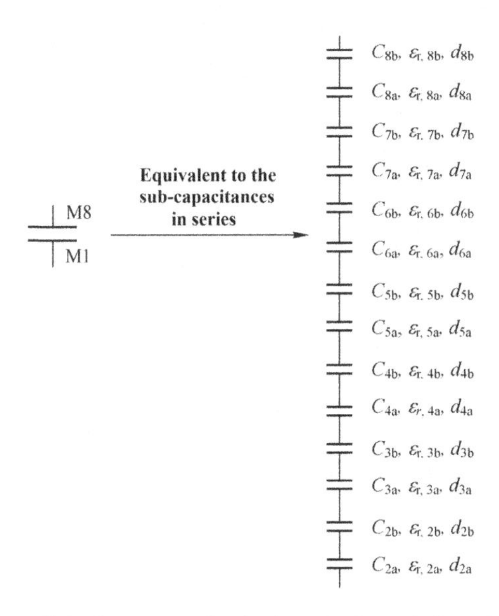

Figure 4.11. Calculation method of equivalent dielectric constant.

After simplified,

$$\varepsilon_{r,\text{total}} = \frac{d_{\text{total}}}{\sum\limits_{i=2}^{8} \frac{d_{ia}}{\varepsilon_{r,ia}} + \sum\limits_{j=2}^{8} \frac{d_{jb}}{\varepsilon_{r,jb}}} \tag{4.19}$$

Based on Equations (4.17)–(4.19), the equivalent dielectric constant $\varepsilon_{r,\text{total}}$ can be readily calculated and the capacitance of the oxide layer can be also estimated.

The equivalent circuit model parameters can be determined by electromagnetic simulation. Moreover, the model parameters can be also obtained by S parameters measurements. The admittance Y-parameter matrix of the equivalent circuit model for open test structure can be expressed as

$$Y^{\text{o}} = \begin{bmatrix} \frac{j\omega C_{\text{ox}g}}{1+j\omega C_{\text{ox}g} R_{\text{pg}}} + j\omega C_{\text{pgd}} & -j\omega C_{\text{pgd}} \\ -j\omega C_{\text{pgd}} & \frac{j\omega C_{\text{oxd}}}{1+j\omega C_{\text{ox}d} R_{\text{pd}}} + j\omega C_{\text{pgd}} \end{bmatrix} \tag{4.20}$$

By utilizing the Y-parameter matrix, the extrinsic elements can be directly determined:

$$C_{\text{ox}g} = -\frac{1}{\omega \text{Im}\left(\frac{1}{Y_{11}^{\text{o}}+Y_{12}^{\text{o}}}\right)} \tag{4.21}$$

$$C_{\text{ox}d} = -\frac{1}{\omega \text{Im}\left(\frac{1}{Y_{22}^{\text{o}}+Y_{12}^{\text{o}}}\right)} \tag{4.22}$$

$$C_{\text{pdg}} = -\frac{\text{Im}(Y_{12}^{\text{o}})}{\omega} \tag{4.23}$$

$$R_{\text{pg}} = \text{Re}\left(\frac{1}{Y_{11}^{\text{o}}+Y_{12}^{\text{o}}}\right) \tag{4.24}$$

$$R_{\text{pd}} = \text{Re}\left(\frac{1}{Y_{22}^{\text{o}}+Y_{12}^{\text{o}}}\right) \tag{4.25}$$

Based on the 130 nm standard CMOS process, the open test structure is fabricated. The extracted capacitances and resistances versus frequency are illustrated in Figure 4.12. As observed, the oxide capacitances $C_{\text{ox}g}$ and C_{oxd} touch at about 20 fF, and the coupling capacitance $C_{\text{pg}d}$ is roughly 1fF. While the extrinsic resistances R_{pd} and R_{pg} appear stable and touch at about 30 Ω in the high-frequency ranges. The extracted capacitances and resistances versus frequency based on the 90 nm standard CMOS process are depicted in Figure 4.13. As seen, the gate oxide capacitance C_{oxg} and the drain oxide capacitance $C_{\text{ox}d}$ are roughly equal to 120 fF and 115 fF, respectively. The extracted value of the coupling capacitance $C_{\text{pg}d}$ is 1 fF, the extrinsic resistances R_{pd} and R_{pg} touch at 8–12 Ω, and the arithmetic mean can be used as the extracted value. Based on 90 nm standard CMOS process, the comparison between the measured and the modeled S parameters of open test structure in the frequency range of 0.1–40 GHz is plotted in Figure 4.14. It can be obviously seen that the modeled data are in good agreement with the measured data, which indicates reasonable good accuracy and robustness of this extraction technique.

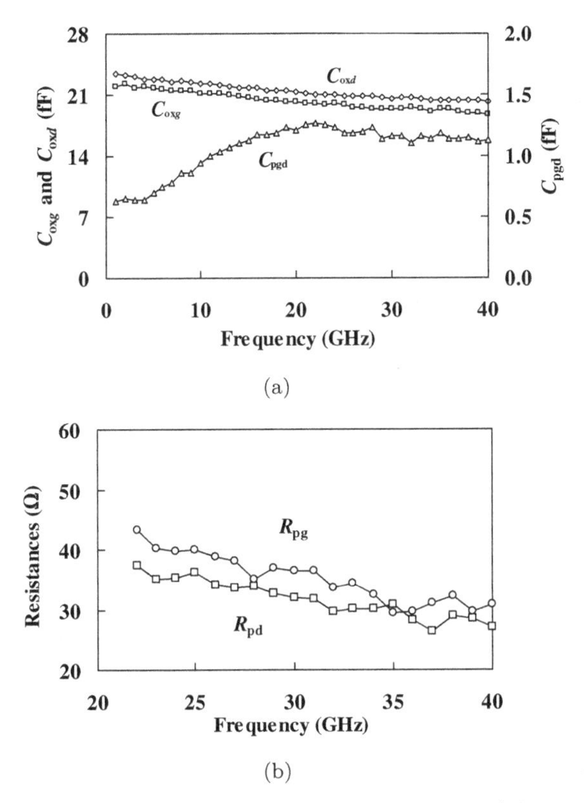

Figure 4.12. Extracted capacitances (a) and resistances (b) versus frequency (130 nm standard CMOS process).

4.2.3 *Short test structure*

The 3D view and plan view of the short test structure are given in Figures 4.15(a) and 4.15(b), respectively. It is worth noting that the short test structure does not include the device under test, and the three interconnects connecting the gate, source, and drain of the transistor can be short-circuited. Figure 4.15(c) shows the corresponding short test structure equivalent circuit model, where R_{lg} and L_g represent the gate feedline resistance and inductance, R_{ld} and L_d represent the drain feedline resistance and inductance, and R_{ls} and L_s represent the source feedline resistance and inductance.

First, de-embedding the open extrinsic network:

$$Y^{\text{short}} = Y^{\text{shortm}} - Y^{\text{open}} \tag{4.26}$$

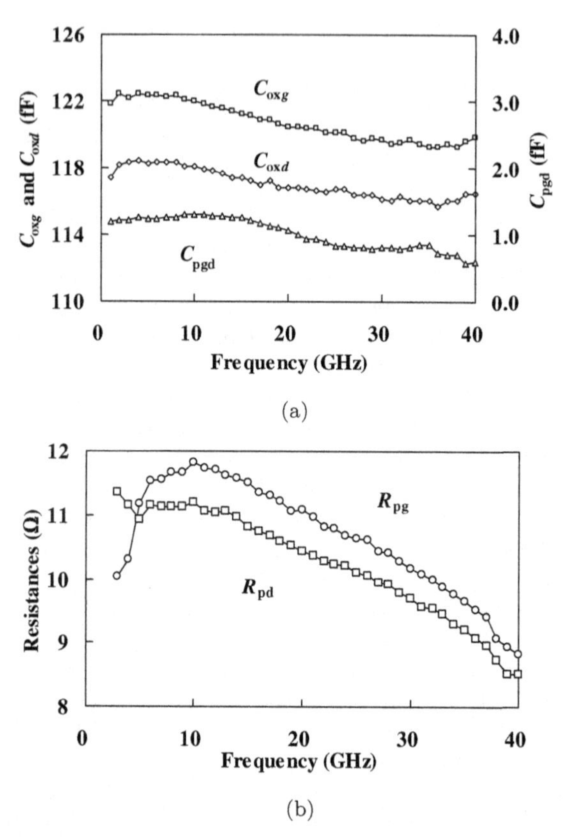

Figure 4.13. Extracted capacitances (a) and resistances (b) versus frequency (90 nm standard CMOS process).

where Y^{shortm} represents the admittance matrix transformed from the measured S parameters of short test structure and Y^{open} represents the admittance matrix transformed from the measured S parameters for the open test structure. Y^{short} represents the Y parameters of the network of the dashed box in Figure 4.15(c). The corresponding short matrix Z^{short} can be expressed as

$$Z_{11}^{\text{short}} = R_{\text{lg}} + R_{\text{ls}} + j\omega(L_g + L_s) \tag{4.27}$$

$$Z_{12}^{\text{short}} = Z_{21}^{\text{short1}} = R_{\text{ls}} + j\omega L_s \tag{4.28}$$

$$Z_{22}^{\text{short}} = R_{\text{ld}} + R_{\text{ls}} + j\omega(L_d + L_s) \tag{4.29}$$

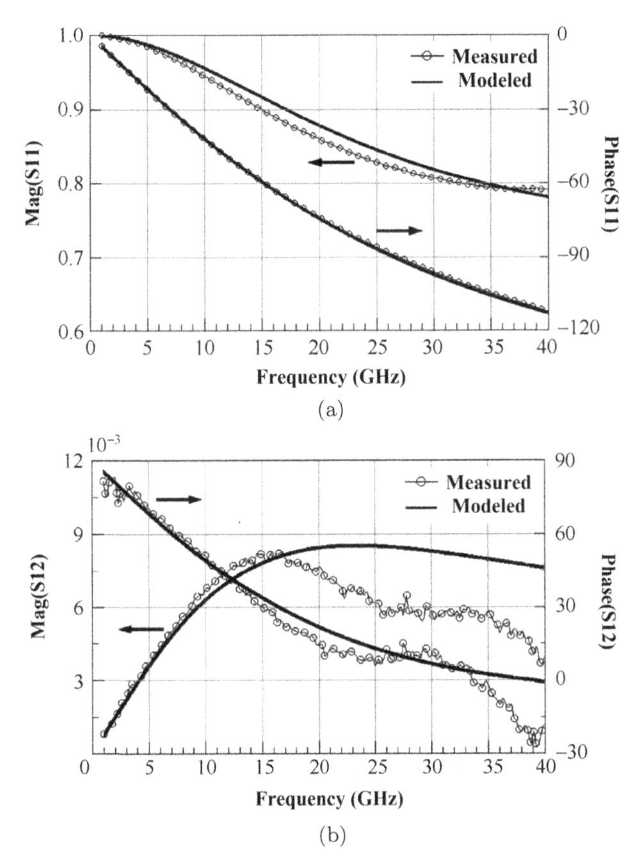

Figure 4.14. Comparison between the measured and the modeled S parameters of open test structure based on 90 nm standard CMOS process. (a) S_{11}; (b) S_{12}.

According to the above equations, the expression of the model parameters can be directly obtained:

$$R_{\mathrm{l}g} = \mathrm{Re}(Z_{11}^{\mathrm{short}} - Z_{12}^{\mathrm{short}}) \tag{4.30}$$

$$R_{\mathrm{l}d} = \mathrm{Re}(Z_{22}^{\mathrm{short}} - Z_{21}^{\mathrm{short}}) \tag{4.31}$$

$$R_{\mathrm{l}s} = \mathrm{Re}(Z_{12}^{\mathrm{short}}) \tag{4.32}$$

$$L_g = \frac{\mathrm{Im}(Z_{11}^{\mathrm{short}} - Z_{12}^{\mathrm{short}})}{\omega} \tag{4.33}$$

$$L_d = \frac{\mathrm{Im}(Z_{22}^{\mathrm{short}} - Z_{21}^{\mathrm{short}})}{\omega} \tag{4.34}$$

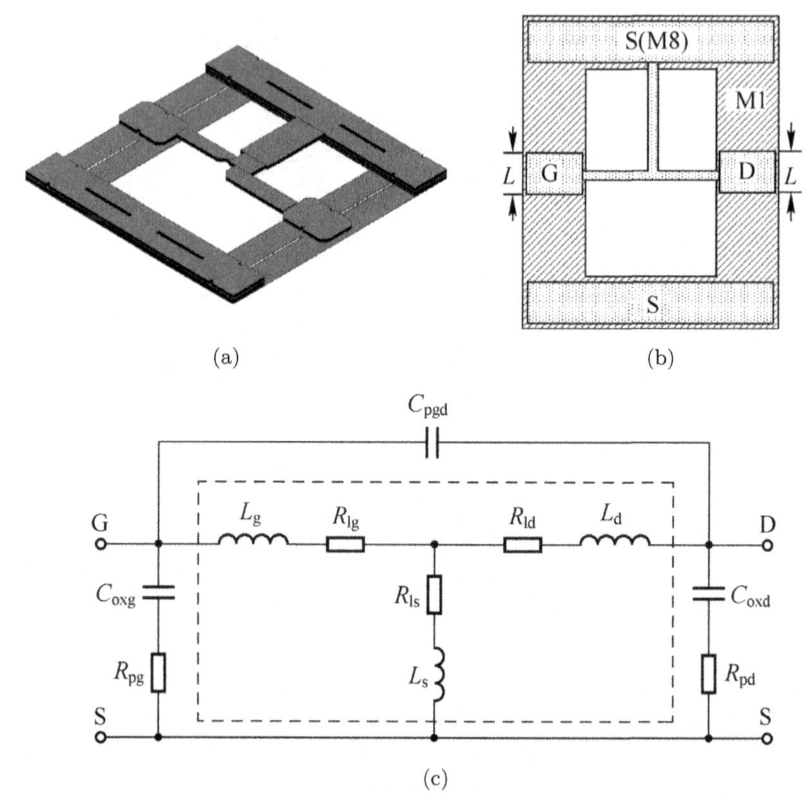

Figure 4.15. (a) 3D structure; (b) plane structure; (c) equivalent circuit model of short test structure.

$$L_s = \frac{\mathrm{Im}(Z_{12}^{\mathrm{short}})}{\omega} \tag{4.35}$$

The short test structure based on the 90 nm and 130 nm standard CMOS process was measured [10]. The corresponding extracted feedline inductances and resistances versus frequency based on the 130 nm standard CMOS process are given in Figures 4.16 and 4.17, respectively. Figure 4.18 plots the comparison between the measured and modeled S parameters of the short test structure in the frequency range of 0.1–40 GHz. It can be clearly observed that the measured data and the modeled S-parameters are in good agreement.

Figures 4.19 and 4.20 illustrate the extracted feedline inductances and resistances versus frequency based on the 90 nm standard CMOS

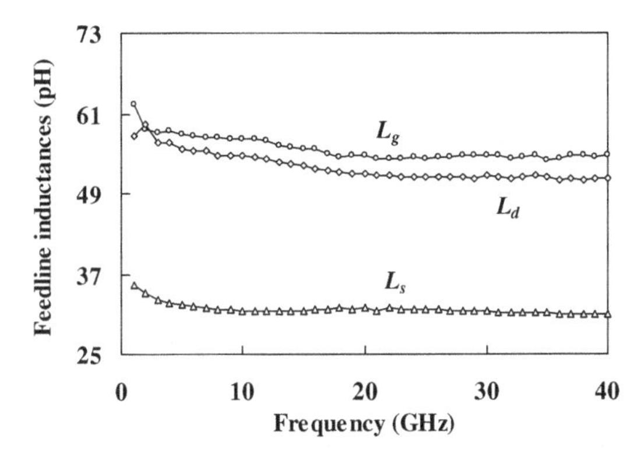

Figure 4.16. Extracted feedline inductances versus frequency (130 nm process).

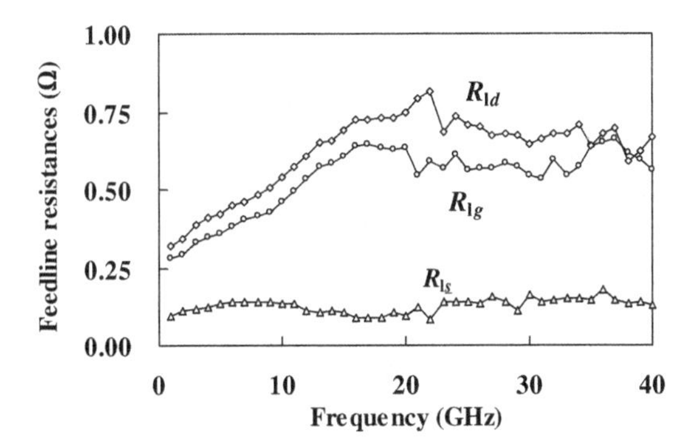

Figure 4.17. Extracted feedline resistances versus frequency (130 nm process).

process, respectively. The extrinsic parameters of the pad and feedline are tabulated in Table 4.4.

4.2.4 *Skin effect*

The skin effect is a well-known physical phenomenon. When the high-frequency current passes through the wire, the skin effect will cause the current to be unevenly distributed in the cross-section of the conductor. The current mainly flows on the surface of the conductor, and the closer it is to the surface of the conductor, the larger the current density, which is equivalent that the cross-sectional area of the conductor decreased and the resistance increased. The conventional short test structure model does not take into account the skin

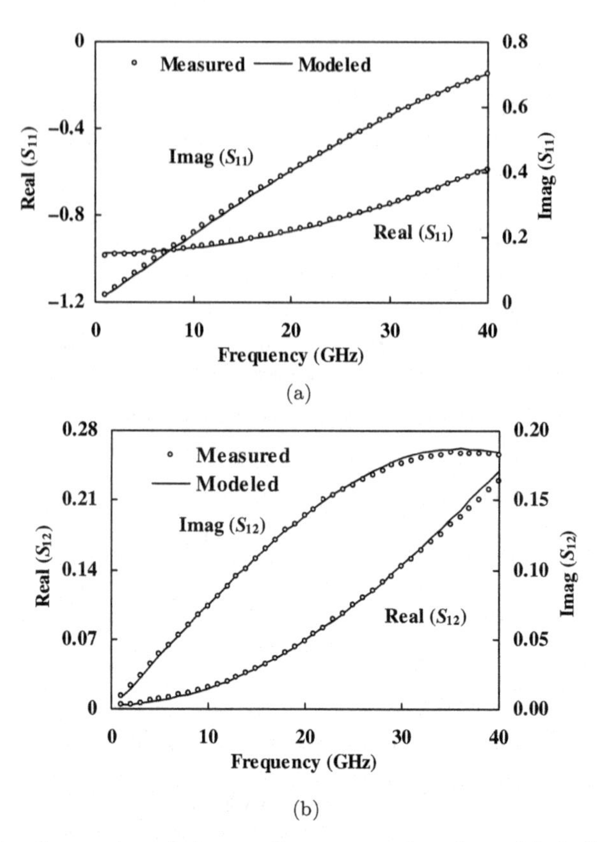

Figure 4.18. Comparison between the measured and modeled S-parameters (130 nm process): (a) S_{11}; (b) S_{12}.

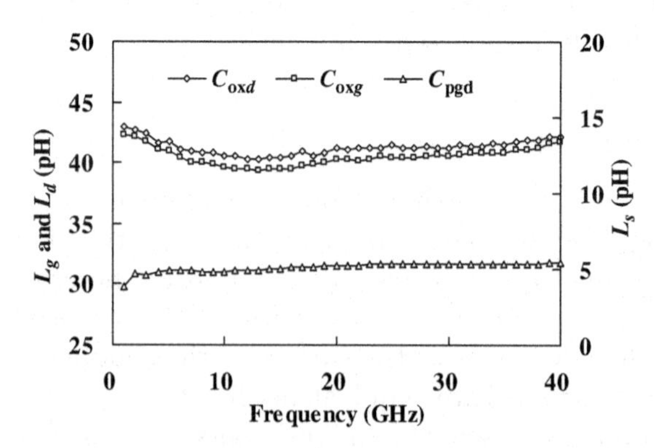

Figure 4.19. Extracted feedline inductances versus frequency (90 nm process).

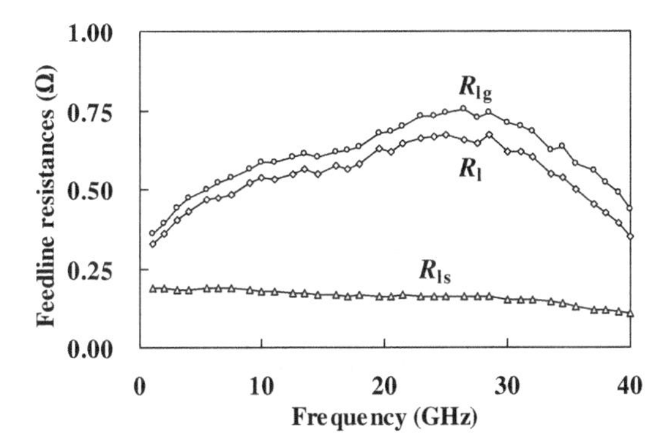

Figure 4.20. Extracted feedline resistances versus frequency (90 nm process).

Table 4.4. Pad and feedline extrinsic parameters.

	Parameters	Value	Parameters	Value
	$C_{\mathrm{ox}g}$(fF)	116.6	$C_{\mathrm{ox}d}$(fF)	114.5
Pad	C_{pgd}(fF)	0.8	$R_{\mathrm{pg}}(\Omega)$	10
	$R_{\mathrm{pd}}(\Omega)$	10.3		
	L_g(pH)	40	L_d(pH)	40
Feedline	L_s(pH)	5	$R_{\mathrm{l}g}(\Omega)$	0.6
	$R_{\mathrm{l}d}(\Omega)$	0.5	$R_{\mathrm{l}s}(\Omega)$	0.2

effect caused by high-frequency current. The extracted feedline resistances versus frequency is illustrated in Figure 4.21. As seen, the loss of feedline resistances increases with frequency. Hence, it is necessary to establish the equivalent circuit model including skin effect [13–15].

Figure 4.22 depicts the equivalent circuit model of short test structure considering the skin effect, where R_{lga}, R_{lda}, and R_{lsa} are the losses of the feedlines. L_{lga}, L_{lda}, and L_{lsa} are the corresponding inductances. Three pairs of parallel inductances and resistances R_{lgb}, L_{lgb}, R_{ldb}, L_{ldb}, R_{lsb}, and L_{lsb} are added to model both the drop of inductances and the increase of line resistances due to the skin effect in the feedlines.

In the proposed model, the Z-parameter matrix can be represented by gate feedline impedance Z_{lg}, drain feedline impedance Z_{ld},

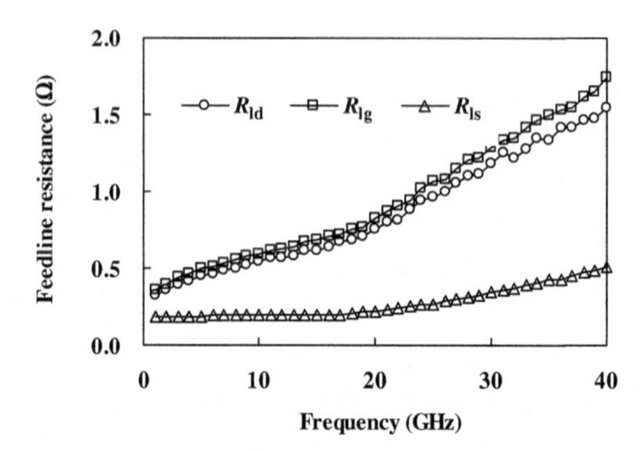

Figure 4.21. Extracted feedline resistances versus frequency.

and source feedline impedance $Z_{\rm ls}$:

$$Z^{\rm short} = \begin{bmatrix} Z_{\rm lg} + Z_{\rm ls} & Z_{\rm ls} \\ Z_{\rm ls} & Z_{\rm ld} + Z_{\rm ls} \end{bmatrix} \tag{4.36}$$

where

$$Z_{ld} = R_{\rm lda} + j\omega L_{\rm lda} + j\omega R_{\rm ldb} L_{\rm ldb}/(j\omega L_{\rm ldb} + R_{\rm ldb})$$
$$Z_{lg} = R_{\rm lga} + j\omega L_{\rm lga} + j\omega R_{\rm lgb} L_{\rm lgb}/(j\omega L_{\rm lgb} + R_{\rm lgb})$$
$$Z_{ls} = R_{\rm lsa} + j\omega L_{\rm lsa} + j\omega R_{\rm lsb} L_{\rm lsb}/(j\omega L_{\rm lsb} + R_{\rm lsb})$$

At low frequencies, $R_{\rm lga}$, $R_{\rm lda}$, and $R_{\rm lsa}$ can be directly obtained from the real part of the impedance, and the low-frequency inductances $L_{\rm lga}$, $L_{\rm lda}$, and $L_{\rm lsa}$ can also be directly calculated from the imaginary part of the impedance, and other parameters need to be obtained by utilizing semi-analytical method.

The extracted extrinsic parameters are tabulated in Table 4.5. Meanwhile, the comparison between the modeled and measured S parameter for short test structure is given in Figure 4.23, where 4.23(a)–4.23(f) are the comparison between the modeled and measured data of the magnitude and phase of S_{11}, S_{12}, and S_{22} in order.

The comparison of S-parameters accuracy between the proposed model and conventional model is illustrated in Figure 4.24. It can be clearly seen that the proposed model is more accurate than the conventional model. For S_{11} and S_{22}, the accuracy of the proposed

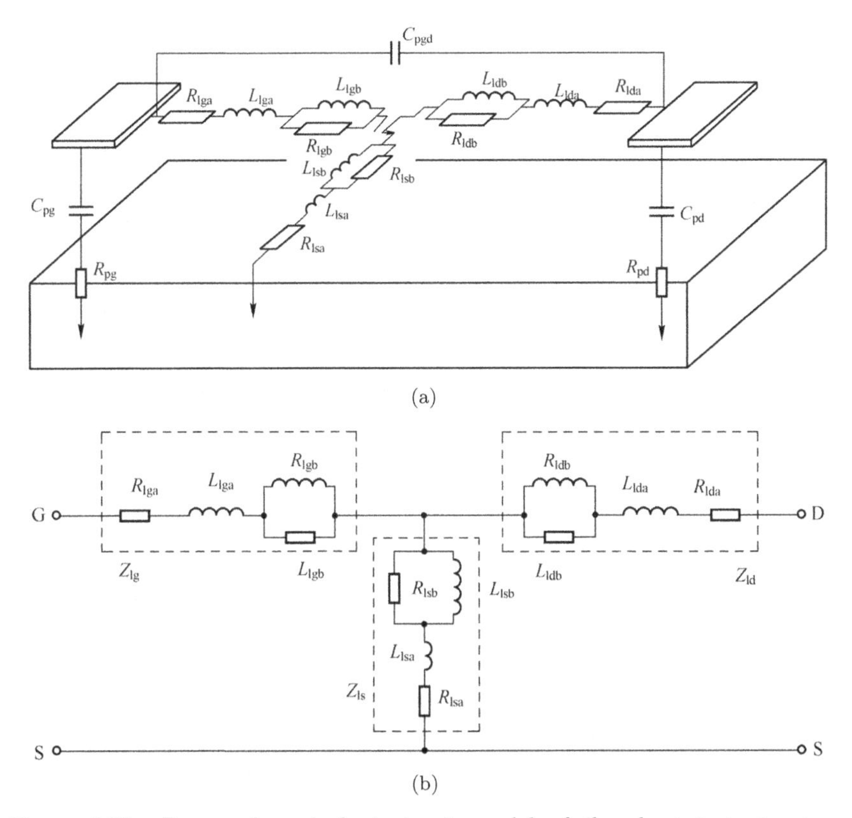

Figure 4.22. Proposed equivalent circuit model of the short test structure: (a) three-dimensional view; (b) plan view.

model is improved by 1% over the conventional model, while the accuracy of S_{12} (S_{21} is equal to S_{12}) is improved by nearly 8%. Note that the accuracy of S-parameters is defined as

$$\Delta S_{pq}^{N} = \left|S_{pq}^{N} - S_{pq}^{M}\right| / \left|S_{pq}^{M}\right| \tag{4.37}$$

$$\Delta S_{pq}^{C} = \left|S_{pq}^{C} - S_{pq}^{M}\right| / \left|S_{pq}^{M}\right| \tag{4.38}$$

where ΔS_{pq}^{N} (p, q = 1, 2) represents the accuracy of modeled S parameters in the proposed model and measured S parameters by utilizing short test structure. ΔS_{pq}^{C} (p, q = 1, 2) represents the accuracy of modeled S parameters in the conventional model and measured S parameters by utilizing short test structure. S_{pq}^{M} represents the measured S-parameter obtained by utilizing short

Table 4.5. Extrinsic parameters.

Type of parameters	Parameter	Value
Pad extrinsic parameters	C_{pg}(fF)	121
	C_{pd}(fF)	117
	C_{pgd}(fF)	1.2
	$R_{\text{pg}}(\Omega)$	9.8
	$R_{\text{pd}}(\Omega)$	10.5
Feedline parameters	$R_{\text{lga}}(\Omega)$	0.4
	$R_{\text{lda}}(\Omega)$	0.34
	$R_{\text{lsa}}(\Omega)$	0.18
	$R_{\text{lgb}}(\Omega)$	3
	$R_{\text{ldb}}(\Omega)$	3
	$R_{\text{lsb}}(\Omega)$	6.5
	L_{lga}(pH)	30
	L_{lda}(pH)	31
	L_{lsa}(pH)	0.1
	L_{lgb}(pH)	10
	L_{ldb}(pH)	10
	L_{lsb}(pH)	4.5

test structure. S_{pq}^{N} and S_{pq}^{C} represent the modeled S parameters in the proposed model and modeled S parameters in the conventional model, respectively.

4.2.5 *Feedline inductance*

Under the cut-off condition ($V_{\text{gs}} = 0$ V or V_{gs} is far less than the threshold voltage), the gate-source capacitance C_{gsp}, gate-drain capacitance C_{gdp}, and drain-source capacitance C_{dsp} can be obtained by utilizing the imaginary part of the Y parameter $\text{Im}(Y_{ij}^{\text{cl}})$:

$$C_{\text{gsp}} = \frac{\text{Im}(Y_{11}^{\text{cl}} + Y_{12}^{\text{cl}})}{\omega} \tag{4.39}$$

$$C_{\text{gdp}} = -\frac{\text{Im}(Y_{12}^{\text{cl}})}{\omega} \tag{4.40}$$

$$C_{\text{jd}} + C_{\text{dsp}} = \frac{\text{Im}(Y_{22}^{\text{cl}} + Y_{12}^{\text{cl}})}{\omega} \tag{4.41}$$

The extracted C_{gsp}, C_{gdp}, and $C_{\text{dsp}} + C_{\text{jd}}$ at low frequencies are illustrated in Figure 4.25. The feedline inductances can be directly

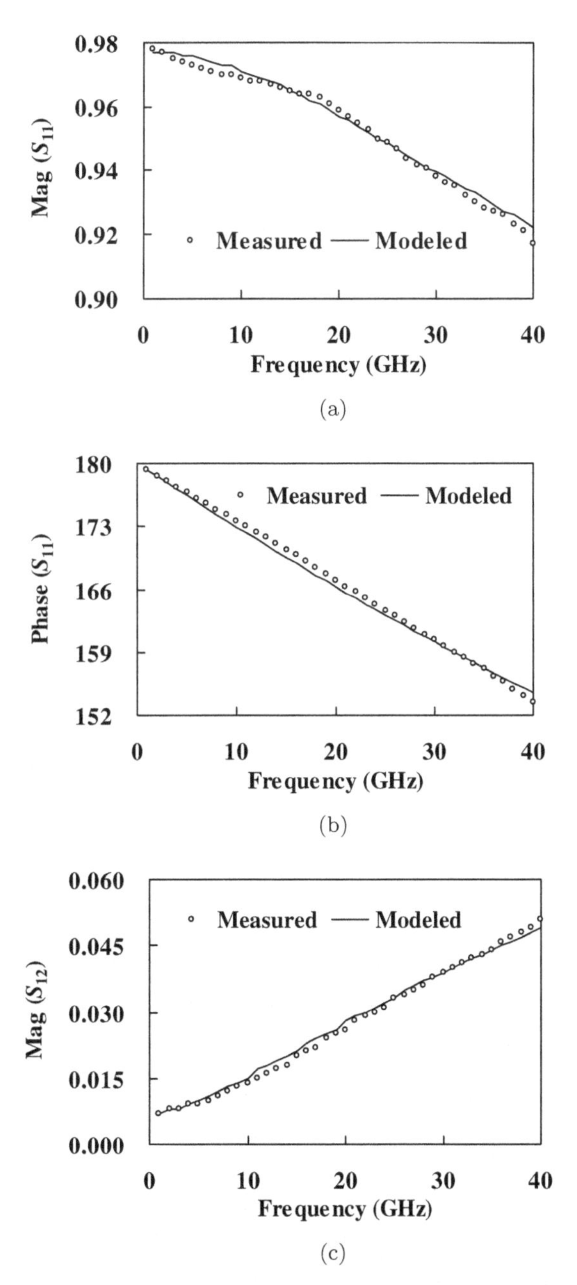

Figure 4.23. Comparison of the modeled and measured S parameter for the short test structure: (a) the magnitude of S_{11}; (b) the phase of S_{11}; (c) the magnitude of S_{12}; (d) the phase of S_{12}; (e) the magnitude of S_{22}; (f) the phase of S_{22}.

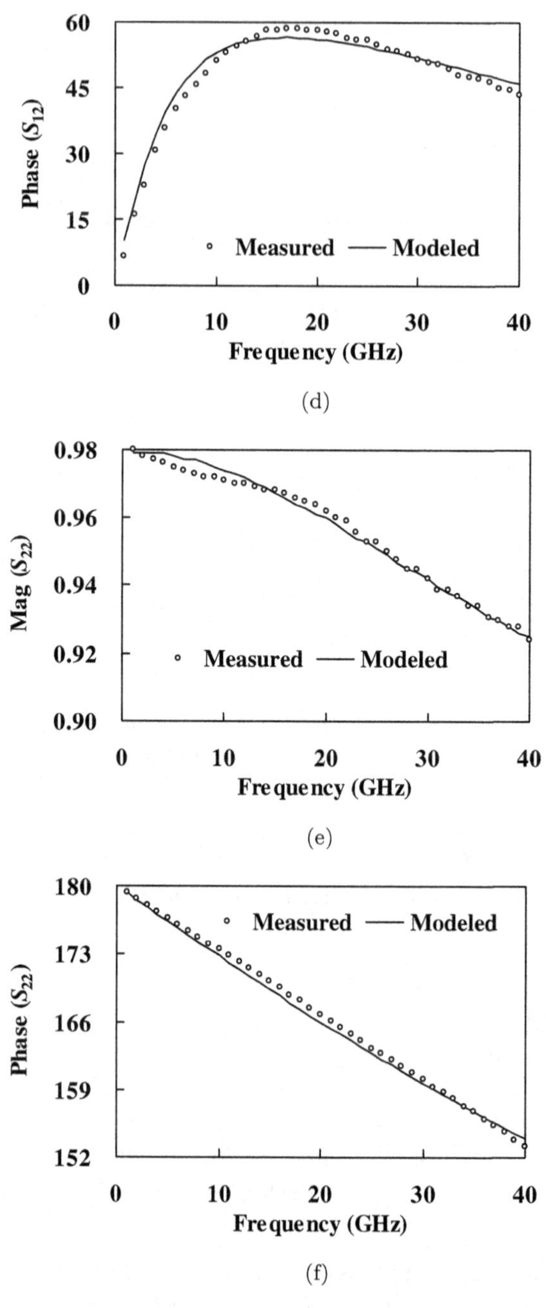

Figure 4.23. *(Continued)*

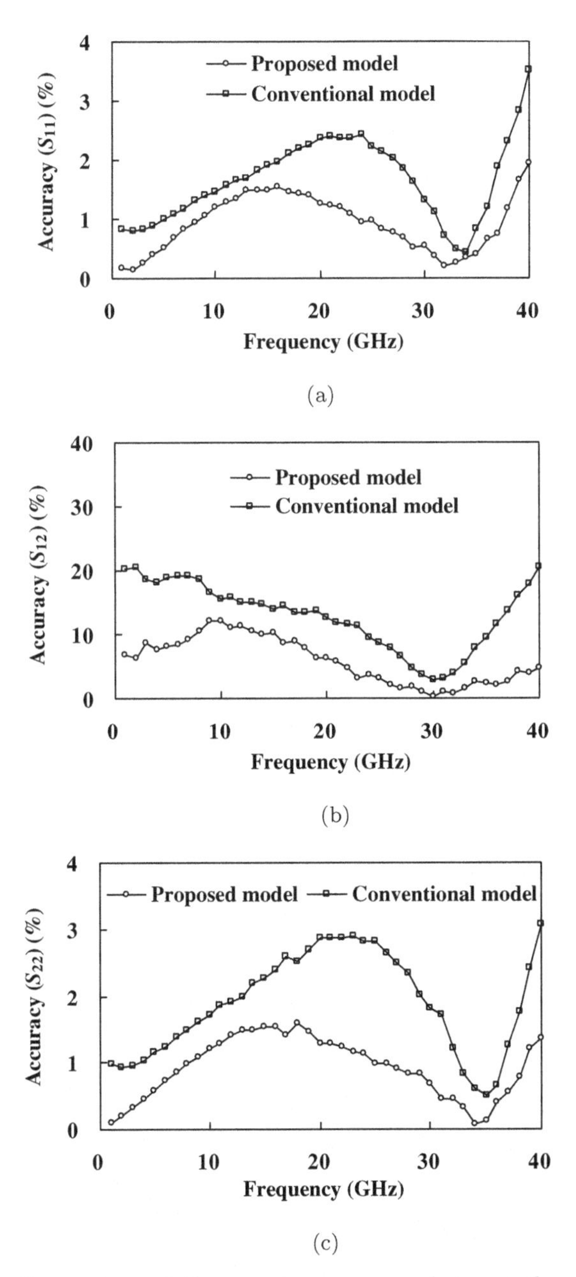

Figure 4.24. Comparison of S-parameters accuracy between the proposed and conventional models: (a) the accuracy of S_{11}; (b) the accuracy of S_{12}; (c) the accuracy of S_{22}.

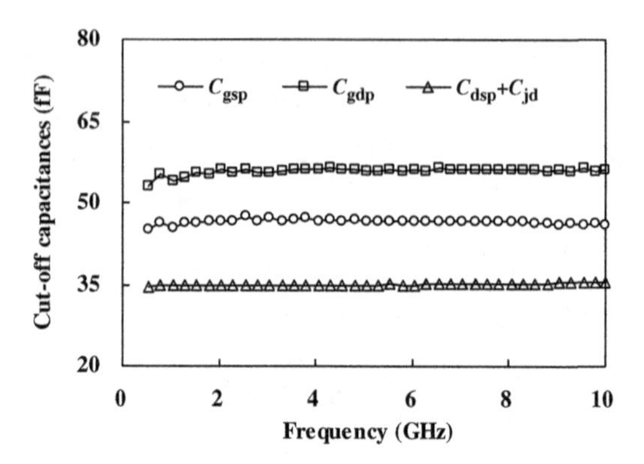

Figure 4.25. Extracted intrinsic capacitances at low frequencies.

calculated by ignoring high-order terms:

$$L_g \approx \frac{\mathrm{Im}(Z_{11}^c - Z_{12}^c) + (C_{\mathrm{dsp}} + C_{\mathrm{jd}})/\omega B}{\omega} \tag{4.42}$$

$$L_d \approx \frac{\mathrm{Im}(Z_{22}^c - Z_{12}^c) + C_{\mathrm{gsp}}/\omega B}{\omega} \tag{4.43}$$

$$L_s \approx \frac{\mathrm{Im}(Z_{12}^c) + C_{\mathrm{gdp}}/\omega B}{\omega} \tag{4.44}$$

wherein

$$B = C_{\mathrm{gsp}}C_{\mathrm{gdp}} + C_{\mathrm{gsp}}(C_{\mathrm{dsp}} + C_{\mathrm{jd}}) + C_{\mathrm{gdp}}(C_{\mathrm{dsp}} + C_{\mathrm{jd}})$$

The extracted feedline inductances L_g, L_d, and L_s at low frequencies are illustrated in Figure 4.26.

4.3 Extraction of Extrinsic Resistances

The gate resistance can greatly affect the performance of the circuit. The gate resistance will introduce thermal noise and increase the noise figure of the circuit, as well as affect the switching speed and maximum oscillation frequency of the transistor. Thence, the gate resistance should be reduced as much as possible in the layout design. The extrinsic resistances mainly caused by factors, such as layout design, process, and device degradation, have a great impact on input and output impedance matching, noise performance, and oscillation frequency. The accuracy of extrinsic resistance will directly influence

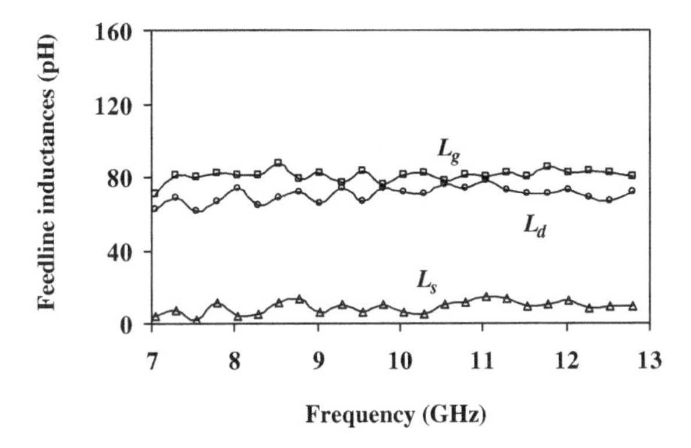

Figure 4.26. Extracted feedline inductances.

the accuracy of the model. Therefore, accurately extracting extrinsic resistances have always been one of the hot topics in modeling.

Currently, the commonly used methods for determining extrinsic resistances mainly include the Cold-FET method, the normal bias method, and the cut-off method [4,16,17]. The following part describes the extracted extrinsic resistance method.

4.3.1 *COLD-FET method*

The Cold-FET method refers to the case that the transistor is working under the strong inversion region ($V_{\mathrm{gs}} = 1.2\mathrm{V}$ and $V_{\mathrm{ds}} = 0\mathrm{V}$). Under such bias conditions, the transconductance g_m is approximately equal to zero and can be ignored. Since the extrinsic impedance of the substrate caused by the reverse PN junction is much greater than the source extrinsic resistance R_s and channel resistance R_{ch}, the equivalent circuit can be simplified to Figure 4.27 [16].

According to the small-signal equivalent circuit model, we have

$$\mathrm{Re}\left(Z_{11}\right) = R_g + R_s + \frac{A}{4} \tag{4.45}$$

$$\mathrm{Re}\left(Z_{12}\right) = \mathrm{Re}\left(Z_{21}\right) = \frac{A}{2} \tag{4.46}$$

$$\mathrm{Re}\left(Z_{22}\right) = R_d + R_s + A \tag{4.47}$$

where

$$A = \frac{R_{\text{ch}}}{1 + \omega^2 C_x R_{\text{ch}}^2}$$

$$C_x = C_{\text{ds}} + \frac{C_{\text{gs}} C_{\text{gd}}}{C_{\text{gs}} + C_{\text{gd}}}$$

The imaginary part of Z_{22} is

$$\text{Im}\,(Z_{22}) = -\frac{\omega C_x R_{\text{ch}}^2}{1 + \omega^2 C_x^2 R_{\text{ch}}^2} \tag{4.48}$$

namely,

$$-\frac{\omega}{\text{Im}\,(Z_{22})} = C_x \omega^2 + \frac{1}{R_{\text{ch}}^2 C_x} \tag{4.49}$$

Thus, C_x and R_{ch} can be obtained from the slope and intercept of $-\omega/\text{Im}(\text{Z}_{22})$ versus ω^2.

After C_x and R_{ch} are determined, R_s, R_d, and R_g can be directly calculated. The extracted extrinsic resistances versus frequency for 130 nm and 90 nm process MOSFET devices are illustrated in Figure 4.28.

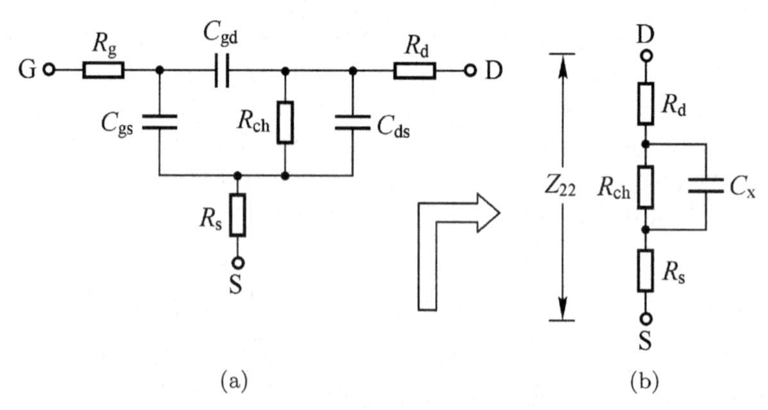

Figure 4.27. Equivalent circuit model of strong inversion region under bias condition of $V_{\text{gs}} = 1.2\,\text{V}$ and $V_{\text{ds}} = 0\,\text{V}$.

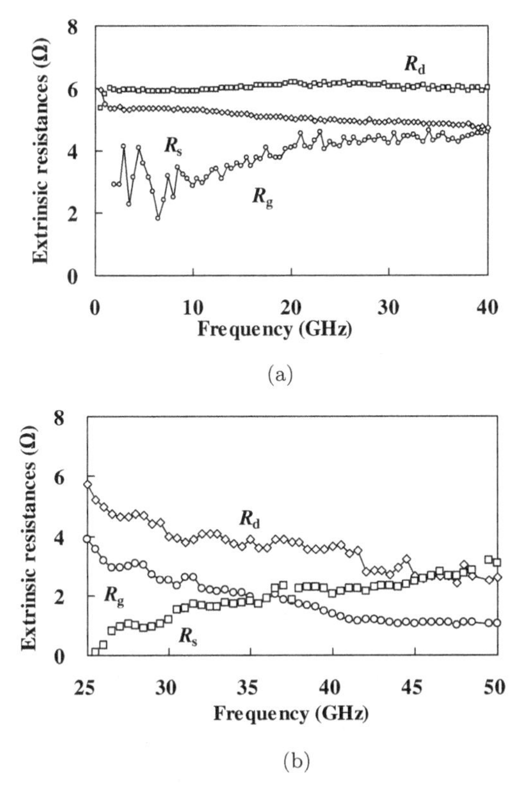

Figure 4.28. Extracted extrinsic resistances versus frequency: (a) 130 nm process; (b) 90 nm process.

4.3.2 *Hot-FET method*

Under normal operating bias conditions, when the parasitic effect of the substrate is ignored, the small signal equivalent circuit model is depicted in Figure 4.29 [17], and the open circuit Z parameter can be described as follows:

$$Z_{11} = R_g + R_s + j\omega(L_g + L_s) + \frac{g_{\text{ds}} - j\omega(C_{\text{gd}} + C_{\text{ds}})}{D} \tag{4.50}$$

$$Z_{12} = R_s + j\omega L_s + \frac{j\omega C_{\text{gd}}}{D} \tag{4.51}$$

$$Z_{21} = R_s + j\omega L_s - \frac{g_m - j\omega C_{\text{gd}}}{D} \tag{4.52}$$

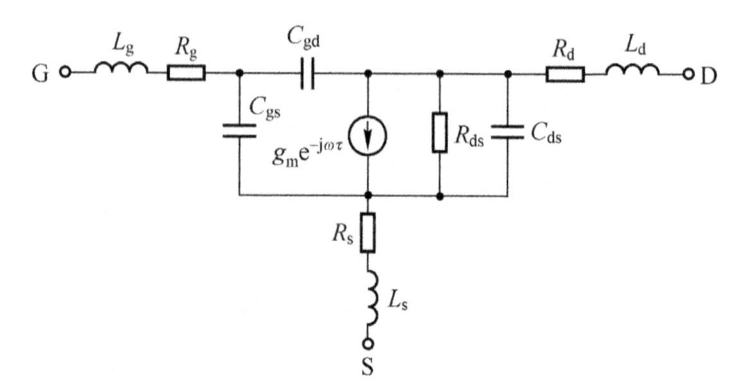

Figure 4.29. Small-signal equivalent circuit model by neglecting the substrate parasitic effect.

$$Z_{22} = R_d + R_s + j\omega(L_d + L_s) + \frac{j\omega(C_{gs} + C_{gd})}{D} \tag{4.53}$$

where

$$D = -\omega^2(C_{gs}C_{ds} + C_{gs}C_{gd} + C_{gd}C_{ds}) + j\omega[g_m C_{gd} + g_{ds}(C_{gs} + C_{gd})]$$

The real part of the Z parameter can be expressed as

$$\text{Re}(Z_{12}) = R_s + \frac{A_s}{\omega^2 + B} \tag{4.54}$$

$$\text{Re}(Z_{22} - Z_{12}) = R_d + \frac{A_d}{\omega^2 + B} \tag{4.55}$$

$$\text{Re}(Z_{11} - Z_{12}) = R_g + \frac{A_g}{\omega^2 + B} \tag{4.56}$$

where A_s, A_d, A_g, and B are constants related to intrinsic parameters but independent of frequency. Then three series extrinsic resistances R_s, R_d, and R_g are extracted by utilizing the curve fitting method. Assuming $\omega^2 \gg B$ at high frequencies, the above expressions can be simplified as

$$\text{Re}\,(Z_{12}) = R_s + A_s\omega^{-2} \tag{4.57}$$

$$\text{Re}\,(Z_{22} - Z_{12}) = R_d + A_d\omega^{-2} \tag{4.58}$$

$$\text{Re}\,(Z_{11} - Z_{12}) = R_g + A_g\omega^{-2} \tag{4.59}$$

Draw the linear curve of $\text{Re}(Z_{ij})(i = 1, 2,\ j = 1, 2)$ versus ω^{-2}, the extrinsic series resistance can be obtained according to the intercept on the vertical axis.

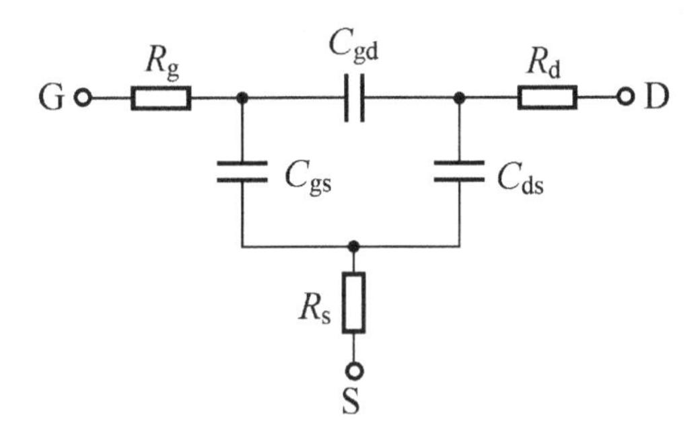

Figure 4.30. Small-signal equivalent circuit model under conventional zero-bias condition.

4.3.3 *Cut-off method*

Under cut-off condition ($V_{gs} = V_{ds} = 0$ V), the transconductance of the transistor is close to zero. When the influence of substrate parasitic effect is neglected, the equivalent circuit after de-embedding the pad and the feedline elements is shown in Figure 4.30 [4].

The extrinsic resistances can be directly determined utilizing the open circuit Z parameter of the equivalent circuit in Figure 4.30:

$$R_g = \mathrm{Re}(Z_{11}) - \mathrm{Re}(Z_{12}) \tag{4.60}$$

$$R_s = \mathrm{Re}(Z_{12}) \tag{4.61}$$

$$R_d = \mathrm{Re}(Z_{22}) - \mathrm{Re}\left(Z_{12}\right) \tag{4.62}$$

The extracted extrinsic resistances versus frequency under cut-off condition is illustrated in Figure 4.31.

Nevertheless, the parameter extraction neglects the influence of substrate parasitics including substrate resistance and capacitance, which will cause the extrinsic resistance to fluctuate with frequency at high frequencies. Since silicon is a semiconductor and has low resistance properties, it cannot completely isolate the transistor from the substrate. Hence, the substrate coupling effects occur. When the device operates at high frequencies, part of the signal flows through the substrate. At low frequencies, the substrate can be ignored in the small signal model of MOSFET devices, however the substrate effect cannot be neglected at high frequencies [18,19].

The small-signal equivalent circuit model after de-embedding the pad extrinsic elements and the feedline inductances under cut-off

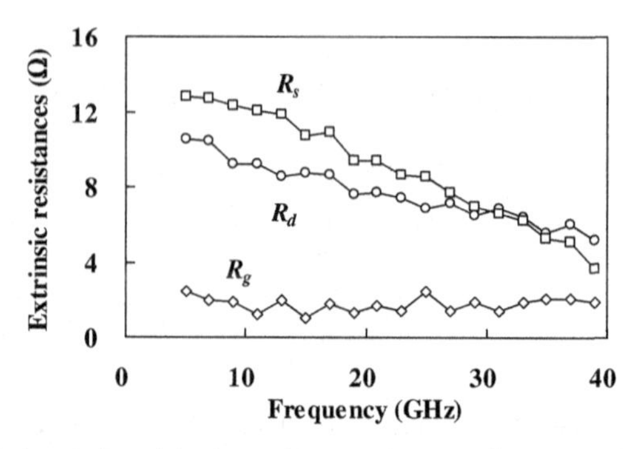

Figure 4.31. Extracted extrinsic resistances versus frequency under the cut-off condition.

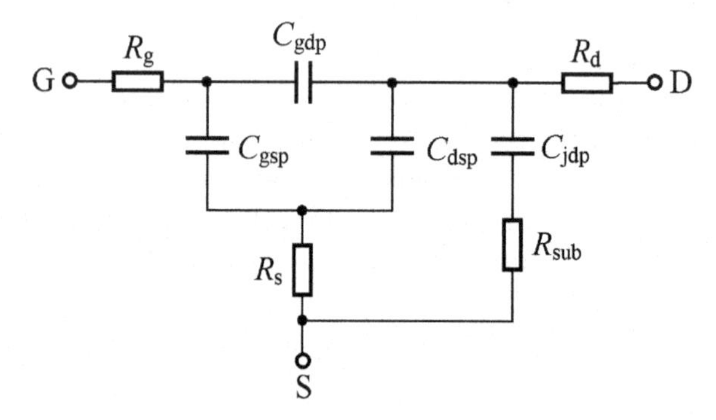

Figure 4.32. Equivalent circuit model after de-embedding the pad extrinsic elements and the feedline inductances under cut-off condition.

condition of the MOSFET device is depicted in Figure 4.32. The Z parameter of the equivalent circuit model can be expressed as

$$Z_{11}^{c} = \frac{j\omega(C_{\mathrm{gdp}} + C_{\mathrm{dsp}}) + M_c R_s + Y_{\mathrm{jd}}[1 + j\omega R_s(C_{\mathrm{gsp}} + C_{\mathrm{dsp}})]}{M_c + Y_{\mathrm{jd}}[j\omega(C_{\mathrm{gdp}} + C_{\mathrm{gsp}}) + M_c R_s]} \tag{4.63}$$

$$Z_{12}^{c} = Z_{21}^{c} = \frac{j\omega C_{\mathrm{gdp}} + M_c R_s}{M_c + Y_{\mathrm{jd}}[j\omega(C_{\mathrm{gdp}} + C_{\mathrm{gsp}}) + M_c R_s]} \tag{4.64}$$

$$Z_{22}^{c} = \frac{j\omega(C_{\mathrm{gdp}} + C_{\mathrm{gsp}}) + M_c R_s}{M_c + Y_{\mathrm{jd}}[j\omega(C_{\mathrm{gdp}} + C_{\mathrm{gsp}}) + M_c R_s]} \tag{4.65}$$

where

$$M_c = -\omega^2(C_{\mathrm{gsp}}C_{\mathrm{dsp}} + C_{\mathrm{gsp}}C_{\mathrm{gdp}} + C_{\mathrm{gdp}}C_{\mathrm{dsp}})$$

Ignoring the higher-order terms of ω^2, we have

$$\mathrm{Re}\left(\frac{1}{Z_{22}^{\mathrm{c}}}\right) \approx \omega^2 C_x^2 (R_s + R_d) + \frac{\omega^2 \left[C_{\mathrm{jdp}}^2 R_{\mathrm{sub}} + C_x C_{\mathrm{jdp}} (R_s + R_d)\right]}{1 + \omega^2 C_{\mathrm{jdp}}^2 R_{\mathrm{sub}}^2} \tag{4.66}$$

$$\mathrm{Im}\left(\frac{1}{Z_{22}^{\mathrm{c}}}\right) \approx \omega C_x + \frac{\omega C_{\mathrm{jdp}}}{1 + \omega^2 C_{\mathrm{jdp}}^2 R_{\mathrm{sub}}^2} \tag{4.67}$$

where

$$C_x = C_{\mathrm{dsp}} + C_{\mathrm{gsp}} C_{\mathrm{gdp}} / (C_{\mathrm{gsp}} + C_{\mathrm{gdp}})$$

After transformation by utilizing the above equation, we have

$$\left(-\frac{d\left[\mathrm{Re}(Z_{22}^{\mathrm{c}})^{-1}/\omega^2\right]}{d\omega^2}\right)^{-\frac{1}{2}} = \frac{C_{\mathrm{jdp}}^2 R_{\mathrm{sub}}^2}{\sqrt{A}} \omega^2 + \frac{1}{\sqrt{A}} = f(\omega) = a\omega^2 + b \tag{4.68}$$

where A, a, and b are constants independent of frequency. According to $f(\omega)$ versus ω^2, as shown in Figure 4.33, it is assumed that the slope and intercept of the fitted straight line are a and b, respectively, so that C_{jdp} multiply R_{sub} can be directly obtained by the following equation:

$$C_{\mathrm{jdp}} R_{\mathrm{sub}} = \sqrt{\frac{a}{b}} \tag{4.69}$$

Combined with other equations, the substrate extrinsic elements C_{jdp} and R_{sub} can be immediately obtained.

For the CMOS process, the gate, source, and drain extrinsic resistances are much less than the substrate extrinsic resistance. Hence, the extrinsic resistances can be approximately expressed as

$$R_s \approx \frac{\omega C_{\mathrm{jdp}} R_{\mathrm{sub}} (B_1^2 + B_2^2) \mathrm{Re}(Z_{12}) - \omega^3 C_{\mathrm{jdp}}^3 C_{\mathrm{gdp}} (C_{\mathrm{gdp}} + C_{\mathrm{gsp}}) R_{\mathrm{sub}}^2}{\omega C_{\mathrm{jdp}} R_{\mathrm{sub}} B_1^2 - B_1 B_2} \tag{4.70}$$

$$R_g \approx \mathrm{Re}(Z_{11}) - \frac{\omega^2 C_{\mathrm{jdp}}^2 C_{\mathrm{gdp}}^2 R_{\mathrm{sub}} + (B_1^2 + B_2^2 - \omega C_{\mathrm{gdp}} C_{\mathrm{jdp}} B_2) R_s}{B_1^2 + B_2^2} \tag{4.71}$$

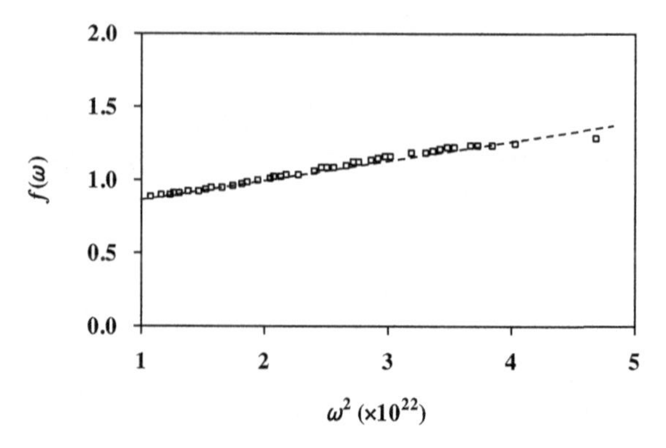

Figure 4.33. $f(\omega)$ versus ω^2.

$$R_d \approx \mathrm{Re}(Z_{22}) - \mathrm{Re}(Z_{12}) - \frac{\omega^2 C_{\mathrm{jdp}}^2 C_{\mathrm{gdp}}(C_{\mathrm{gdp}} + C_{\mathrm{gsp}})R_{\mathrm{sub}}}{B_1^2 + B_2^2} \tag{4.72}$$

where

$$B_1 = -\omega^2 C_r C_{\mathrm{jdp}} R_{\mathrm{sub}}$$

$$B_2 = \omega[C_{\mathrm{gsp}}C_{\mathrm{gdp}} + C_{\mathrm{gsp}}(C_{\mathrm{dsp}} + C_{\mathrm{jdp}}) + C_{\mathrm{gdp}}(C_{\mathrm{dsp}} + C_{\mathrm{jdp}})]$$

$$C_r = C_{\mathrm{gsp}}C_{\mathrm{gdp}} + C_{\mathrm{gsp}}C_{\mathrm{dsp}} + C_{\mathrm{gdp}}C_{\mathrm{dsp}}$$

The MOSFET device with 90 nm gate length, 1 μm cell gate width, 16 gate fingers, and two cells is utilized as an example. The extrinsic resistances versus frequency is shown in Figure 4.34. It can be observed that relatively stable resistances can be achieved at middle frequencies.

The comparison of extracted extrinsic resistances and variation range for different methods is summarized in Table 4.6. As seen, compared with the conventional method, the fluctuation range of the data obtained by the proposed method becomes smaller versus frequency.

4.4 Extraction of Intrinsic Parameters

The small-signal equivalent circuit model can be divided into two parts: the extrinsic network and the intrinsic network. The intrinsic part is the core area of the small-signal model for a MOSFET device.

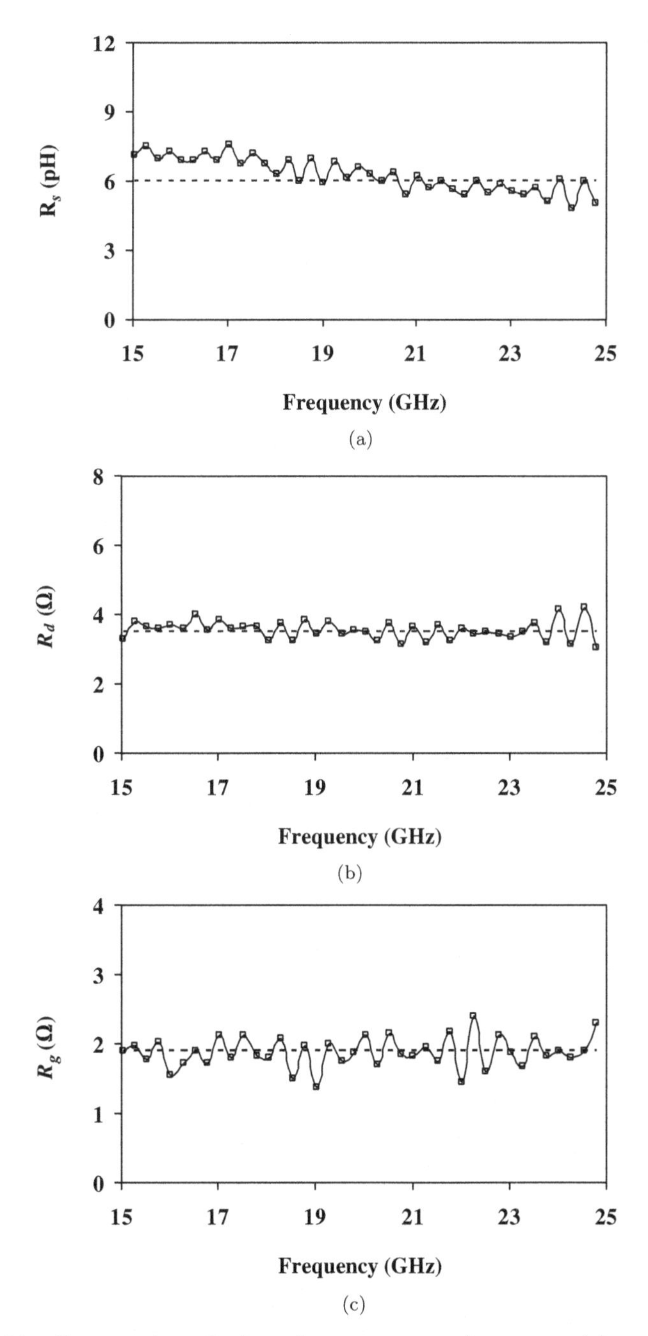

Figure 4.34. Extracted extrinsic resistances versus frequency: (a) source resistance; (b) gate resistance; (c) drain resistance.

Table 4.6. Comparison of extracted extrinsic resistances for different methods.

Type of the extraction method	R_g (Ω)	Variation (Ω)	R_d (Ω)	Variation (Ω)	R_s (Ω)	Variation (Ω)
Conventional cut-off method	6.9	±3.8	1.8	±1.0	7.4	±4.3
Cold-FET	1.9	±1.2	3.6	±2.0	1.9	±1.3
Normal bias	2.5	—	2.3	—	1.2	—
Proposed cut-off method	3.5	±0.3	1.9	±0.3	6.0	±0.7

After the pad and feedline extrinsic elements and extrinsic resistances are determined, the next step is to determine the parameters of the drain substrate network, and finally, the intrinsic parameters can be obtained.

4.4.1 *Substrate elements*

Usually, zero-bias method is utilized to extract the extrinsic substrate parameters R_{sub} and C_{jd} of a MOSFET device. The zero-bias method is the test method for devices without any bias ($V_{\text{gs}} = V_{\text{ds}} = 0\text{V}$). The equivalent circuit model for calculating Z_{22} under zero-bias condition is depicted in Figure 4.35 [9].

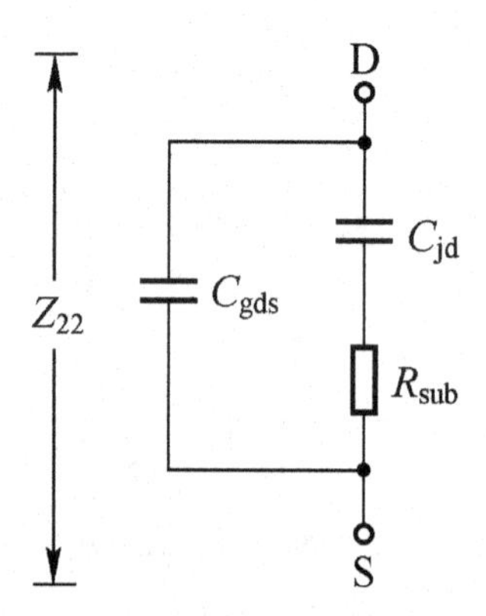

Figure 4.35. Equivalent circuit model for calculating Z_{22} under zero-bias condition.

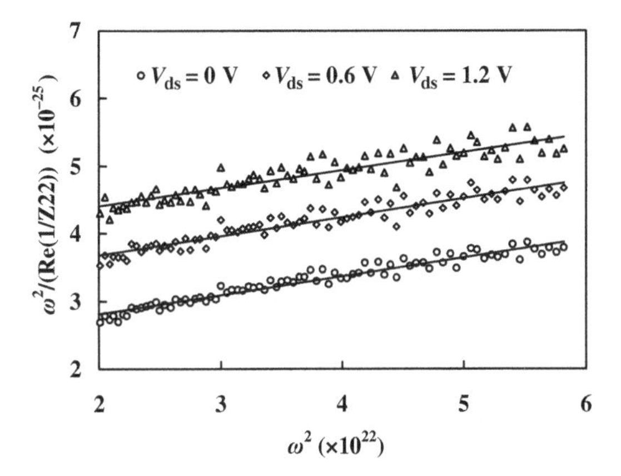

Figure 4.36. $\omega^2/\mathrm{Re}(1/Z_{22})$ versus ω^2 under bias condition of $V_{\mathrm{ds}} = 0$ V, 0.6 V, and 1.2 V.

The impedance of Z_{22} can be expressed as

$$\frac{1}{Z_{22}} = \frac{\omega^2 R_{\mathrm{sub}} C_{\mathrm{jd}}^2}{1 + \omega^2 C_{\mathrm{jd}}^2 R_{\mathrm{sub}}^2} + j\left(\omega C_{\mathrm{gds}} + \frac{\omega C_{\mathrm{jd}}}{1 + \omega^2 C_{jd}^2 R_{\mathrm{sub}}^2}\right) \tag{4.73}$$

After transformation, we have

$$\frac{\omega^2}{\mathrm{Re}(1/Z_{22})} = \omega^2 R_{\mathrm{sub}} + \frac{1}{R_{\mathrm{sub}} C_{\mathrm{jd}}^2} \tag{4.74}$$

$\omega^2/\mathrm{Re}(1/Z_{22})$ versus ω^2 under different bias conditions is depicted in Figure 4.36. Utilizing the linear fitting method, R_{sub} can be obtained from the slope of the curve, and C_{jd} can be obtained by the following expression:

$$C_{\mathrm{jd}} = \frac{1}{\sqrt{b_0 R_{\mathrm{sub}}}} \tag{4.75}$$

where b_0 represents the intercept of the fitted curve on the vertical axis.

Note that the above extraction method assumes that the extrinsic capacitance C_{jd} and extrinsic resistance R_{sub} are independent of the drain voltage. Actually, the extrinsic capacitance C_{jd} represents the extrinsic diode junction capacitance between the drain and substrate, and the extrinsic resistance R_{sub} represents the loss from the drain to substrate. According to the basic principle of diode, the diode junction capacitance C_{sub} is related to the drain voltage, when the drain voltage varies, the junction capacitance C_{jd} also varies.

4.4.2 *Direct-extraction method*

Since the extrinsic parameters have been determined, the intrinsic parameters versus frequency can be directly determined by the following steps:

(1) Convert the measured S parameters to Y parameters:

$$S_{\text{m}} = \begin{bmatrix} S_{11}^{\text{m}} & S_{12}^{\text{m}} \\ S_{21}^{\text{m}} & S_{22}^{\text{m}} \end{bmatrix} \Rightarrow Y_{\text{m}} = \begin{bmatrix} Y_{11}^{\text{m}} & Y_{12}^{\text{m}} \\ Y_{21}^{\text{m}} & Y_{22}^{\text{m}} \end{bmatrix} \tag{4.76}$$

(2) De-embedding the extrinsic pad open structure:

$$Y_{\text{m1}} = Y_{\text{m}} - Y_{\text{open}} \tag{4.77}$$

(3) Convert Y parameters to Z parameters:

$$Y_{\text{m1}} = \begin{bmatrix} Y_{11}^{\text{m1}} & Y_{12}^{\text{m1}} \\ Y_{21}^{\text{m1}} & Y_{22}^{\text{m1}} \end{bmatrix} \Rightarrow Z_{\text{m1}} = \begin{bmatrix} Z_{11}^{\text{m1}} & Z_{12}^{\text{m1}} \\ Z_{21}^{\text{m1}} & Z_{22}^{\text{m1}} \end{bmatrix} \tag{4.78}$$

(4) De-embedding the feedline extrinsic inductances and the gate and drain extrinsic resistances:

$$Z_{\text{m2}} = Z_{\text{m1}} - \begin{bmatrix} R_g + j\omega\left(L_g + L_s\right) & j\omega L_s \\ j\omega L_s & R_d + j\omega\left(L_d + L_s\right) \end{bmatrix} \tag{4.79}$$

(5) Convert Z parameters to Y parameters:

$$Z_{\text{m2}} = \begin{bmatrix} Z_{11}^{\text{m2}} & Z_{12}^{\text{m2}} \\ Z_{21}^{\text{m2}} & Z_{22}^{\text{m2}} \end{bmatrix} \Rightarrow Y_{\text{m2}} = \begin{bmatrix} Y_{11}^{\text{m2}} & Y_{12}^{\text{m2}} \\ Y_{21}^{\text{m2}} & Y_{22}^{\text{m2}} \end{bmatrix} \tag{4.80}$$

(6) De-embedding the extrinsic substrate network:

$$Y_{\text{m3}} = Y_{\text{m2}} - \begin{bmatrix} 0 & 0 \\ 0 & \frac{\omega^2 C_{\text{jd}}^2 R_{\text{sub}}}{1+\omega^2 C_{\text{jd}}^2 R_{\text{sub}}^2} + j\frac{\omega C_{\text{jd}}}{1+\omega^2 C_{\text{jd}}^2 R_{\text{sub}}^2} \end{bmatrix} \tag{4.81}$$

(7) Convert Y parameters to Z parameters:

$$Y_{\mathrm{m3}} = \begin{bmatrix} Y_{11}^{\mathrm{m3}} & Y_{12}^{\mathrm{m3}} \\ Y_{21}^{\mathrm{m3}} & Y_{22}^{\mathrm{m3}} \end{bmatrix} \Rightarrow Z_{\mathrm{m3}} = \begin{bmatrix} Z_{11}^{\mathrm{m3}} & Z_{12}^{\mathrm{m3}} \\ Z_{21}^{\mathrm{m3}} & Z_{22}^{\mathrm{m3}} \end{bmatrix} \tag{4.82}$$

(8) De-embedding the source extrinsic resistance R_s:

$$Z_{\mathrm{m4}} = Z_{\mathrm{m3}} - \begin{bmatrix} R_s & R_s \\ R_s & R_s \end{bmatrix} \tag{4.83}$$

(9) Convert Z parameters to Y parameters. Here, Y parameters are the measured data of the intrinsic network:

$$Z_{\mathrm{m4}} = \begin{bmatrix} Z_{11}^{\mathrm{m4}} & Z_{12}^{\mathrm{m4}} \\ Z_{21}^{\mathrm{m4}} & Z_{22}^{\mathrm{m4}} \end{bmatrix} \Rightarrow Y_{\mathrm{int}} = \begin{bmatrix} Y_{11}^{\mathrm{int}} & Y_{12}^{\mathrm{int}} \\ Y_{21}^{\mathrm{int}} & Y_{22}^{\mathrm{int}} \end{bmatrix} \tag{4.84}$$

(10) According to the Y parameter of the intrinsic network:

$$Y_{\mathrm{int}} = \begin{bmatrix} j\omega\left(C_{\mathrm{gs}} + C_{\mathrm{gd}}\right) & -j\omega C_{\mathrm{gd}} \\ g_{\mathrm{m}}e^{-j\omega\tau} - j\omega C_{\mathrm{gd}} & g_{\mathrm{ds}} + j\omega\left(C_{\mathrm{ds}} + C_{\mathrm{gd}}\right) \end{bmatrix} \tag{4.85}$$

The intrinsic elements can be achieved by the following equation:

$$C_{\mathrm{gd}} = -\frac{\mathrm{Im}\left(Y_{11}^{\mathrm{int}}\right)}{\omega} \tag{4.86}$$

$$C_{\mathrm{gs}} = \frac{\mathrm{Im}\left(Y_{11}^{\mathrm{int}} + Y_{12}^{\mathrm{int}}\right)}{\omega} \tag{4.87}$$

$$C_{\mathrm{gs}} = \frac{\mathrm{Im}\left(Y_{22}^{\mathrm{int}} + Y_{12}^{\mathrm{int}}\right)}{\omega} \tag{4.88}$$

$$g_{\mathrm{ds}} = \mathrm{Re}\left(Y_{22}^{\mathrm{int}}\right) \tag{4.89}$$

$$g_m = \left|Y_{21}^{\mathrm{int}} - Y_{12}^{\mathrm{int}}\right| \tag{4.90}$$

$$\tau = -\frac{1}{\omega}\arctan\left[\frac{\mathrm{Im}\left(Y_{21}^{\mathrm{int}} - Y_{12}^{\mathrm{int}}\right)}{\mathrm{Re}\left(Y_{21}^{\mathrm{int}} - Y_{12}^{\mathrm{int}}\right)}\right] \tag{4.91}$$

In order to validate and assess the accuracy of the extraction, the MOSFET device with 90 nm gate length, 16 gate fingers, and two cells was investigated. Figures 4.37(a)–4.37(f) provide the intrinsic

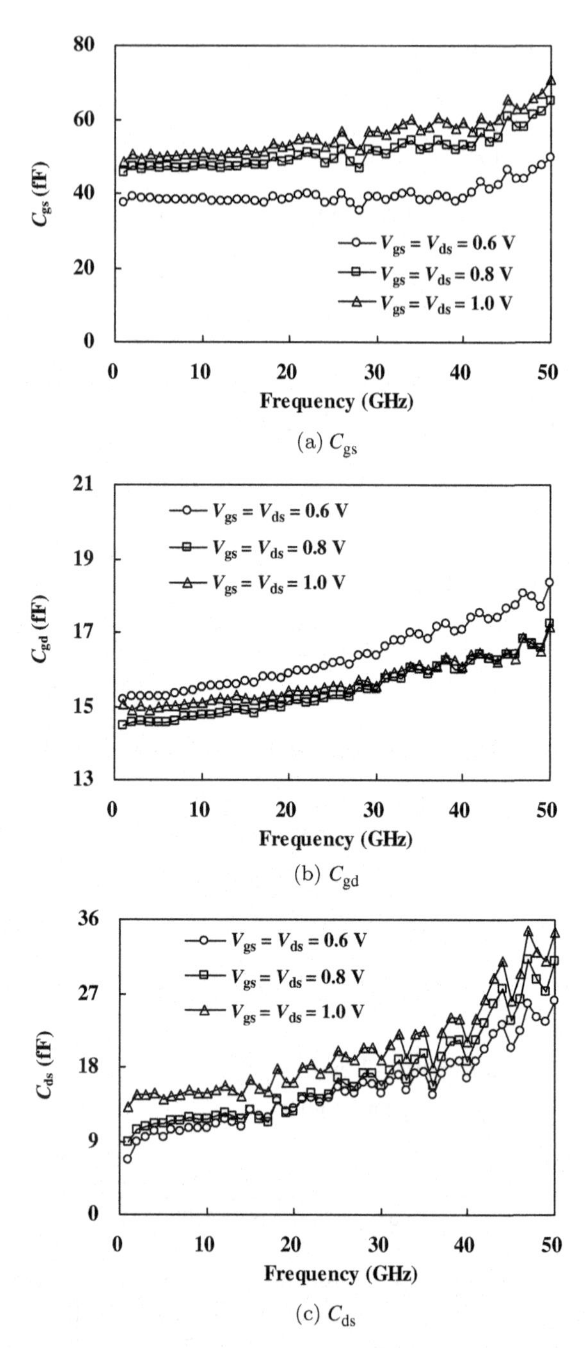

(a) C_{gs}

(b) C_{gd}

(c) C_{ds}

Figure 4.37. Intrinsic parameters versus frequency. Bias: $V_{gs} = V_{ds} = 0.6\,\text{V}$, $V_{gs} = V_{ds} = 0.8\,\text{V}$, and $V_{gs} = V_{ds} = 1.0\,\text{V}$.

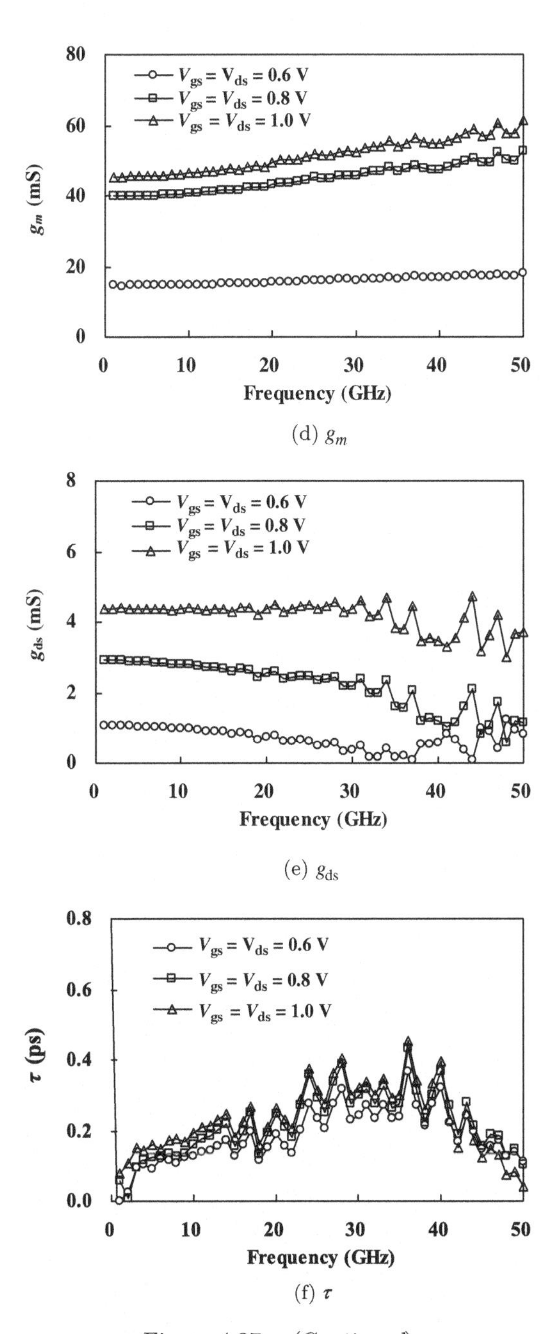

(d) g_m

(e) g_{ds}

(f) τ

Figure 4.37. *(Continued)*

Table 4.7. Intrinsic parameters of MOSFET device. The cell gate width is 1 μm, the number of gate fingers is 16, and the cell number is 2.

Parameter	$V_{gs} = V_{ds} = 0.6$V	$V_{gs} = V_{ds} = 0.8$V	$V_{gs} = V_{ds} = 1.0$V
g_m(mS)	16.2	44.3	51.4
τ(ps)	0.15	0.23	0.31
C_{gs}(fF)	38.6	46.7	54.3
C_{gd}(fF)	14.9	14.4	14.2
C_{ds}(fF)	11.6	12.5	14.6
g_{ds}(mS)	4.34	2.70	0.82

parameters versus frequency under three different bias conditions ($V_{gs} = V_{ds} = 0.6$ V, $V_{gs} = V_{ds} = 0.8$ V, and $V_{gs} = V_{ds} = 1.0$ V). It is significant to observe that in the frequency range of 1–50 GHz, the intrinsic parameters fluctuate versus frequency, hence the average is used as the final extracted result. The extracted intrinsic parameters are tabulated in Table 4.7.

In the frequency range of 0.5–50 GHz, the comparison between the measured and the modeled S parameters for MOSFET devices under three different bias conditions ($V_{ds} = 1.0$ V and $V_{gs} = 0$ V, $V_{ds} = 0.6$ V and $V_{gs} = 0.6$ V, $V_{ds} = 0.8$ V and $V_{gs} = 1.2$ V) are illustrated in Figure 4.38. Good agreements are obtained over the entire measured frequency range to verify the validity of the model parameter extraction method.

The MOSFET device with 90 nm gate length, 4 fingers $\times$ 0.6 μm (gate finger $\times$ gate width), and 18 cells was investigated to demonstrate the accuracy of the intrinsic parameter extraction method. Figure 4.39 illustrates the extracted intrinsic parameters C_{gs}, C_{gd}, g_m, and R_{ds} versus bias voltage, where the gate-source voltage V_{gs} and the drain-source voltage V_{ds} range are from 0.6 V to 1.2 V, step is 0.2 V, and bias points are 16 in total. As seen in Figure 4.39, the gate-source capacitance C_{gs} is mostly independent of the drain-source voltage V_{ds}, only controlled by the gate-source voltage V_{gs}. With increased V_{gs}, the capacitance C_{gs} increases. Moreover, with the increase in V_{gs}, the gate-drain capacitance C_{gd} increases and with increased V_{ds}, C_{gd} decreases. As observed, the output resistance R_{ds} increases slightly with the increase in V_{ds}, and decreases dramatically with increased V_{gs}. The transconductance g_m increases

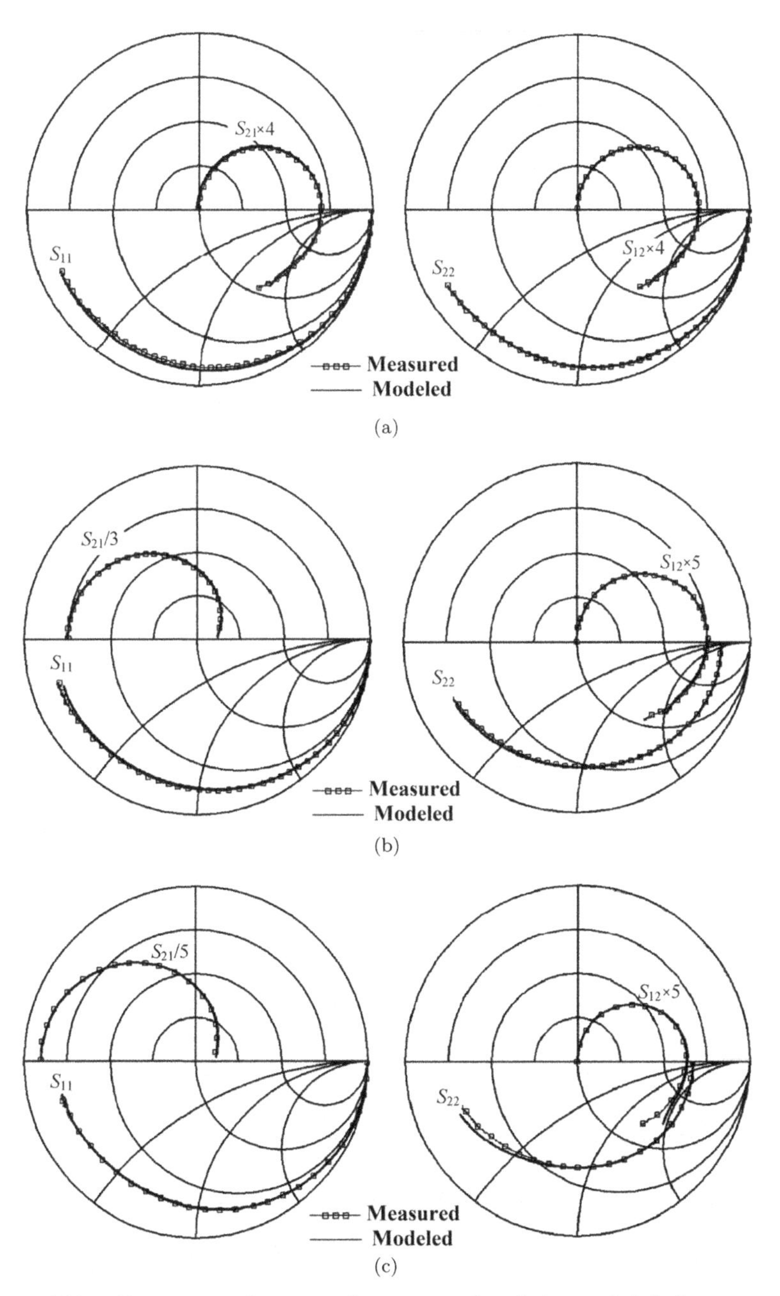

Figure 4.38. Comparison between the measured and the modeled S parameters for MOSFET devices in the frequency range of 0.5–50 GHz. Bias: (a) $V_{ds} = 1.0\,V$, $V_{gs} = 0\,V$; (b) $V_{ds} = 0.6\,V$, $V_{gs} = 0.6\,V$; (c) $V_{ds} = 0.8\,V$, $V_{gs} = 1.2V$.

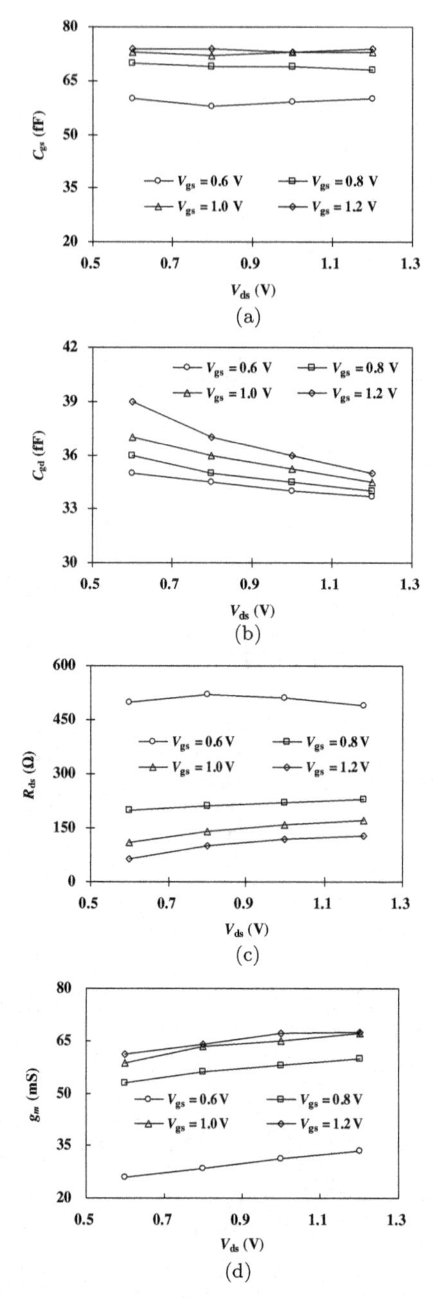

Figure 4.39. Intrinsic parameters versus frequency: (a) Gate-source capacitance C_{gs}; (b) Gate-drain capacitance C_{gd}; (c) Output resistance R_{ds}; (d) Transconductance g_m.

gradually with increased V_{ds}, and increases rapidly with the increase in V_{gs}.

4.4.3 *Optimization method*

If the extrinsic elements of the device are determined, it is ready to obtain the intrinsic elements utilizing de-embedding techniques. However, if only the S parameters of the device are provided without the required test conditions, it will be difficult to extract the intrinsic elements according to the direct extraction method. Thence, the semi-analytical method combining analysis and optimization is a good option here. The basic principle of the semi-analytical method is to optimize the extrinsic elements as unknown variables, and the intrinsic elements are regarded as the function of extrinsic elements. The specific extraction steps are presented as follows:

(1) Set initial values of extrinsic resistances R_g, R_d, and R_s and substrate extrinsic elements C_{jd} and R_{sub}.
(2) Calculate the intrinsic elements utilizing the analytical expressions, and the intrinsic elements can be regarded as functions of five extrinsic elements:

$$C_{gs} = f_1(\omega_i, R_g, R_d, R_s, C_{jd}, R_{sub}) \tag{4.92}$$

$$C_{gd} = f_2(\omega_i, R_g, R_d, R_s, C_{jd}, R_{sub}) \tag{4.93}$$

$$C_{ds} = f_3(\omega_i, R_g, R_d, R_s, C_{jd}, R_{sub}) \tag{4.94}$$

$$g_m = f_4(\omega_i, R_g, R_d, R_s, C_{jd}, R_{sub}) \tag{4.95}$$

$$g_{ds} = f_5(\omega_i, R_g, R_d, R_s, C_{jd}, R_{sub}) \tag{4.96}$$

$$\tau = f_6(\omega_i, R_g, R_d, R_s, C_{jd}, R_{sub}) \tag{4.97}$$

(3) The error ε between the modeled and measured S-parameters is used as the optimization standard. The five extrinsic elements R_g, R_d and R_s, C_{jd} and R_{sub} are the optimization variables, and the iteration ends when the optimization standard is satisfied.

The MOSFET device with 40 nm gate length, 5 μm cell gate width, and four gate fingers is utilized as an example. First, the pad extrinsic elements C_{oxg}, C_{oxd}, C_{pgd}, R_{pg}, and R_{pd} and feedline extrinsic elements L_g, L_d, and L_s are extracted by using an open and short test structure. Then, multiple sets of data are selected from the measured data for optimization, respectively, and the obtained extracted elements are averaged to extract the final model parameters. The comparison between the measured and modeled S parameters under three different bias conditions is shown in Figure 4.40.

4.4.4 *Scaling rules*

Scaling rules mainly focus on the total gate width of the device (gate width of single finger × number of gate finger). In order to extract the small-signal equivalent circuit model of the device versus the gate width, the detail of MOSFET devices fabricated by 130 nm standard CMOS process is tabulated in Table 4.8 [11].

It is worth noting that if the gate width of a single finger for the device is too large, it will be limited by process conditions. Hence, it is necessary to adopt a multi-gate finger structure (with the same total gate width) in the layout design process, as shown in Figure 4.41, where Figure 4.41(a) shows a single-finger structure device with a total gate width of $1 \times W$, and Figure 4.41(b) shows an interdigital structure device with a total gate width of $2 \times \frac{W}{2}$.

Figure 4.42 illustrates the extracted extrinsic resistances R_{g}, R_{s}, and R_{d} versus the total gate width of the device, where the device dimensions are 0.13 μm × 5 μm × 8 fingers, 0.13 μm × 5 μm × 16 fingers, 0.13 μm × 5 μm × 32 fingers, and 0.13 μm × 5 μm × 48 fingers (gate length × gate width × number of gate finger), respectively. It can be seen from the figure that the source resistance R_{s} and the drain resistance R_d are inversely proportional to the gate width, and the proportional model expression can be written as follows:

$$R_s = \frac{R_{s0}}{W} \tag{4.98}$$

$$R_d = \frac{R_{d0}}{W} \tag{4.99}$$

The gate extrinsic resistance R_g is the linear function of gate width:

$$R_g = \frac{R_{g0}}{W} + R_{g1} \tag{4.100}$$

where W represents the total gate width of the transistor. The data in Figure 4.42 are linearly fitted by the least square method, and the

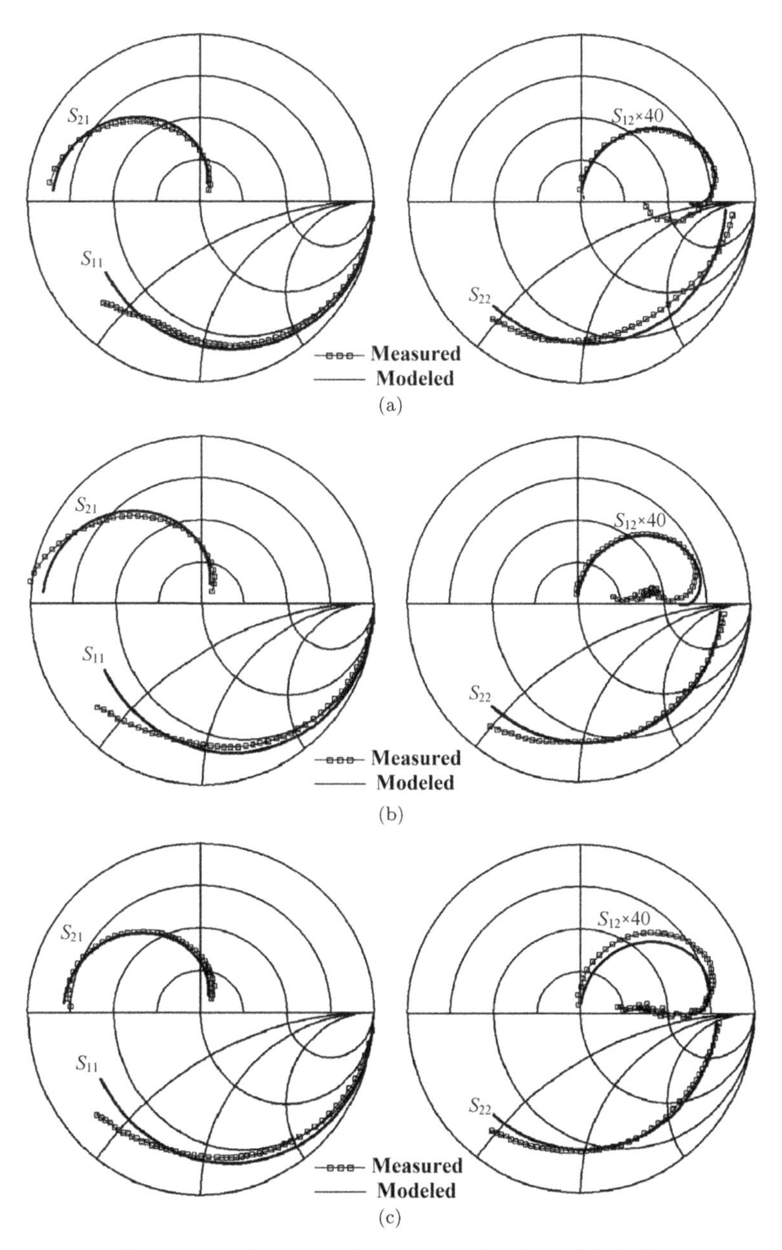

Figure 4.40. Comparison between the measured and modeled S parameters for MOSFET devices. Bias: (a) $V_{\mathrm{ds}} = 0.6\,\mathrm{V}$, $V_{\mathrm{gs}} = 0.8\,\mathrm{V}$; (b) $V_{\mathrm{ds}} = 0.7\,\mathrm{V}$, $V_{\mathrm{gs}} = 0.8\,\mathrm{V}$; (c) $V_{\mathrm{gs}} = 0.8\,\mathrm{V}$, $V_{\mathrm{ds}} = 0.8\,\mathrm{V}$.

Table 4.8. The detail of MOSFET devices.

Gate length (nm)	Single finger gate width (μm)	Gate finger	Effective gate width (μm)
130	5	8	40
130	5	16	80
130	5	32	160
130	5	48	240

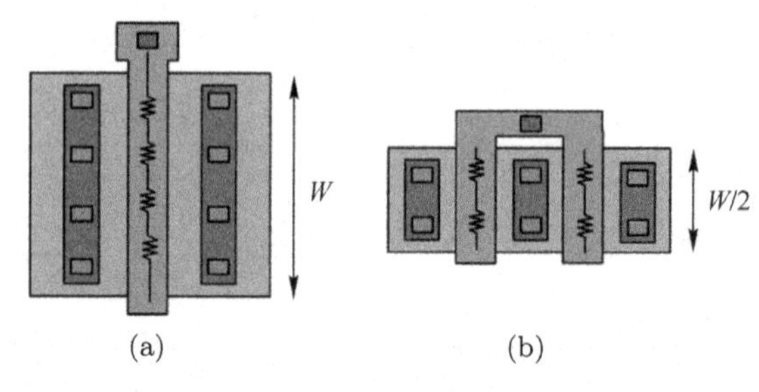

Figure 4.41. Layout of MOSFET devices: (a) single finger structure; (b) interdigital structure.

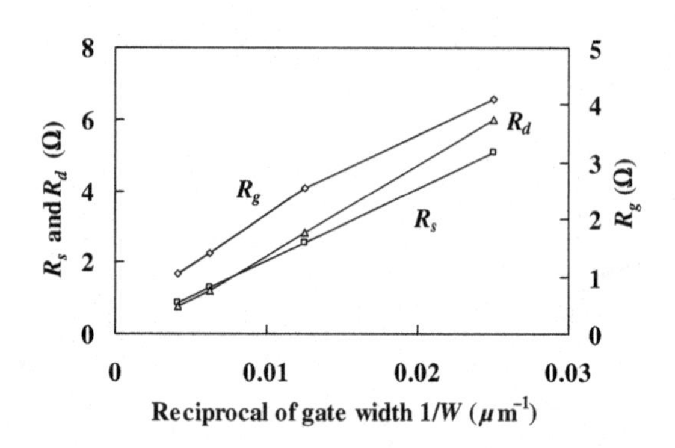

Figure 4.42. Extrinsic resistances versus gate width.

corresponding model parameters can be obtained: $R_{g0} = 145\ \Omega \cdot \mu\text{m}$, $R_{g1} = 0.53\ \Omega$, $R_{s0} = 204.5\ \Omega \cdot \mu\text{m}$, and $R_{d0} = 234.5\ \Omega \cdot \mu\text{m}$.

The extrinsic capacitance C_{jd} and resistance R_{sub} versus the drain bias voltage V_{ds} is depicted in Figures 4.43(a) and 4.43(b). With the drain bias voltage increased, the substrate capacitance C_{jd} sightly decreases. Thence, the dependence of the substrate capacitance C_{jd}

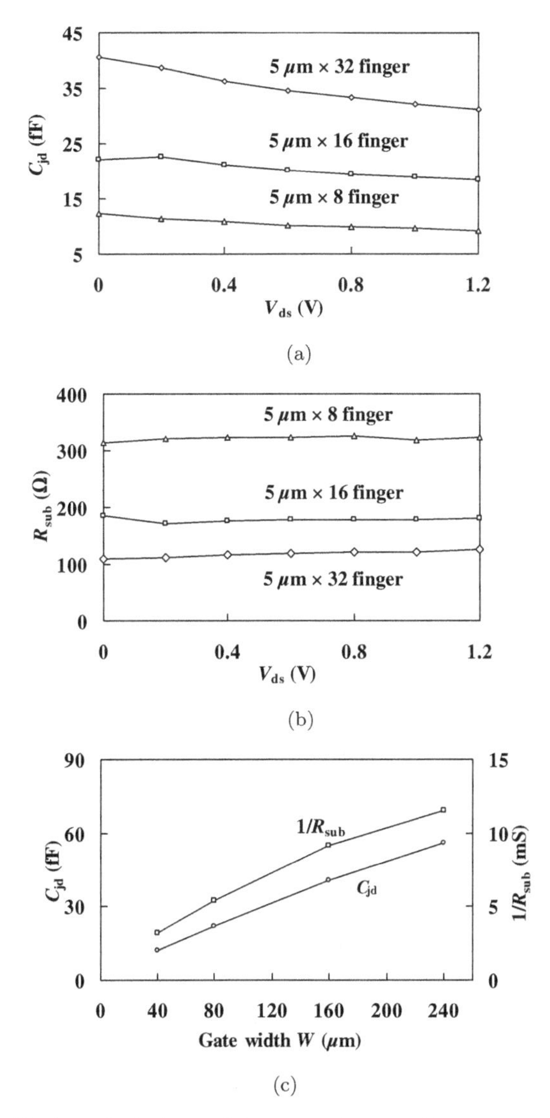

Figure 4.43. Drain substrate resistance and capacitance versus gate width: (a) $C_{\rm jd}$ versus $V_{\rm ds}$; (b) $R_{\rm sub}$ versus $V_{\rm ds}$; (c) $C_{\rm jd}$ and $R_{\rm sub}$ versus W.

on the source-drain voltage needs to be considered. The conventional extraction method only extracts the capacitance $C_{\rm jd}$ when the drain voltage is equal to zero, ignoring the dependence of the extrinsic junction capacitance on the drain bias, while it can be observed from Figure 4.43(b) that the substrate resistance R_{sub} is substantially independent of the drain bias voltage.

The substrate capacitance C_{jd} and resistance R_{sub} versus the gate width of the device is illustrated in Figure 4.43(c). It is noted that C_{jd} is proportional to the gate width of the device, while the resistance R_{sub} is inversely proportional to the gate width of the device. The proportional relationships of the parameters are shown as follows:

$$C_{\text{jd}} = C_{\text{jd0}} W \tag{4.101}$$

$$R_{\text{sub}} = \frac{R_{\text{sub0}}}{W} \tag{4.102}$$

Utilizing the linear fitting method in Figure 4.43, we have $C_{\text{sub}} = 0.243\ \text{fF}/\mu\text{m}$ and $R_{\text{sub0}} = 1.4\ \text{k}\Omega \cdot \mu\text{m}$.

The transconductance g_{m} and the output resistance R_{ds} versus the gate width W are illustrated in Figure 4.44. As seen, g_m is proportional to the gate width, and R_{ds} is inversely proportional to the gate width. The scale model is

$$g_m = g_{m0} W \tag{4.103}$$

$$1/R_{\text{ds}} = g_{\text{ds0}} W \tag{4.104}$$

Using linear fitting to the data in Figure 4.44, we can achieve $g_{m0} = 0.5885\ \text{mS}/\mu\text{m}$ and $g_{\text{ds0}} = 0.028\ \text{mS}/\mu\text{m}$.

The intrinsic capacitances C_{gd}, C_{gs}, and C_{ds} versus gate width W is depicted in Figure 4.45. It is significant to observe that the intrinsic capacitance is proportional to gate width, and the proportional

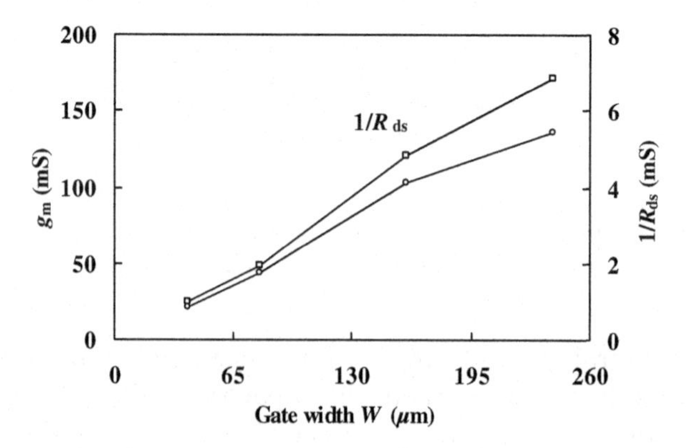

Figure 4.44. g_m and R_{ds} versus gate width W.

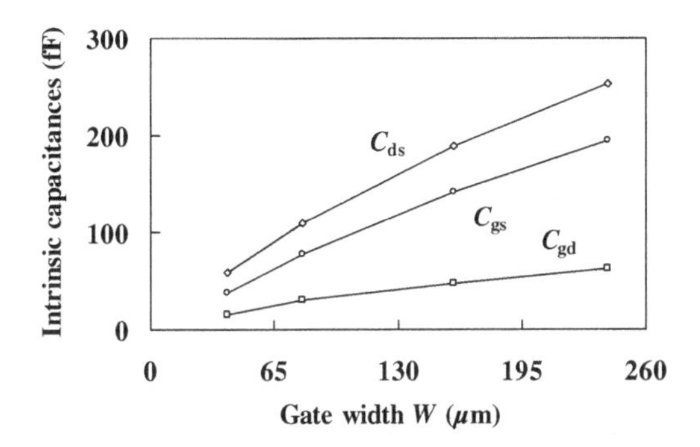

Figure 4.45. Intrinsic capacitances $C_{\rm gd}$, $C_{\rm gs}$, and $C_{\rm ds}$ versus gate width W.

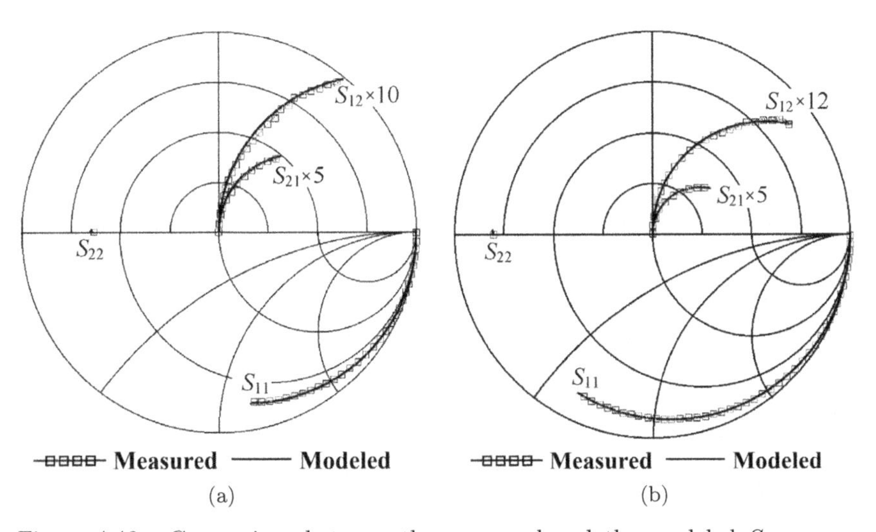

Figure 4.46. Comparison between the measured and the modeled S parameters under the bias condition of $V_{\rm gs} = 1.2$ V and $V_{\rm ds} = 0$ V. MOSFET device: (a) 0.13 μm × 5 μm × 8 finger; (b) 0.13 μm × 5 μm × 16 finger.

model parameters can be written as follows:

$$C_{\rm gs} = C_{\rm gs0} W \tag{4.105}$$

$$C_{\rm gd} = C_{\rm gd0} W \tag{4.106}$$

$$C_{\rm ds} = C_{\rm ds0} W \tag{4.107}$$

Utilizing the linear fitting method in Figure 4.45, we can get $C_{\rm gs0} = 0.8403$ fF/μm, $C_{\rm gd0} = 0.2829$ fF/μm, and $C_{\rm ds0} = 1.173$ fF/μm.

Figures 4.46–4.48 depict the comparison between the measured and modeled S parameters for MOSFET devices under different bias conditions. The frequency range is 100 MHz–40 GHz, the step is 100 MHz, and the input signal power is –20 dBm. It can be clearly observed from the figure that the measured data are in good agreement with the modeled data.

4.5 Sensitivity Analysis

RF and microwave computer-aided circuit design relies on the small-signal equivalent circuit model of the device, and the accuracy of the model directly affects the electrical characteristics of the device under test. Small-signal equivalent circuit model parameters are usually utilized by semiconductor fabrication to monitor the changes in process parameters. Generally, the model parameter extraction and validation are based on the device measurement. Nevertheless, the measurement data usually contain the measurement errors, which will ultimately lead to the imprecise of the model. In order to improve the model accuracy, the parameter extraction methods combined with the uncertainty estimation caused by the microwave and RF measurement can be utilized to quantify the uncertainty of the model. In this section, the sensitivity of the device equivalent circuit model parameters is analyzed, and the fluctuation range of the device model parameters is given. Finally, the influence of the model parameter fluctuations on the device characteristics is discussed.

4.5.1 *Uncertainty of S parameters*

In order to quantify the variation of model parameters in the small-signal equivalent circuit caused by the change in port network characteristics, sensitivity analysis is introduced to represent the sensitivity of the S-parameters to the model parameters [3–6]. In circuit design, the sensitivity analysis method is normally utilized to weigh the influence of parameters on circuit stability. Based on this method,

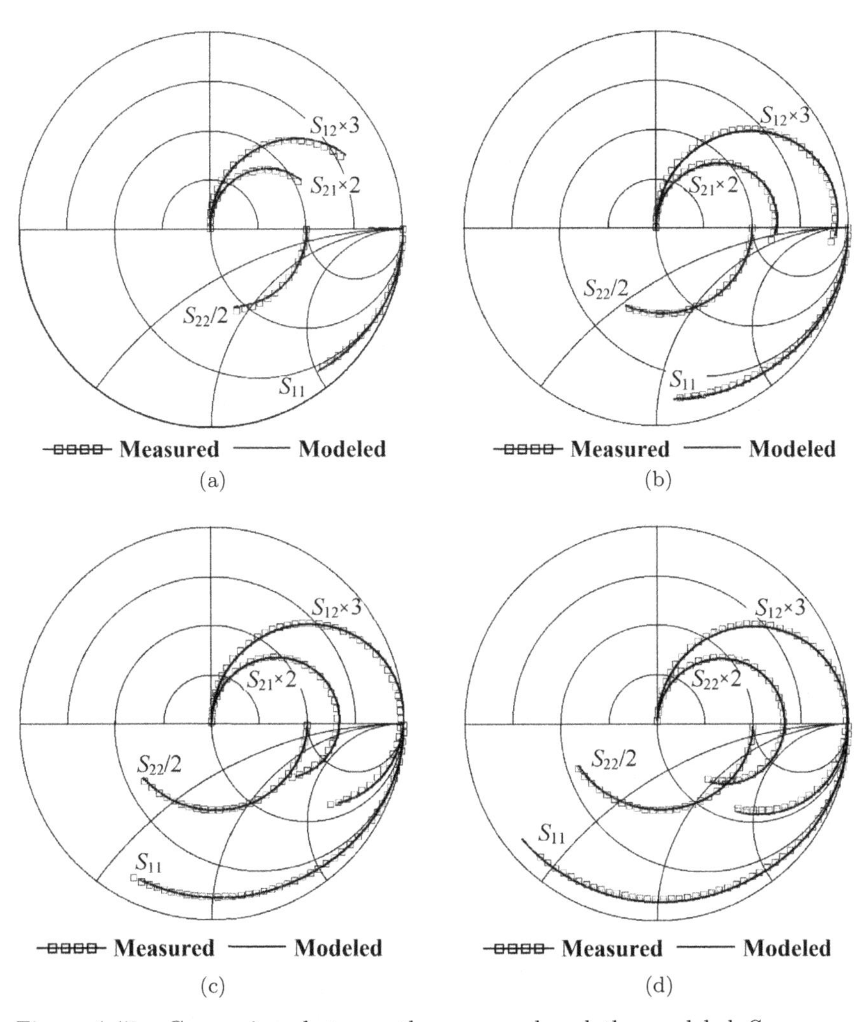

Figure 4.47. Comparison between the measured and the modeled S parameters under the bias condition of $V_{\text{gs}} = 0\text{V}$ and $V_{\text{ds}} = 0\,\text{V}$. MOSFET device: (a) 0.13 μm × 5 μm × 8 finger; (b) 0.13 μm × 5 μm × 16 finger; (c) 0.13 μm × 5 μm × 32 finger; (d) 0.13 μm × 5 μm × 48 finger.

the stability of analog circuits with the same characteristics but different structures can be obtained, so as to optimize the circuit. In the process of device modeling, the accuracy of the vector network analyzer, the selection of the measurement reference plane, the limited dynamic generation range, and the calibration method will bring

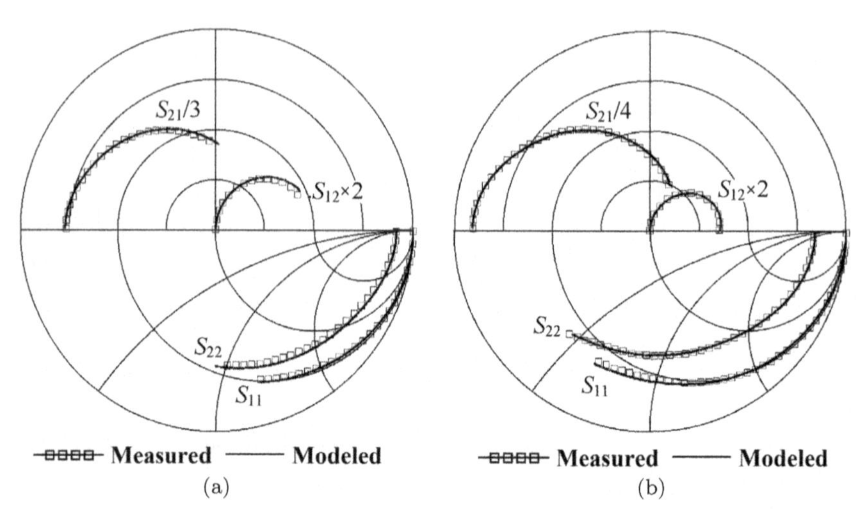

Figure 4.48. Comparison between the measured and modeled S-parameters. under the bias condition of $V_{gs} = 0.6\,V$ and $V_{ds} = 0.6\,V$. MOSFET device: (a) 0.13 μm $\times$ 5 μm $\times$ 8 finger; (b) 0.13 μm $\times$ 5 μm $\times$ 16 finger.

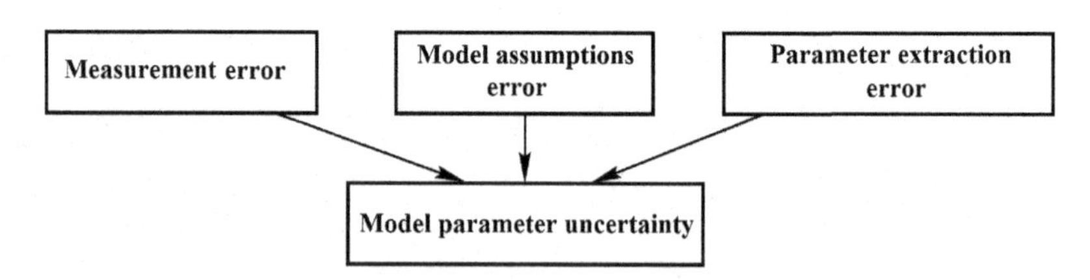

Figure 4.49. Model parameter uncertainty analysis.

measurement errors, resulting in the imprecise of the S-parameter measurement results. In order to quantify the error, the sensitivity of intrinsic parameters to S parameters is given. Nevertheless, it is difficult to illustrate the uncertainty estimation of the model parameters and the optimal frequency range of the parameter extraction clearly by using sensitivity. Hence, combining with the uncertainty of the measurement equipment to obtain the uncertainty of the model parameters is necessary.

The analysis of the model parameter uncertainty is depicted in Figure 4.49. As apparent from Figure 4.49, the measurement error, model assumptions error, and parameter extraction error are the main factors affecting the uncertainty of model parameters which leads to the model inaccuracy. During the measurement process, the vector network analyzer (VNA) is inevitably affected by the environment and the measurement error of the equipment, resulting in the deviation between the measured S-parameters and the actual S-parameters of the device.

4.5.2 *Circuit sensitivity analysis*

The small-signal equivalent circuit model of the MOSFET device is quite complicated, and uncertainty estimation is required to determine the optimal solution of the model parameters. Hence, this part mainly discusses the uncertainty of the model parameters caused by the measurement errors of the equipment, and the optimal frequency extraction range of the model parameters is also given to improve the accuracy of model.

In computer-aided design software, sensitivity analysis is utilized to quantify the sensitivity of internal circuit parameters to the output signal. Here, sensitivity analysis is utilized to calculate the influence of the deviation of the output signal (such as S-parameters) on the intrinsic model parameters [20–24].

When the network S-parameters change, the percentage change in the model parameter x can be expressed as

$$\frac{\Delta x}{x} \cong \frac{\partial x}{\partial |S_{ij}|} \frac{1/x}{1/|S_{ij}|} \frac{\Delta |S_{ij}|}{|S_{ij}|} = K^{x}_{|S_{ij}|} \frac{\Delta |S_{ij}|}{|S_{ij}|} \tag{4.108}$$

Thence, the relative sensitivity of the model parameter x to the magnitude of the S-parameter is defined as

$$K^{x}_{|S_{ij}|} = \frac{\partial x}{x} / \frac{\partial |S_{ij}|}{|S_{ij}|} \tag{4.109}$$

The absolute phase deviation is written as

$$\frac{\Delta x}{x} \cong \frac{\partial x}{\partial \angle S_{ij}} \frac{1}{x} \Delta \angle S_{ij} = K^{x}_{\angle S_{ij}} \Delta \angle S \tag{4.110}$$

The absolute sensitivity of the model parameter x to the phases of S-parameter is defined as

$$K^{x}_{\angle S_{ij}} = \frac{\partial x}{\partial \angle S_{ij}} \frac{1}{x} \tag{4.111}$$

Based on the small-signal equivalent circuit model of the measured data, the extraction of the model parameters depends on the measured S-parameters of the MOSFET device. Considering the S-parameters consist of magnitude and phase, the deviation of the model parameter x can be expressed in terms of S-parameter measurement deviation and sensitivity as

$$\frac{\Delta x}{x} \cong \sum_{\forall i,j \in \{1,2\}} K^{x}_{|S_{ij}|} \frac{\Delta |S_{ij}|}{|S_{ij}|} + K^{x}_{\angle S_{ij}} \Delta \angle S_{ij} \tag{4.112}$$

with

$$K^{x}_{|S_{ij}|} = K^{x}_{S_{ij}}$$
$$K^{x}_{\angle S_{ij}} = jK^{x}_{S_{ij}}$$

wherein $K^{x}_{|S_{ij}|}$ is the relative sensitivity of the model parameter x to the magnitude of the S parameter, $\Delta|S_{ij}|/|S_{ij}|$ is the relative change in the magnitude of the S parameter, $K^{x}_{\angle S_{ij}}$ is the absolute sensitivity of the model parameter x to the phase of the S parameter, and ΔS_{ij} is the absolute change in the magnitude of the S parameter.

Figure 4.50 depicts the process to calculate the corresponding deviation of the model parameters according to the S parameter deviation and the sensitivity expressions. First, the sensitivity of the Y parameter to the S parameter is calculated, and then the calculation of the sensitivity of the model parameter to the Y parameter is given. Finally, the sensitivity of the model parameter to the S parameter is obtained.

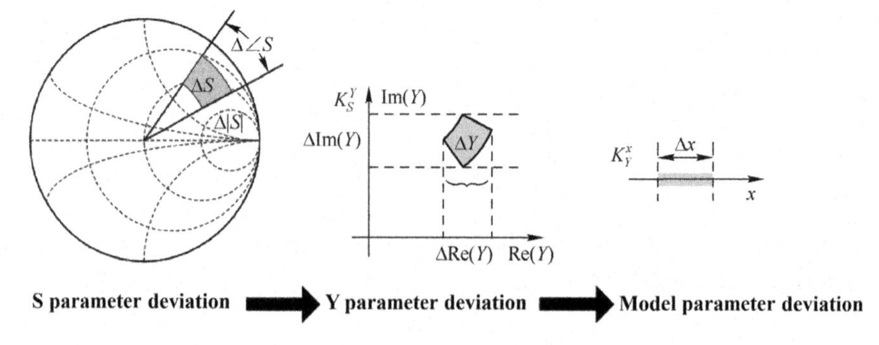

Figure 4.50. Calculation of model parameter deviation.

4.5.3 Sensitivity of intrinsic model parameters

Figure 4.51 illustrates the flowchart for calculating the uncertainty of intrinsic model parameters. It is worth noting that the influence of extrinsic elements on the uncertainty of intrinsic parameters is ignored. Thence, it is assumed that the sensitivity of the extrinsic Y parameter to the measured S parameter is approximately equal to the sensitivity of the intrinsic Y parameter to the measured S parameter, namely:

$$K_{S_{ij}}^{Y_{\text{ext}}} = K_{S_{ij}}^{Y_{\text{int}}} \tag{4.113}$$

In the calculation of model parameters sensitivity, the first step is to convert the measured S-parameters into Y-parameters:

$$Y_{\text{ext}} = \frac{1}{\Delta_3}\begin{bmatrix} \Delta_1 & -2S_{12} \\ -2S_{21} & \Delta_2 \end{bmatrix} \tag{4.114}$$

wherein

$$\Delta_1 = (S_{11} - 1)(1 + S_{22}) - S_{12}S_{21}$$

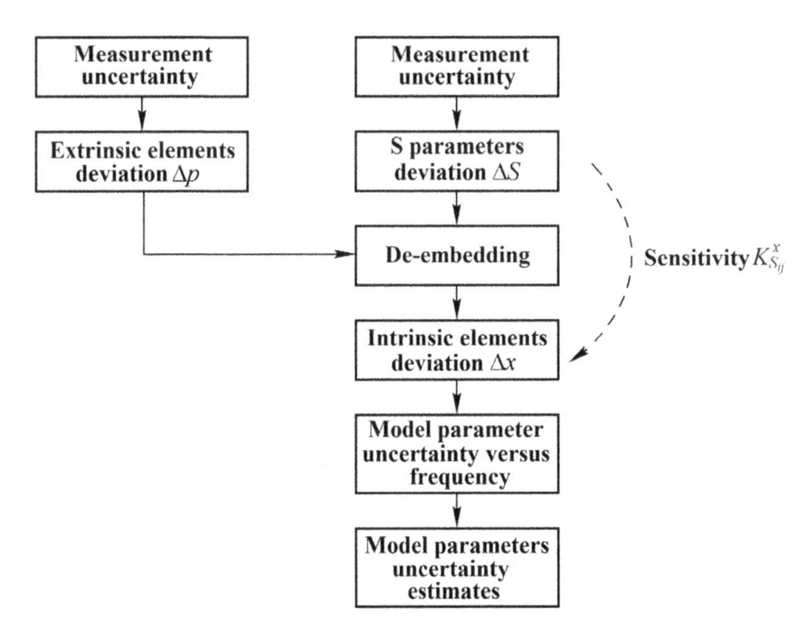

Figure 4.51. Flowchart for calculating the uncertainty of intrinsic model parameters.

$$\Delta_2 = (1 + S_{11})(S_{22} - 1) - S_{12}S_{21}$$
$$\Delta_3 = (1 + S_{11})(1 + S_{22}) - S_{12}S_{21}$$

The expressions for calculating the sensitivity of the Y parameters to the S parameters are tabulated in Table 4.9.

In order to simplify the calculation, the transformations on Y parameters are given as follows:

$$Y_m = Y_{21} - Y_{12} \tag{4.115}$$
$$Y_{\text{gs}} = Y_{11} + Y_{12} \tag{4.116}$$
$$Y_{\text{gd}} = -Y_{12} \tag{4.117}$$
$$Y_{\text{ds}} = Y_{22} + Y_{12} \tag{4.118}$$

According to the above sensitivity expressions, the sensitivity of Y_{gs}, Y_{gd}, Y_{ds}, and Y_m to the S parameters can be written as

$$K_{S_{ij}}^{Y_m} = \frac{K_{S_{ij}}^{Y_{21}} Y_{21}}{Y_m} - \frac{K_{S_{ij}}^{Y_{12}} Y_{12}}{Y_m} \tag{4.119}$$
$$K_{S_{ij}}^{Y_{\text{gs}}} = \frac{K_{S_{ij}}^{Y_{11}} Y_{11}}{Y_{\text{gs}}} + \frac{K_{S_{ij}}^{Y_{12}} Y_{12}}{Y_{\text{gs}}} \tag{4.120}$$
$$K_{S_{ij}}^{Y_{\text{gd}}} = K_{S_{ij}}^{Y_{12}} \tag{4.121}$$

Table 4.9. Sensitivity of Y parameters to S parameter.

$K_{S_{ij}}^{Y_{ij}}$	S_{11}	S_{12}	S_{21}	S_{22}
Y_{11}	$\frac{2(1+S_{22})^2 S_{11}}{\Delta_3\Delta_1}$	$-\frac{2(1+S_{22})S_{12}S_{21}}{\Delta_3\Delta_1}$	$-\frac{2(1+S_{22})S_{12}S_{21}}{\Delta_3\Delta_1}$	$\frac{2S_{12}S_{21}S_{22}}{\Delta_3\Delta_1}$
Y_{12}	$-\frac{(1+S_{22})S_{11}}{\Delta_3}$	$\frac{(1+S_{11})(1+S_{22})}{\Delta_3}$	$\frac{S_{21}S_{12}}{\Delta_3}$	$-\frac{(1+S_{11})S_{22}}{\Delta_3}$
Y_{21}	$-\frac{(1+S_{22})S_{11}}{\Delta_3}$	$\frac{S_{21}S_{12}}{\Delta_3}$	$\frac{(1+S_{11})(1+S_{22})}{\Delta_3}$	$-\frac{(1+S_{11})S_{22}}{\Delta_3}$
Y_{22}	$\frac{2S_{11}S_{12}S_{21}}{\Delta_2\Delta_3}$	$-\frac{2(1+S_{11})S_{12}S_{21}}{\Delta_2\Delta_3}$	$-\frac{2(1+S_{11})S_{12}S_{21}}{\Delta_2\Delta_3}$	$\frac{2(1+S_{11})^2 S_{22}}{\Delta_2\Delta_3}$

$$K_{S_{ij}}^{Y_{\rm ds}} = \frac{K_{S_{ij}}^{Y_{12}} Y_{12}}{Y_{\rm ds}} + \frac{K_{S_{ij}}^{Y_{22}} Y_{22}}{Y_{\rm ds}} \tag{4.122}$$

Moreover, based on the small-signal equivalent circuit model, the transconductance for the MOSFET device can be also given as

$$Y_m = g_m {\rm e}^{-{\rm j}\omega\tau} \tag{4.123}$$

In this way, the sensitivity of Y_m to the S parameters can be expressed by the sensitivity of transconductance g_m and time delay τ to S parameters, so we have

$$\begin{aligned} K_{S_{ij}}^{Y_m} &= \frac{\partial Y_m}{\partial S_{ij}}\frac{S_{ij}}{Y_m} = \frac{\partial Y_m}{\partial g_m}\frac{\partial g_m}{\partial S_{ij}}\frac{S_{ij}}{g_m}\frac{g_m}{Y_m} + \frac{\partial Y_m}{\partial \tau}\frac{\partial \tau}{\partial S_{ij}}\frac{S_{ij}}{\tau}\frac{\tau}{Y_m} \\ &= e^{-j\omega\tau} K_{S_{ij}}^{g_m}\frac{g_m}{Y_m} - j\omega g_m e^{-j\omega\tau} K_{S_{ij}}^{\tau}\frac{\tau}{Y_m} = K_{S_{ij}}^{g_m} - j\omega\tau K_{S_{ij}}^{\tau} \end{aligned} \tag{4.124}$$

Hence, the relative magnitude sensitivity and the absolute phase sensitivity of the g_m and τ to the S parameters can be written as

$$K_{|S_{ij}|}^{g_m} = \mathrm{Re}\left(K_{S_{ij}}^{Y_m}\right) \tag{4.125}$$

$$K_{|S_{ij}|}^{\tau} = -\mathrm{Im}\left(K_{S_{ij}}^{Y_m}\right) / (\omega\tau) \tag{4.126}$$

$$K_{\angle S_{ij}}^{g_m} = \mathrm{Im}\left(K_{S_{ij}}^{Y_m}\right) \tag{4.127}$$

$$K_{\angle S_{ij}}^{\tau} = -\mathrm{Re}\left(K_{S_{ij}}^{Y_m}\right) / (\omega\tau) \tag{4.128}$$

Similarly, the sensitivity of other intrinsic parameters can be obtained immediately. In the small-signal equivalent circuit, the sensitivity expressions of intrinsic parameters are tabulated in Table 4.10.

Wherein, Re represents the real part of the number, and Im represents the imaginary part of the number.

4.5.4 *Uncertainty estimation*

On account of the uncertainty of the measurement equipment varying with the frequency, it is difficult to illustrate the uncertainty estimation of the model parameters and the optimal frequency range of the

Table 4.10. Intrinsic parameters sensitivity.

Intrinsic parameter x	Relative magnitude sensitivity $K^{x}_{\lvert S_{ij}\rvert}$	Absolute phase sensitivity $K^{x}_{\angle S_{ij}}$
C_{gs}	$\mathrm{Re}\left(K^{Y_{\mathrm{gs}}}_{S_{ij}}\right)$	$-\mathrm{Im}\left(K^{Y_{\mathrm{gs}}}_{S_{ij}}\right)$
C_{gd}	$\mathrm{Re}\left(K^{Y_{\mathrm{gd}}}_{S_{ij}}\right)$	$-\mathrm{Im}\left(K^{Y_{\mathrm{gd}}}_{S_{ij}}\right)$
g_{m}	$\mathrm{Re}(K^{Y_{\mathrm{m}}}_{S_{ij}})$	$-\mathrm{Im}(K^{Y_{\mathrm{m}}}_{S_{ij}})$
τ	$-\mathrm{Im}(K^{Y_{\mathrm{m}}}_{S_{ij}})/\omega\tau$	$\mathrm{Re}(K^{Y_{\mathrm{m}}}_{S_{ij}})/\omega\tau$
g_{ds}	$-\mathrm{Re}(K^{Y_{\mathrm{ds}}}_{S_{ij}} Y_{\mathrm{ds}})/g_{\mathrm{ds}}$	$\mathrm{Im}(K^{Y_{\mathrm{ds}}}_{S_{\mathrm{ij}}} Y_{\mathrm{ds}})/g_{\mathrm{ds}}$
C_{ds}	$-\mathrm{Im}(K^{Y_{\mathrm{ds}}}_{S_{ij}} Y_{\mathrm{ds}})/\omega C_{\mathrm{ds}}$	$\mathrm{Re}(K^{Y_{\mathrm{ds}}}_{S_{ij}} Y_{\mathrm{ds}})/\omega C_{\mathrm{ds}}$

parameter extraction clearly by using sensitivity. Hence, combining the sensitivity with the uncertainty of the measurement equipment to obtain the uncertainty of the model parameters is necessary.

To calculate the parameter uncertainty versus frequency, the uncertainties in the measurement S parameters should be known first. Figure 4.52 gives the S parameter uncertainty specified for Agilent 8510C VNA [25], showing the worst-case uncertainty in reflection and transmission measurement. It is interesting to note that the relative magnitude and absolute phase uncertainties are specified equally. To describe approximately the S parameter measurement uncertainty, the empirical models are developed based on the equipment specifications. The relative magnitude and absolute phase uncertainties can be expressed as

$$\begin{aligned}\sigma_{\lvert S_{11,22}\rvert} = \sigma_{\angle S_{11,22}} &= \{(k_1\lvert S_{11,22}\rvert + k_2)\left[\ln\left(k_3\lvert S_{11,22}\rvert + k_4\right)\right] \\ &\quad \times + k_5\}\left(b_1 f^2 + b_2 f + b_3\right) \end{aligned} \tag{4.129}$$

$$\begin{aligned}\sigma_{\mathrm{dB}\lvert S_{12,21}\rvert} = \sigma_{\angle S_{12,21}} &= \{(t_1\mathrm{dB}\lvert S_{12,21}\rvert + t_2)\left[\ln\left(t_3\mathrm{dB}\lvert S_{12,21}\rvert + t_4\right)\right] \\ &\quad \times + t_5\}\left(a_1 f^2 + a_2 f + a_3\right) \end{aligned} \tag{4.130}$$

wherein $\lvert S_{11,22}\rvert$ and $\mathrm{dB}\lvert S_{12,21}\rvert$ represent the measured relative magnitude of S parameters. $\angle S_{11,22}$ and $\angle S_{12,21}$ represent the measurement absolute phase given in radians. Besides, k_n,t_n, b_n, and a_n are

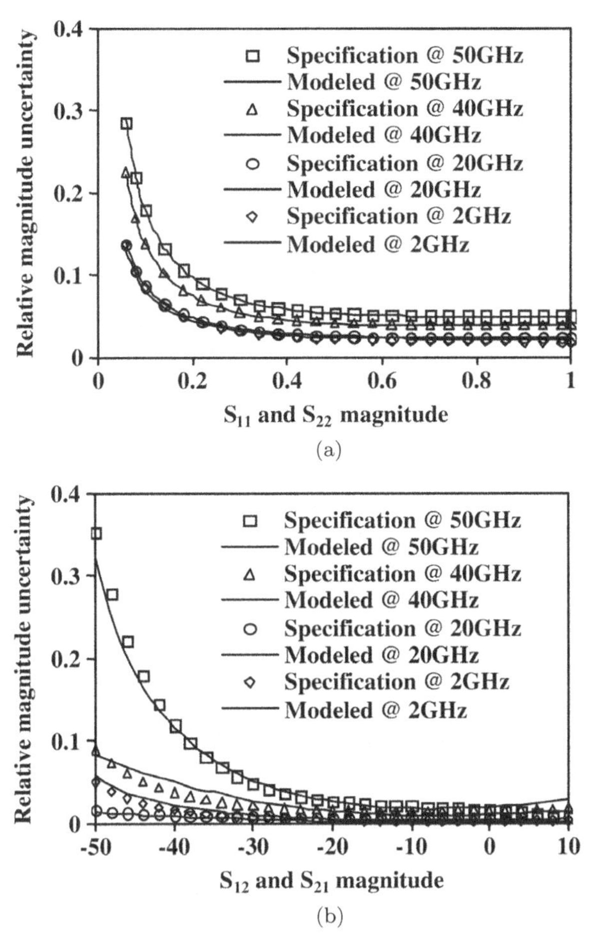

Figure 4.52. Specified and modeled uncertainty in S parameters: (a) S_{11} and S_{22}; (b) S_{12} and S_{21}.

model parameters. The comparison of the VNA specification and modeled results is also shown in Figure 4.52.

As mentioned, the measurement uncertainty is heavily determined by the performance of the VNA used, calibrations performed, and measurement reference plane. Hence, it is not ready to predict in general. In order to quantify the measurement uncertainty, the S parameter deviation is assumed to be normally distributed and uncorrelated. The uncertainty of x can be characterized by the sensitivity of the x to S parameter and the uncertainties of the

S parameter. The expression is expressed as follows:

$$\sigma_x^2 = \sum_{\forall i,j \in \{1,2\}} [(K^x_{|S_{ij}|})^2 \sigma^2_{|S_{ij}|} + (K^x_{\angle S_{ij}})^2 \sigma^2_{\angle S_{ij}}] \tag{4.131}$$

By utilizing the above expressions, the uncertainty of intrinsic parameters can be readily determined. Figure 4.53 illustrates the estimated relative uncertainty of intrinsic parameters C_{gs}, C_{gd}, g_m, g_{ds}, and C_{ds} versus frequency under the bias condition of $V_{\mathrm{gs}} = V_{\mathrm{ds}} = 0.6\,\mathrm{V}$, $V_{\mathrm{gs}} = V_{\mathrm{ds}} = 0.8\,\mathrm{V}$, and $V_{\mathrm{gs}} = V_{\mathrm{ds}} = 1.0\,\mathrm{V}$. It can be observed that the optimal frequency ranges of C_{gs}, C_{gd}, g_{ds}, and g_{m} are concentrated in the low- and middle-frequency band (5–20 GHz), and the uncertainty has remained within 10% at the optimal extraction frequencies. While the optimal frequency range of the source-drain capacitance C_{ds} is concentrated in the intermediate frequency (15–35 GHz), and the uncertainty is about 7% at the optimal extraction frequencies. In order to further illustrate the model parameters and the corresponding uncertainty, the extracted intrinsic parameters and uncertainty range versus frequency under the bias condition of $V_{\mathrm{gs}} = V_{\mathrm{ds}} = 1.0\,\mathrm{V}$ are provided in Figure 4.54. Besides, the extracted intrinsic parameters, minimum uncertainty, and optimal extraction frequency are summarized in Table 4.11.

To further analyze the relationship between the uncertainties and bias voltage, Figure 4.55 depicts the extracted intrinsic parameters and estimated uncertainty versus bias voltage. The results illustrate that the smooth and well-behaved extraction can be performed without any *a priori* knowledge about the device characteristics. It is clear to see that when the bias voltage steps from 0 to 1.2 V, the estimated uncertainty of capacitances C_{gs}, C_{gd}, and C_{ds} and output conductance g_{ds} has little ripples versus bias voltage and the uncertainty is all within 10%.

4.5.5 *Monte Carlo analysis*

Monte Carlo method is one of the significant methods to analyze the tolerance of design parameters. Monte Carlo methods applied in electronic circuits are mainly based on probability and statistics, which can predict the yield of batch products before going into production. Utilizing Monte Carlo method to simulate the circuit can provide reliable basis for the actual production of the circuit so that the yield of

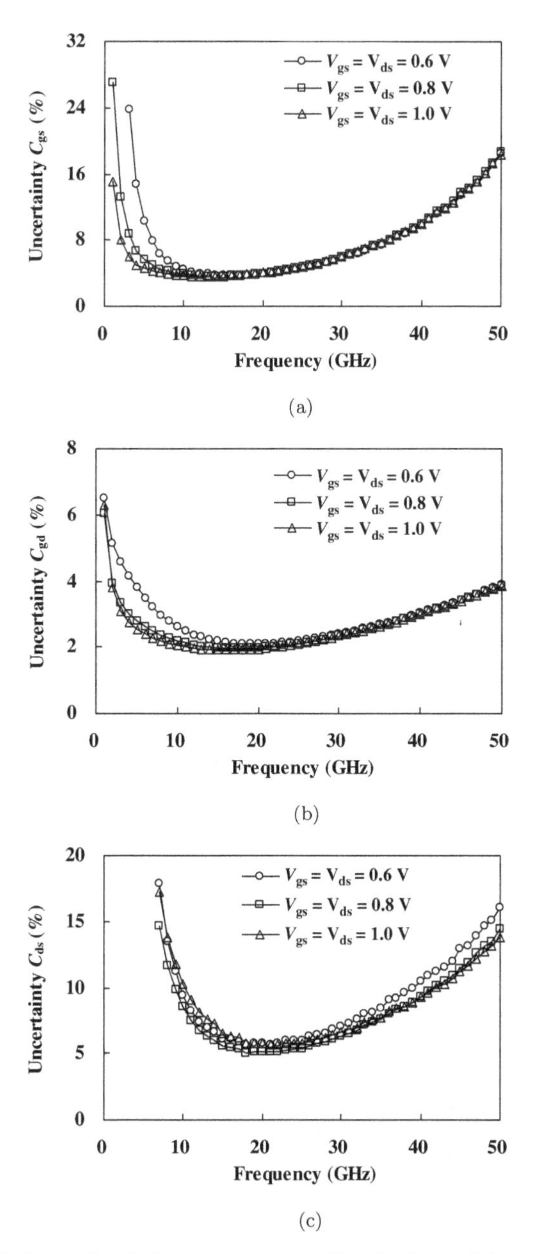

Figure 4.53. Estimated relative uncertainty of intrinsic model parameters versus frequency. Bias: $V_{\rm gs} = V_{\rm ds} = 0.6\,\rm V$, $V_{\rm gs} = V_{\rm ds} = 0.8\,\rm V$, and $V_{\rm gs} = V_{\rm ds} = 1.0\,\rm V$: (a) Gate-source capacitance $C_{\rm gs}$; (b) Gate-drain capacitance $C_{\rm gd}$; (c) Source-drain capacitance $C_{\rm ds}$; (d) Transconductance $g_{\rm m}$; (e) Outputconductance $g_{\rm ds}$.

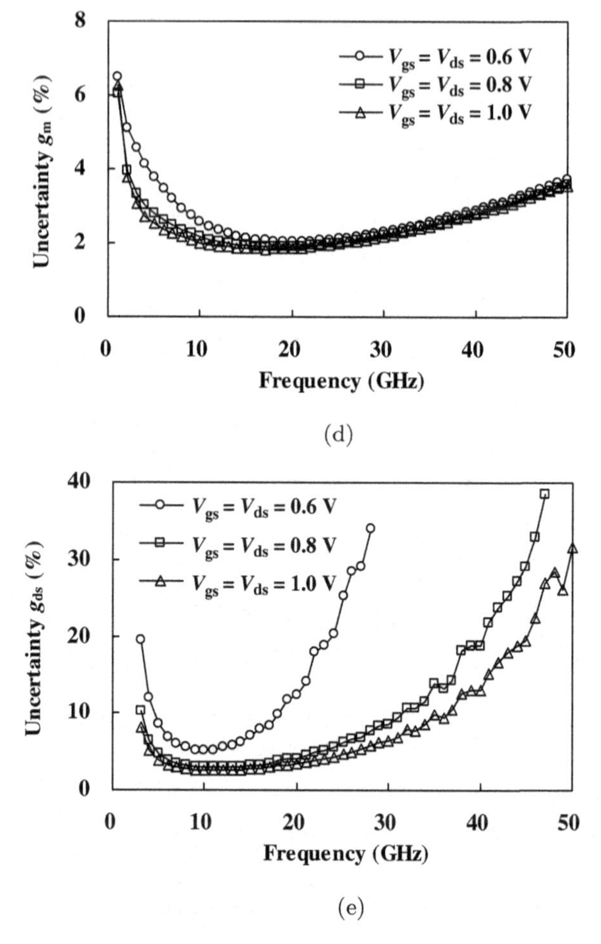

Figure 4.53. (*Continued*)

Table 4.11. Extracted intrinsic parameters.

Parameter	Value	Uncertainty $\sigma_{\min}$ (%)	Optimal extraction frequency $f_{\rm opt}$ (GHz)
$C_{\rm gs}$(fF)	52.7	2.6	12.0
$C_{\rm gd}$(fF)	14.8	1.9	14.3
$C_{\rm ds}$(fF)	18.4	4.6	22.0
g_m(mS)	47.4	1.8	7.5
$g_{\rm ds}$(mS)	4.3	2.6	11.3

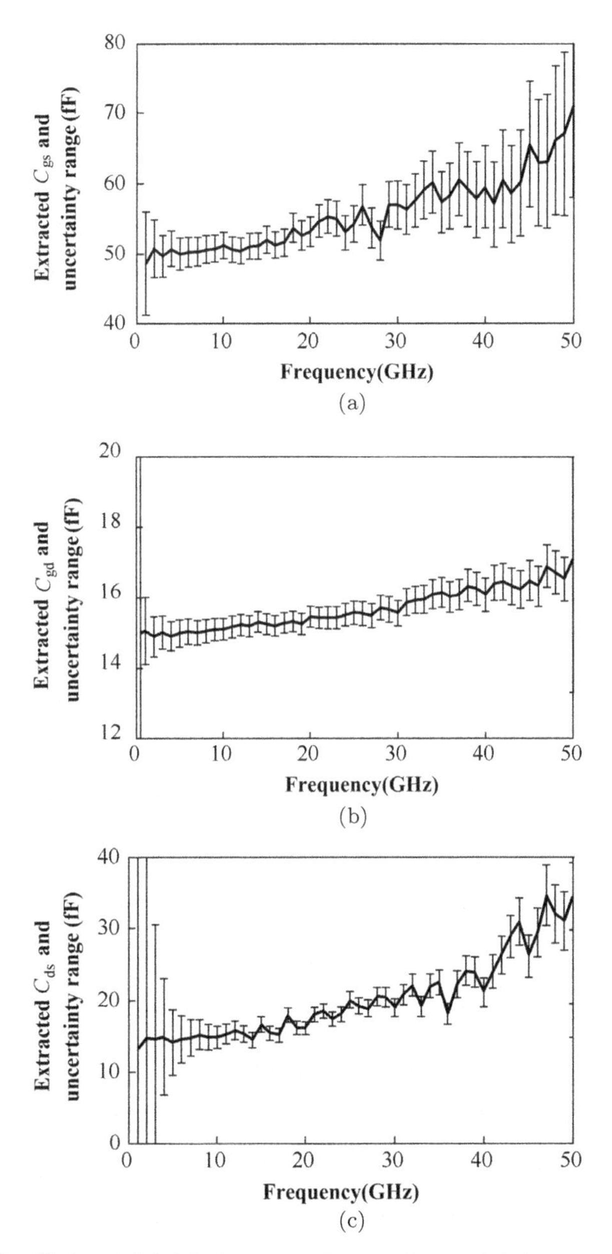

Figure 4.54. Extracted intrinsic parameters and uncertainty range versus frequency: (a) C_{gs}; (b) C_{gd}; (c) C_{ds}; (d) g_m; (e) g_{ds}.

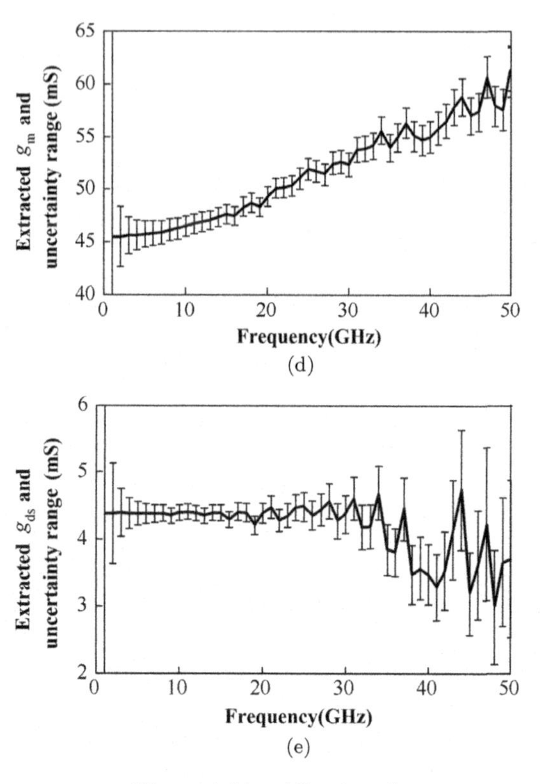

Figure 4.54. (*Continued*)

products is staying in control. The obtained model parameters are analyzed by Monte Carlo analysis method to demonstrate that small ripples in model parameters will not have a significant impact on the circuit design. In order to verify the sensitivity of the model parameters obtained, the tolerance range of each model parameter in the equivalent circuit model of the MOSFET device is given, and then the S parameters of the small signal circuit model of the MOSFET device are analyzed to verify the variation range of S parameters. Here, the stability of S parameters represents the stability of the output characteristics of the circuit model.

It is of great significance to investigate the influence of intrinsic parameters simultaneously rippled on model accuracy. Obviously, the S parameters can be regarded as a function of intrinsic elements.

$$S_{ij} = f(C_{\mathrm{gs}}, C_{\mathrm{gd}}, C_{\mathrm{ds}}, g_m, \tau, R_{\mathrm{ds}}) \tag{4.132}$$

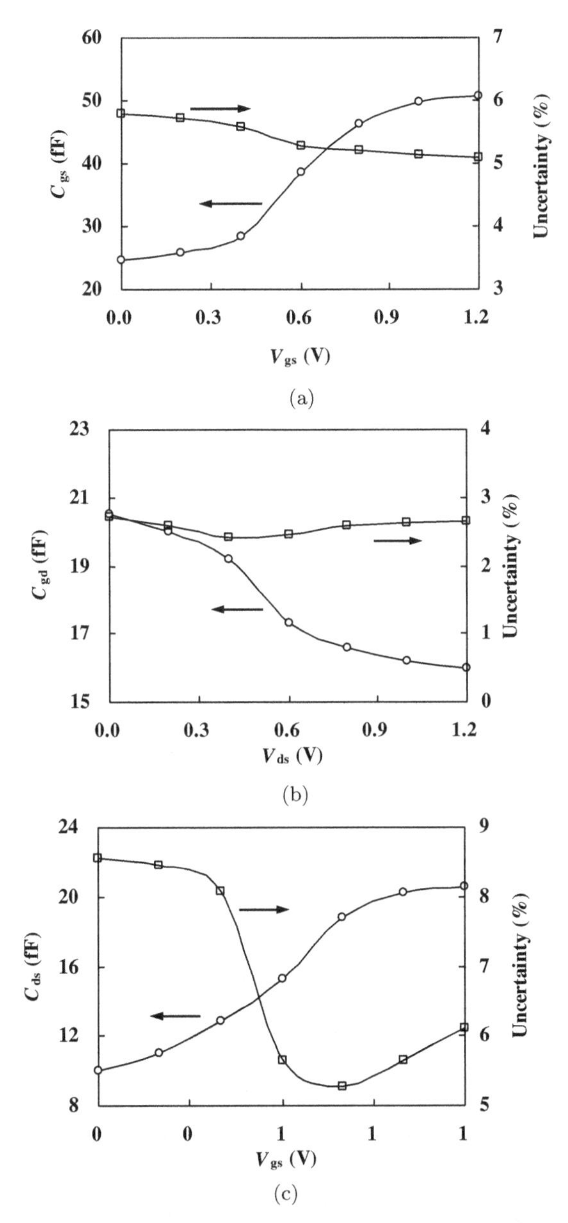

Figure 4.55. Extracted intrinsic parameters and estimated uncertainty versus bias voltage: (a) $C_{\rm gs}$ and estimated uncertainty versus $V_{\rm gs}(V_{\rm ds} = 1.0\,{\rm V})$; (b) $C_{\rm gd}$ and estimated uncertainty versus $V_{\rm ds}(V_{\rm gs} = 1.0\,{\rm V})$; (c) $C_{\rm ds}$ and estimated uncertainty versus $V_{\rm gs}(V_{\rm ds} = 1.0\,{\rm V})$; (d) g_m and estimated uncertainty versus $V_{\rm gs}$ $(V_{\rm ds} = 1.0\,{\rm V})$; (e) $g_{\rm ds}$ and estimated uncertainty versus $V_{\rm ds}(V_{\rm gs} = 1.0\,{\rm V})$.

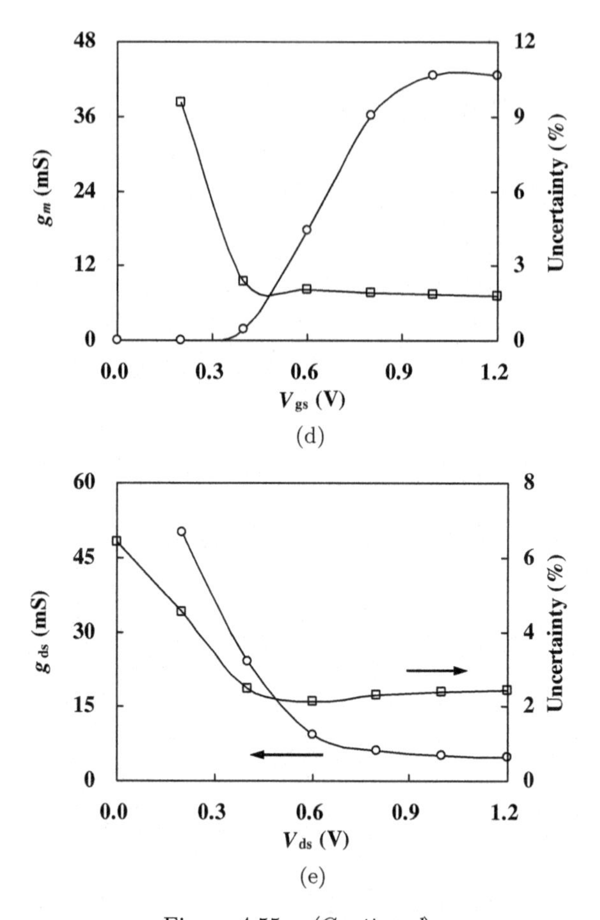

(d)

(e)

Figure 4.55. (*Continued*)

wherein S_{ij} $(i, j = 1, 2)$ represent the S-parameters of network characteristics.

The deviation of S_{ij} due to the deviation of the intrinsic parameters can be obtained by the following expression:

$$\begin{aligned}\frac{\Delta S_{ij}}{S_{ij}} &= K_{C_{\text{gs}}}^{S_{ij}} \frac{\Delta C_{\text{gs}}}{C_{\text{gs}}} + K_{C_{\text{gd}}}^{S_{ij}} \frac{\Delta C_{\text{gd}}}{C_{\text{gd}}} + K_{C_{\text{ds}}}^{S_{ij}} \frac{\Delta C_{\text{ds}}}{C_{\text{ds}}} \\ &\quad + K_{g_{\text{m}}}^{S_{ij}} \frac{\Delta g_{\text{m}}}{g_{\text{m}}} + K_{\tau}^{S_{ij}} \frac{\Delta \tau}{\tau} + K_{R_{\text{ds}}}^{S_{ij}} \frac{\Delta R_{\text{ds}}}{R_{\text{ds}}}\end{aligned} \tag{4.133}$$

wherein $K_x^{S_{ij}}$ represents the sensitivity of S_{ij} to intrinsic parameter x.

The uncertainty of the measurement instrument mainly causes the deviation of the measured S parameters, which leads to the uncertainty of the intrinsic parameters extraction. Monte Carlo analysis is utilized to analyze the worst case of the equivalent circuit model. The worst case is that when the circuit elements are simultaneously rippled, the ripples of S parameters magnitude reach the maximum or minimum. In the frequency range of 0.5–50 GHz, the circuit characteristics of the MOSFET device model are simulated. First, the standard values of intrinsic parameters are determined, respectively, where the standard value is the extracted value under the optimal

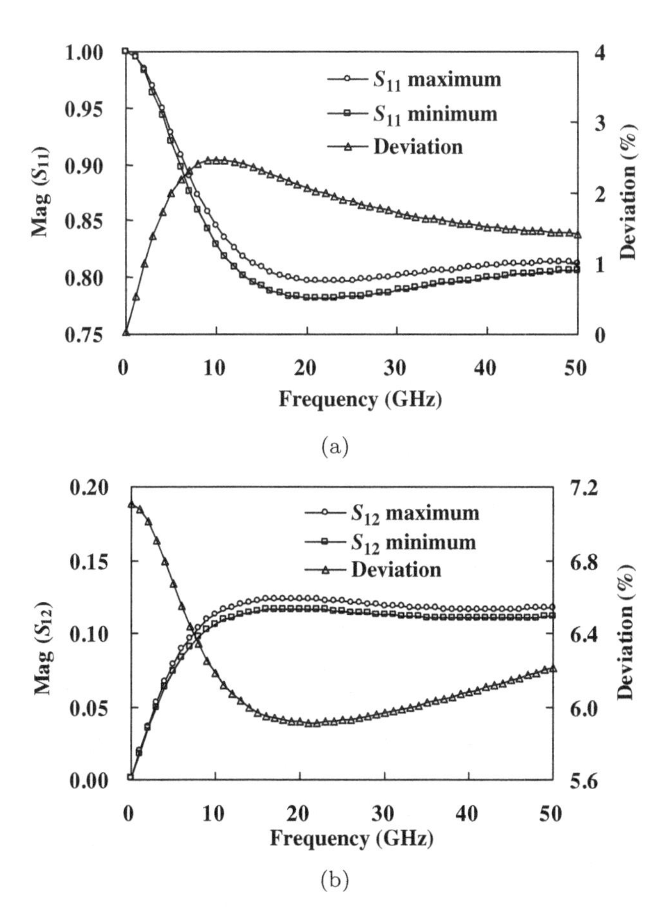

Figure 4.56. Maximum and minimum magnitude of S parameters and corresponding deviation versus frequency: (a) S_{11}; (b) S_{12}; (c) S_{21}; (d) S_{22}.

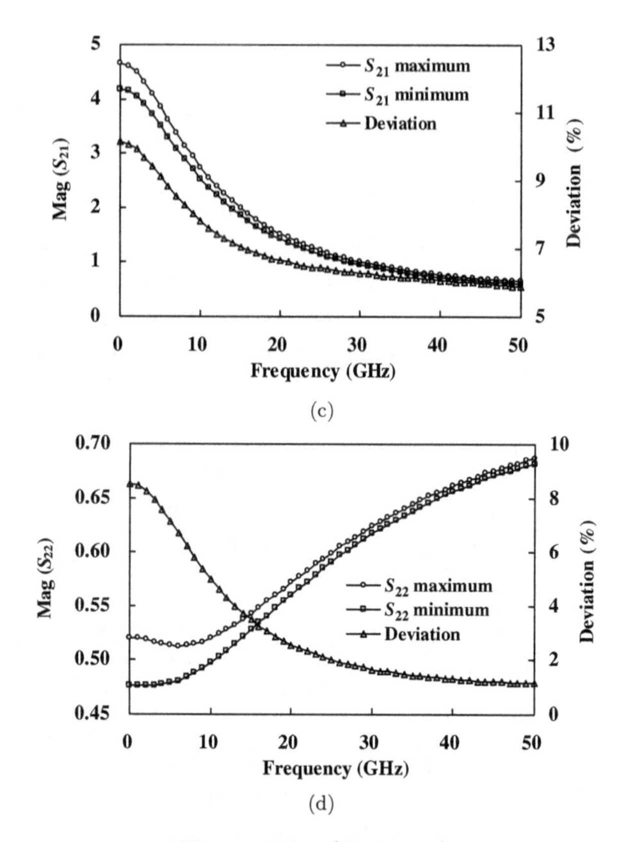

Figure 4.56. *(Continued)*

extraction frequency in Table 4.11, and the ripple range is the uncertainty of the parameters, for example, the standard value of C_{gs} is 52.7 fF, the ripple range is 2.6%, and the number of samples is 200.

In order to clearly illustrate the accuracy of the S-parameters in the two extreme cases, the maximum and minimum magnitude of S parameters in sampled data, as well as the deviation against the standard data versus frequency, are provided in Figure 4.56. The deviation represents the maximum minus the minimum of S parameter magnitude divided by the standard S parameter magnitude when the model parameter fluctuates, which can be expressed as

$$\text{Deviation} = \left| \frac{S_{ij}^{\max} - S_{ij}^{\min}}{S_{ij}} \right| \cdot 100\% \tag{4.134}$$

As can be seen from Figure 4.56, the deviations for $S11$ and $S12$ are within 3% and 7%, respectively. Whereas for the $S12$, the deviation is 10% at low frequencies (0.5–10 GHz) and remains at 6% in the high-frequency range. In the same way, the deviation of S22 is about 9% at low frequency, and the deviation drops to about 1% at medium and high frequencies. However, the swings in the magnitude of S parameter are all within expectations.

4.6 Summary

This chapter discusses the modeling method and parameter extraction of the small-signal equivalent circuit model for MOSFET devices. First, the commonly used RF de-embedding methods are analyzed and the corresponding equivalent circuit model is given. The different methods for extracting extrinsic resistances in series are also compared and analyzed, and a set of closed-form expressions for model parameters are deduced. Since the device model parameter extraction and validation are based on the device measurement, the measurement data usually contain the measurement errors, which will ultimately lead to the imprecise of the model. The influence of small signal equivalent circuit model parameter extraction has been deeply investigated and the sensitivity of intrinsic parameters to S parameters has been deduced. Therefrom, the uncertainty of model parameters to S parameters has been obtained. Finally, the uncertainty of model parameters and optimal frequency extraction range is given.

References

[1] T. Ytterdal, Y. Cheng, and T. A. Fjeldly, *Device Modeling for Analog and RF CMOS Circuit Design*. England: John Wiley & Sons, Ltd., 2003.
[2] J. Gao, *Optoelectronic Integrated Circuit Design and Device Modeling*. Singapore: John Wiley & Sons, Ltd., 2010.
[3] J. Gao, *RF and Microwave Modeling and Measurement Techniques for Field Effect Transistors*. NC, USA: Raleigh, SciTech Publishing, Inc., 2010.

[4] J. Gao and A. Werthof, "Direct parameter extraction method for deep submicrometer MOSFET small signal equivalent circuit," *IET Microwaves, Antennas and Propagation*, 3(4): 564–571, 2009.

[5] A. M. Mangan, S. P. Voinigescu, M. T. Yang, and M. Tazlauanu, "De-embedding transmission line measurements for accurate modeling of IC designs," *IEEE Transactions on Electron Devices*, 53(2): 235–241, 2006.

[6] H. Cho and D. E. Burk, "A three-step method for the de-embedding of high-frequency S-parameter measurements," *IEEE Transactions on Electron Devices*, 38(6): 1371–1374, 1991.

[7] E. P. Vandamme, D. M. M. Schreurs, and C. V. Dinther, "Improved three-step de-embedding method to accurately account for the influence of pad parasitics in silicon on-wafer RF test-structures," *IEEE Transactions on Electron Devices*, 48(4): 737–742, 2001.

[8] A. Aktas and M. Ismail, "Pad de-embedding in RF CMOS," *IEEE Circuits and Devices Magazine*, 17(3): 8–11, 2001.

[9] J. Cheng, "Parameter extraction and modeling technology of RF microwave MOS device," Ph.D Dissertation, East China Normal University, 2012.

[10] J. Cheng and J. Gao, "*Analysis and modeling of the pads for RF CMOS based on EM simulation[J]*," *Microwave Journal*, 53(10): 96–108, 2010.

[11] J. Cheng, B. Han, S. Li, G. Zhai, L. Sun, and J. Gao, "An improved and simple parameter extraction method and scaling model for RF MOSFETs up to 40 GHz," *International Journal of Electronics*, 99(5): 707–718, 2012.

[12] T. E. Kolding, "Shield-based microwave on-wafer device measurements," *IEEE Transaction on Microwave Theory and Techniques*, 49(6): 1039–1044, 2001.

[13] S. Mei and Y. I. Ismail, "Modeling skin and proximity effects with reduced realizable RL circuits," *IEEE Transaction on Very Large Scale Integration System*, 12(4): 437–447, 2004.

[14] Y. Zhou, P. Yu, N. Yan, and J. Gao, "An improved de-embedding procedure for nanometer MOSFET small signal modeling," *Microelectronics Journal*, 57(11): 60–65, 2016.

[15] Y. Zhou, P. Yu, and J. Gao, "Radio-frequency modeling and parameters extraction of multi-cell MOSFET device," *Journal of Infrared and Millimeter Waves*, 36(5): 550–554, 2017.

[16] E. T. Rios, R. T. Torres, G. V. Fierro, and E. A. Gutierrez-D, "A method to determine the gate bias-dependent and gate bias-independent components of MOSFET series resistance from S-parameters," *IEEE Transactions on Electron Devices*, 53(3): 571–573, 2006.

[17] S. Lee and H. K. Yu, "A semi-analytical parameter extraction of a SPICE BSIM3v3 for RF MOSFET's using S-parameters," *IEEE Transactions on Microwave Theory and Techniques*, 48(3): 412–416, 2000.

[18] P. Yu and J. Gao, "A novel approach to extracting extrinsic resistances for equivalent circuit model of nanoscale MOSFET," *International Journal of Numerical Modeling: Electronic Networks, Devices and Fields*, 29(6): 1045–1054, 2016.

[19] P. Yu, "Microwave modeling and parameter extraction for 90-nm gate-length MOSFET devices," Ph.D Dissertation, East China Normal University, 2018.

[20] Y. Zhou, "Parameter extraction and sensitivity analysis of small signal model for multi-cell MOSFET devices," Master Dissertation, East China Normal University, 2017.

[21] R. Anholt, R. Worley, and R. Neidhard, "Statistical analysis of GaAs MESFET S-parameter equivalent-circuit models," *International Journal of Microwave and Millimeter Wave Computer Aided Engineering*, 1(3): 263–227, 1991.

[22] F. D. King, P. Winson, A. D. Snider, L. Dunleavy, and D. P. Levinson, "Math methods in transistor modeling: Condition numbers for parameter extraction," *IEEE Transactions on Microwave Theory and Techniques*, 46(9): 1313–1314, 2002.

[23] J. Cheng, A. C. Cangellaris, A. M. Yaghmour, and J. L. Prince, "Sensitivity analysis of multiconductor transmission lines and optimization for high-speed interconnect circuit design," *IEEE Transactions on Advanced Packaging*, 23(2): 132–141, 2002.

[24] C. Fager, L. J. P. Linner, and J. C. Pedro, "Optimal parameter extraction and uncertainty estimation in intrinsic FET small-signal models," *IEEE Transactions on Microwave Theory and Techniques*, 50(12): 2797–2803, 2002.

[25] "8510C Data Sheet," Agilent Technologies, July, 2006.

Chapter 5

MOSFET Large-Signal Model

Compared with III-V compound semiconductor devices, MOSFET devices based on silicon materials have the advantages of low cost and high integration. With the rapid development of wireless communication technology, the application of silicon CMOS technology in radio frequency and microwave has also been widely promoted. Due to the tight constraints on power consumption and noise, it is critical for designers to be able to accurately predict the performance of the designed circuit, improving first design success.

5.1 Nonlinear Modeling

The small-signal equivalent circuit model of the MOSFET device is of great significance for understanding the physical properties of the device and improving the process. Nonetheless, it cannot reflect the large-signal power harmonic characteristics of the transistor. In order to accurately describe the large-signal physical characteristics of the device and guide the design of integrated circuits such as microwave power amplifier circuits, mixer circuits, and oscillators, establishing a nonlinear equivalent circuit model is an essential part of computer-aided design. The small-signal equivalent circuit model is the model of the device under fixed bias conditions, while the large-signal model needs to determine the small-signal model under different bias conditions, so the large-signal model can be regarded as the collection of numerous small-signal models [1–3].

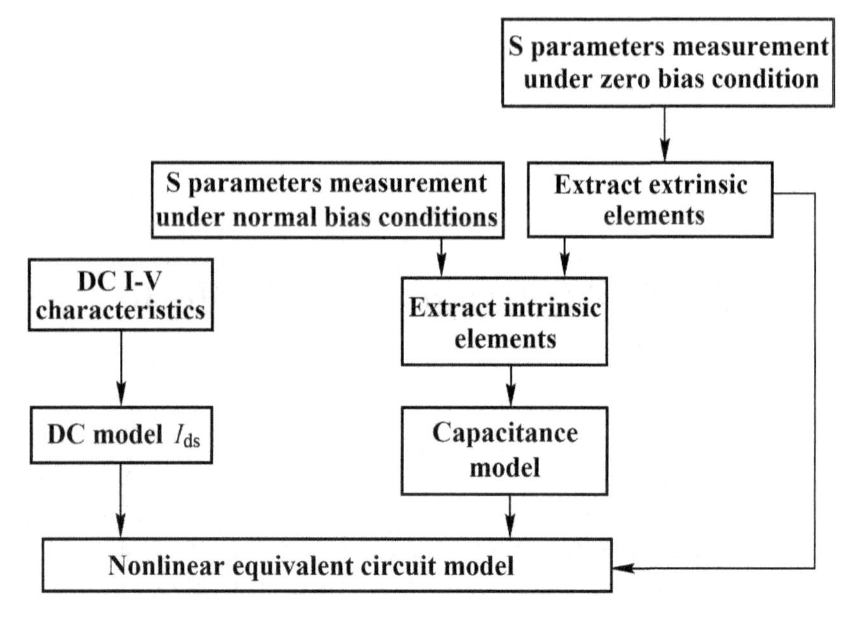

Figure 5.1. Flowchart of the nonlinear modeling for RF MOSFET device.

The flowchart of the nonlinear modeling for the RF MOSFET device is sketched in Figure 5.1. First, the extrinsic elements are extracted by utilizing the measured S parameters under zero bias condition, and then the intrinsic elements are obtained by utilizing the measured S parameters under normal bias conditions. Finally, all elements are determined by optimization in the circuit simulation software. Note that the nonlinear charge model can be established by utilizing the extracted intrinsic capacitances C_{gs} and C_{gd} at each bias point. Meanwhile, the DC model is established by utilizing the measured DC I–V characteristics. At Last, a complete nonlinear equivalent circuit model is established. The ultimate purpose of device modeling is to establish the model that can be applied to the circuit simulation software, so the device model incorporated with the circuit simulation software is an indispensable step in device modeling. Large-signal modeling of a MOSFET device usually utilizes drain output current $I_{ds}(V_{gs}, V_{ds})$, gate-source charge $Q_{gs}(V_{gs}, V_{ds})$, and gate-drain charge $Q_{gd}(V_{gs},V_{ds})$ to describe the large-signal characteristics.

5.2 Commonly Used MOSFET Models

5.2.1 *BSIM model*

At present, the most commonly used MOSFET device model in the industry is the BSIM model [4,5]. Most chip manufacturers use the BSIM model to describe the electrical characteristics of MOSFET devices. There are two different methods for extracting model parameters: the single device-based method and the multi-device-based method. The single device-based extraction method uses the data of one device to extract all parameters of the model. This method can ensure excellent fitting accuracy for one device but not for the devices with other dimensions. Moreover, this method cannot guarantee that the extracted parameters have reasonable physical meaning and cannot achieve the scalable model. BSIM3v3 model adopts the multi-device-based method. This method requires measured data of transistors with different dimensions, and the fitting accuracy for the multi-devices is better than the previous method. The parameter optimization also includes two different methods: global optimization method and local optimization method. The global optimization method regards each parameter as the fitting parameter to optimize, which is ready to appear the optimization results inconsistent with the physical meaning of model parameters, while the local optimization method can ensure that the extracted parameters are consistent with the physical meaning. It is worth noting that the BSIM3v3 model usually uses the local optimization method to optimize the extracted model parameters.

The parameter extraction and optimization of the BSIM model can be done automatically by utilizing the specific model parameter extraction software. Figure 5.2 illustrates the comparison between the measured and modeled DC I–V characteristics of 0.13 μm $\times$ 5 μm $\times$ 4 finger MOSFET device (gate length $\times$ gate width $\times$ gate finger), and the range of gate-source voltage V_{gs} is from 0.5 V to 1.2 V. Figures 5.3–5.5 also exhibit the comparison between the measured and modeled DC I–V characteristics of 0.18 μm $\times$ 5 μm$\times$ 4 finger, 0.24 μm $\times$ 5 μm $\times$ 4 finger, and 0.35 μm $\times$ 5 μm $\times$ 4 finger

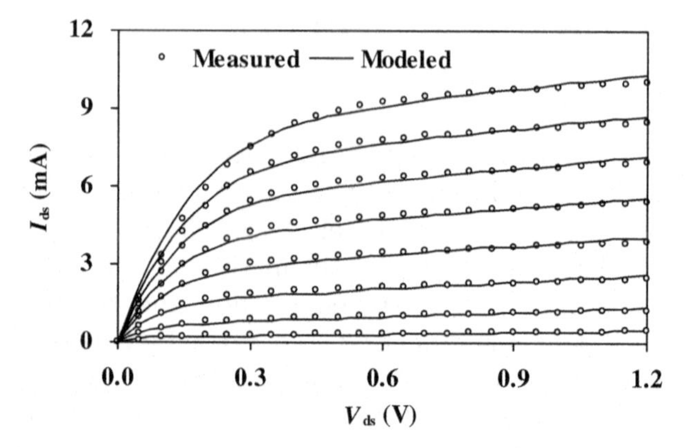

Figure 5.2. Comparison between the measured and modeled DC I–V characteristics. Device: 0.13 μm × 5 μm × 4 finger MOSFET device.

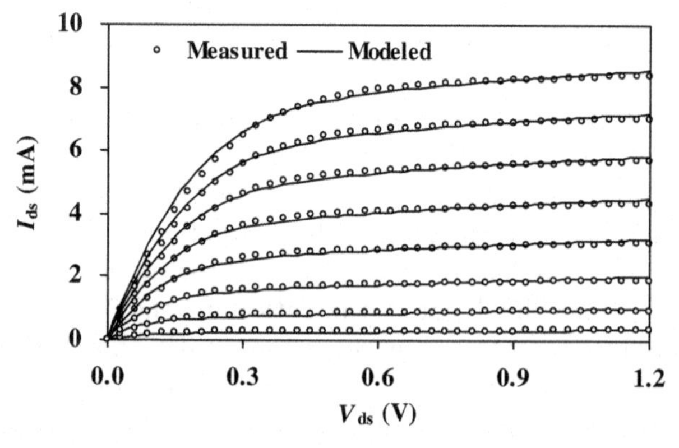

Figure 5.3. Comparison between the measured and modeled DC I–V characteristics. Device: 0.18 μm × 5 μm × 4 finger MOSFET device.

MOSFET devices, respectively. As seen in the above figures, with a decreased gate length of the device, the transconductance increases, and under the same bias condition, the source-drain current also increases. In terms of data fit, the modeled data agree well with the measured data when the device channel length is longer, nonetheless the error increases as the channel size decreases. The reason is that as the channel length becomes shorter, various high-order effects, especially the short-channel effect, have a greater influence on the device [6].

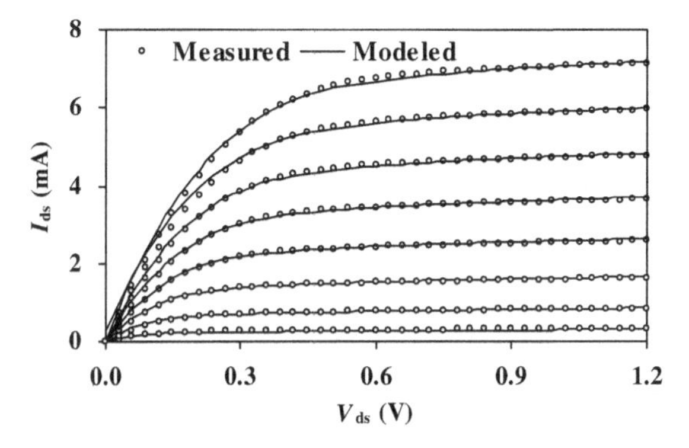

Figure 5.4. Comparison between the measured and modeled DC I–V characteristics. Device: 0.24 μm $\times$ 5 μm $\times$ 4 finger MOSFET device.

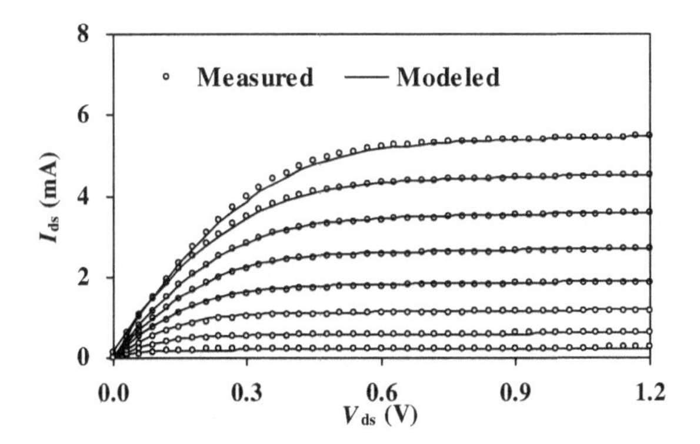

Figure 5.5. Comparison between the measured and modeled DC I–V characteristics. Device: 0.35 μm $\times$ 5 μm $\times$ 4 finger MOSFET device.

5.2.2 *Angelov model*

In 1992, Angelov proposed a unified nonlinear model that can be used for field-effect transistors, including MESFET, MOSFET, and HEMT devices [7]. It is called the Angelov model in commercial circuit simulation software. In this model, the source-drain DC is expressed as

$$I_{\text{ds}} = I_{\text{pk}}(1 + \tan\text{h}(\phi)) \tan\text{h}(\alpha V_{\text{ds}})(1 + \lambda V_{\text{ds}}) \tag{5.1}$$

or

$$I_{\text{ds}} = I_{\text{pk}}(1 + \tan\text{h}(\phi)) \tan\text{h}(\alpha V_{\text{ds}}) \exp(\lambda V_{\text{ds}}) \tag{5.2}$$

And the corresponding transconductance and output conductance expressions are

$$g_m = \frac{\partial I_{\mathrm{ds}}}{\partial V_{\mathrm{gs}}} = I_{\mathrm{pk}} \operatorname{sec h}^2(\phi) \frac{d\phi}{dV_{\mathrm{gs}}} \tan \mathrm{h}(\alpha V_{\mathrm{ds}})(1 + \lambda V_{\mathrm{ds}}) \quad (5.3)$$

$$g_{\mathrm{ds}} = \frac{\partial I_{\mathrm{ds}}}{\partial V_{\mathrm{ds}}} = I_{\mathrm{pk}}[1 + \tan \mathrm{h}(\phi)][\lambda \tan \mathrm{h}(\alpha V_{\mathrm{ds}}) + \alpha \operatorname{sec h}^2(\alpha V_{\mathrm{ds}})(1 + \lambda V_{\mathrm{ds}})] \quad (5.4)$$

The channel length modulation effect can be also represented by a first-order approximation of the exponential function $\exp(\lambda V_{\mathrm{ds}})$:

$$\exp(\lambda V_{\mathrm{ds}}) = 1 + \lambda V_{\mathrm{ds}} + \frac{1}{2}(\lambda V_{\mathrm{ds}})^2 + \cdots \quad (5.5)$$

where ϕ can be expressed as

$$\phi = P_1(V_{\mathrm{gs}} - V_{\mathrm{pk}}) + P_2(V_{\mathrm{gs}} - V_{\mathrm{pk}})^2 + P_3(V_{\mathrm{gs}} - V_{\mathrm{pk}})^3 + \cdots \quad (5.6)$$

Thence, the expression of source-drain DC can be simplified as

$$I_{\mathrm{ds}} = I_{\mathrm{pk}}(1 + \tan \mathrm{h}(P_1(V_{\mathrm{gs}} - V_{\mathrm{pk}}))) \tan \mathrm{h}(\alpha V_{\mathrm{ds}}) \exp(\lambda V_{\mathrm{ds}}) \quad (5.7)$$

where I_{pk} and V_{pk} represent the drain current and gate voltage when the transconductance reaches the maximum, respectively, and the relationship between the drain-source voltage and V_{pk} can be expressed as

$$V_{\mathrm{pk}}(V_{\mathrm{ds}}) = V_{\mathrm{pko}} + (V_{\mathrm{pks}} - V_{\mathrm{pko}}) \tan \mathrm{h}(\alpha V_{\mathrm{ds}}) \quad (5.8)$$

The saturation voltage parameter α can be described as the function of V_{gs}:

$$\alpha = \alpha_{\mathrm{r}} + \alpha_{\mathrm{s}} \exp\left(\frac{V_{\mathrm{gs}}}{nkT}\right) \quad (5.9)$$

The charge model in the Angelov model uses a similar function to I_{ds}:

$$C_{\mathrm{gs}} = C_{\mathrm{gso}}[1 + \tan \mathrm{h}(\phi_1)][1 + \tan \mathrm{h}(\phi_2)] \quad (5.10)$$

$$C_{\mathrm{gd}} = C_{\mathrm{gdo}}[1 + \tan \mathrm{h}(\phi_3)][1 - \tan \mathrm{h}(\phi_4)] \quad (5.11)$$

where

$$\phi_1 = P_{0gsg} + P_{1gsg}V_{gs} + P_{2gsg}V_{gs}^2 + P_{3gsg}V_{gs}^3 + \cdots . \tag{5.12}$$

$$\phi_2 = P_{0gsd} + P_{1gsd}V_{ds} + P_{2gsd}V_{ds}^2 + P_{3gsd}V_{ds}^3 + \cdots . \tag{5.13}$$

$$\phi_3 = P_{0gdg} + P_{1gdg}V_{gs} + P_{2gdg}V_{gs}^2 + P_{3gdg}V_{gs}^3 + \cdots . \tag{5.14}$$

$$\phi_4 = P_{0gdd} + (P_{1gdd} + P_{1cc}V_{gs})V_{ds} + P_{2gdd}V_{ds}^2 + P_{3gdd}V_{ds}^3 + \cdots . \tag{5.15}$$

The above equations satisfy the symmetry:

$$\frac{\partial C_{gs}}{\partial V_{dg}} = \frac{\partial C_{gd}}{\partial V_{gs}} \tag{5.16}$$

The corresponding charge formula can be written as

$$Q_{gs} = C_{gsp}V_{gs} + C_{gso}[V_{gs} + L_{c1} + V_{gs}T_{ch2} + L_{c1}T_{ch2}) \tag{5.17}$$

$$Q_{gd} = C_{gdp}V_{gd} + C_{gdo}[V_{gd} + L_{c4} + V_{gs}T_{ch3} + L_{c4}T_{ch3}) \tag{5.18}$$

where

$$L_{c1} = \frac{\log[\cos\mathrm{h}(P_{10} + P_{11}V_{gs})]}{P_{11}} \tag{5.19}$$

$$T_{h2} = \tan\mathrm{h}[P_{20} + P_{21}V_{ds}] \tag{5.20}$$

$$L_{c4} = \frac{\log[\cos\mathrm{h}(P_{40} + P_{41}V_{gd})]}{P_{41}} \tag{5.21}$$

$$T_{h3} = \tan\mathrm{h}[P_{30} + P_{31}V_{ds}] \tag{5.22}$$

5.3 Effect of Guard Ring

It is essential to analyze the influence of guard ring (GR) on DC and high-frequency performance for deep submicron MOSFET devices. In this section, MOSFET devices with four different guard ring structures were fabricated utilizing 90 nm standard CMOS process and a detailed comparison of the devices' performance is carried out. DC and small-signal equivalent circuit model including guard ring effects is proposed, and a set of simple, but efficient expressions are derived to provide a bidirectional bridge for the S parameters transformation between devices with different guard ring structures [8].

In standard CMOS technology, guard rings have been widely used to improve the performance of MOSFET devices and circuits. The main functions include the following [9–12]:

(1) The guard ring prevents the coupling current, cuts down the crosstalk of the signal from the substrate to the MOSFET device, and reduces the noise injected by the substrate.
(2) The device latch-up effect can be suppressed by guard rings.
(3) The bulk resistance can be minimized by placing the guard ring around the entire device and connecting it to the p-well of the device.

The influence of guard rings on the DC and high-frequency performance of deep submicron MOSFETs is presented in the following, along with the experimental results of guard rings for different test structures. It is worth noting that all MOSFET devices were fabricated on the same wafer and were fabricated using the same process. By establishing circuit models of various test structures, using circuit simulation tools to optimize the performance of MOSFET devices to observe the effect of guard rings to evaluate the influence of guard rings on devices and circuits.

5.3.1 *Structure design*

The NMOSFET devices utilized in this work were fabricated on 90 nm standard CMOS process. To deliberate the effect of guard rings on the DC and high-frequency performance of deep submicron MOSFET devices, four different types of GR test structures (i.e., elementary cell) are employed for NMOSFET, as illustrated in Figure 5.6:

(1) device without guard ring structure (NGR), as shown in Figure 5.6(a),
(2) device with rectangular GR around the device, as shown in Figure 5.6(b),
(3) device with one-sided guard ring structure, including two structures (horizontal or vertical guard ring only, i.e., half of the guard ring), is also called SGR, as shown in Figure 5.6(c).

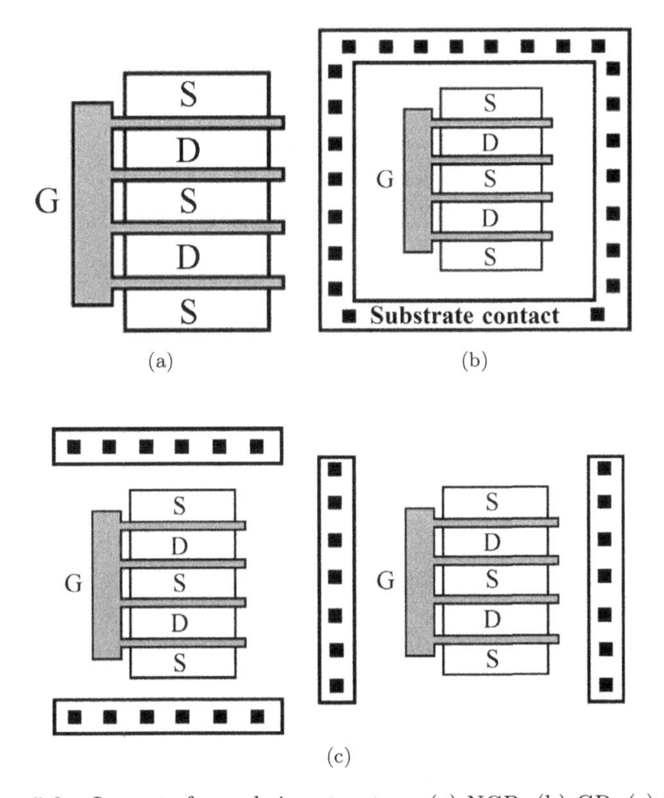

Figure 5.6. Layout of guard ring structure: (a) NGR; (b) GR; (c) SGR.

From a foundry point of view, the chip area and cost will be increased due to the presence of GRs. The corresponding schematic layout of large-size MOSFET that consists of multiple elementary cells is presented in Figure 5.7, and n represents the number of cells included in the large-size MOSFET device.

5.3.2 *DC characteristics*

First, we discuss the large-signal behavior. The cross-section of the n-channel MOSFET device is illustrated in Figure 5.8. For MOSFET devices without GR structure (NGR) (as shown in Figure 5.8(a)), the floating-body effects appear as a slightly higher current in the linear region and early in saturation, a “Kink” later in saturation, and eventually premature breakdown. As is known, the kink effect is due to the impact ionization caused by the high electric field in the

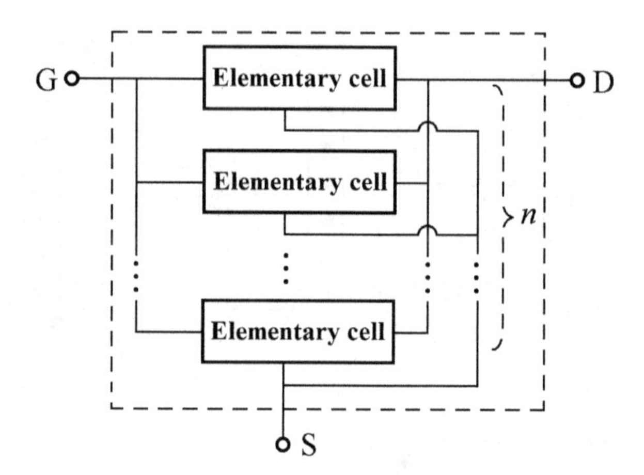

Figure 5.7. Schematic diagram of large-size MOSFET that consists of multiple elementary cells.

pinched-off region of the channel. Hence, it occurs when the MOSFET device is operated in the saturation region, which decreases the threshold voltage and increases the channel current. Since the source-body junction becomes forward-biased, additional electrons are injected from the source to the body. If the channel is short enough, a significant fraction of these electrons do not recombine with the holes in the channel and are collected by the drain. This effect can be represented by a parasitic npn bipolar transistor in parallel with the FET transistor.

For a MOSFET device with GR structure (as shown in Figure 5.8(b)), GR means there is a highly doped substrate ring around the transistor. Therefore, the body-source junction is always reverse biased and the kink effect will be eliminated. However, if the body is grounded by using half guarding (horizontal or vertical GR) only, due to the potentials in the body region being nonuniform distribution, a small kink current will occur in the pinched-off region of the channel (as shown in Figure 5.8(c)).

In order to describe the nonlinear behavior of the drain to source current, an improved compact DC model which is similar to FET devices has been used for elementary cell as follows [15]:

$$I_{\mathrm{ds}} = \frac{\beta(V_{\mathrm{gs}} - V_{\mathrm{t}})^2}{1 + bV_{\mathrm{gs}}}(1 + \lambda_{\mathrm{t}} V_{\mathrm{ds}}) \tan\mathrm{h}(\alpha_{\mathrm{t}} V_{\mathrm{ds}}) \tag{5.23}$$

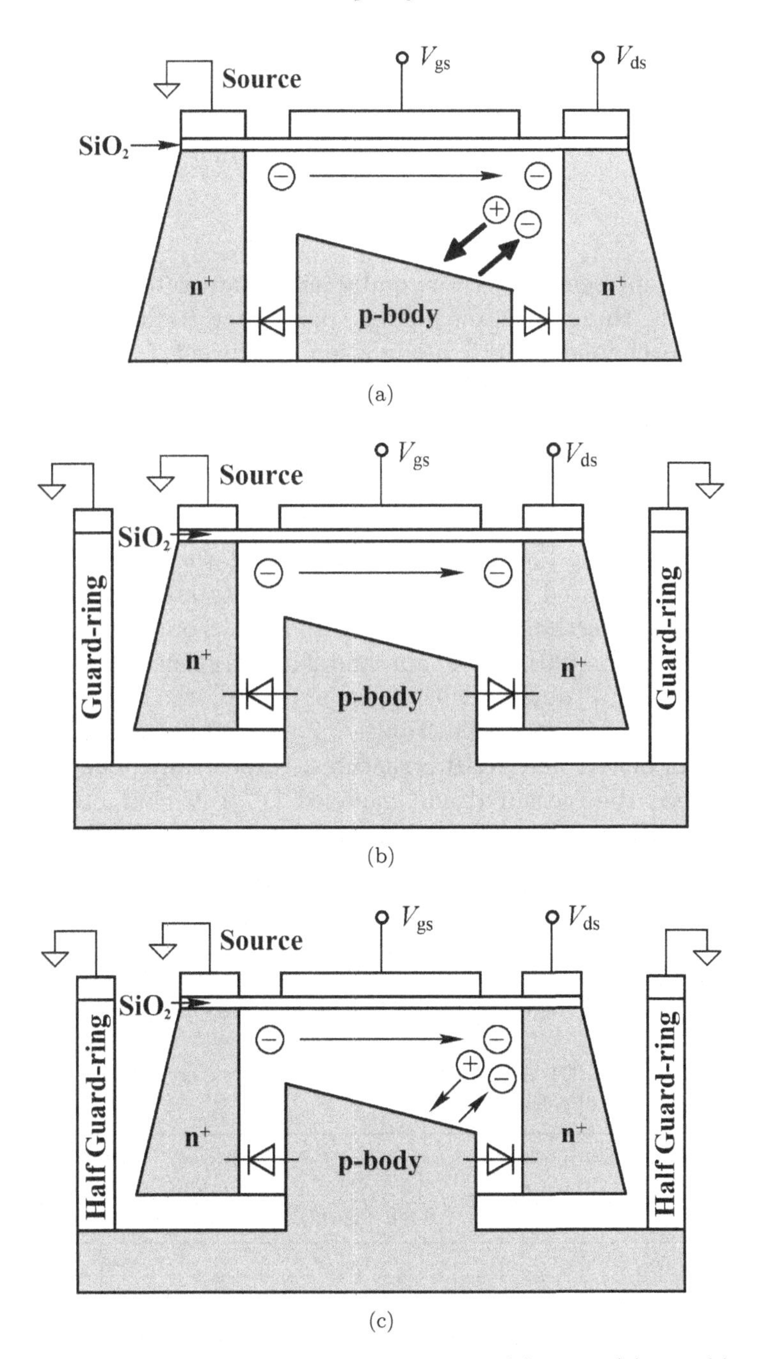

Figure 5.8. Cross-section of n-channel MOSFET: (a) NGR; (b) GR; (c) SGR.

with

$$V_t = V_{to} - k_t V_{ds}$$

$$\lambda_t = \lambda - k_\lambda V_{gs}$$

$$\alpha_t = \alpha + k_\alpha V_{gs}$$

wherein β represents the transconductance parameter (unit A/V^2), α represents the saturation voltage parameter (unit V^{-1}), λ represents the channel length modulation coefficient (unit V^{-1}), V_{to} indicates the threshold voltage (unit V), b defines the doping tail extending parameter (unit V^{-1}), k_t models the threshold voltage changed with V_{ds} (dimensionless), k_λ models the channel length modulation coefficient versus V_{gs} (unit V^{-2}), and k_α models the saturation voltage parameter versus V_{gs} (unit V^{-2}).

For deliberating the effect of the guard ring, the MOSFET device structures fabricated by 90 nm CMOS process are summarized in Table 5.1. It is worth noting that the gate lengths of the four different devices are 90 nm, 90 nm, 240 nm, and 240 nm, respectively.

For MOSFET devices with 4 finger $\times$ 0.6 μm $\times$ 18 cells and 32 finger $\times$ 1.5 μm $\times$ 2 cells, Tables 5.2 and 5.3 give the DC model parameters of GR and NGR structures, respectively. The comparison between the measured and modeled DC I–V characteristics of GR and NGR structures for 90 nm and 240 nm MOSFET devices is illustrated in Figure 5.9. It can be observed from Figure 5.9 that as the gate length decreases, the Kink effect appears when V_{ds} is low, since the shorter gate length leads to a lower saturation voltage V_{sat}, resulting in a higher multiplication factor. A further increase in the drain voltage beyond the kink leads to a quadratic increase in

Table 5.1. MOSFET device structure (gate finger $\times$ gate width of single finger $\times$ number of cells).

Device structure	4 finger $\times$ 0.6 μm$\times$ 18 cell (90 nm)	8 finger $\times$ 1 μm $\times$ 6 cell (90 nm)	32 finger $\times$ 1.5 μm $\times$ 2 cell (240 nm)	32 finger $\times$ 1 μm $\times$ 2 cell (240 nm)
GR	√	√	√	√
SGR	×	√	√	√
NGR	√	√	√	×

Table 5.2. DC model parameters of 90 nm MOSFET device.

	$0.09\ \mu m \times 4$ finger $\times 0.6\ \mu m \times 18$ cell		
		NGR	
Parameter	GR	Without Kink effect	Kink region
$\beta(\times 10^{-3})$	5.35	5.8	5.1
α	8.0	8.6	8.0
V_{to}	0.45	0.45	0.35
λ	0.88	0.88	0.95
b	1.65	1.65	1.65
k_λ	0.55	0.48	0.4

Table 5.3. DC model parameters of 240 nm MOSFET device.

	$0.24\ \mu m \times 32$ finger $\times 1.5\ \mu m \times 2$ cell		
		NGR	
Parameter	GR	Without Kink effect	Kink region
$\beta(\times 10^{-2})$	4.98	5.4	3.42
α	5.8	5.7	5.5
V_{to}	0.4	0.4	0.35
λ	0.13	0.5	1.1
b	1.24	1.24	0.9
k_λ	0.3	0.7	0.59

the drain current, which finally results in a strong parasitic bipolar transistor action. Figure 5.10 illustrates the comparison between the modeled and measured transconductance and output conductance for GR and NGR structures under the bias condition of $V_{ds} = 1.2\,\text{V}$. It is obvious that the kink effect can be eliminated by using GR for both submicrometer and deep submicrometer MOSFET devices.

Table 5.4 summarizes the comparison of DC model parameters between GR and SGR for MOSFET devices $0.09\ \mu m \times 8$ fingers $\times$ $1\ \mu m \times 6$ cells and $0.24\ \mu m \times 32$ fingers $\times 1.5\ \mu m \times 2$ cells (gate length $\times$ number of gate finger $\times$ single-finger gate width $\times$ cell number), and

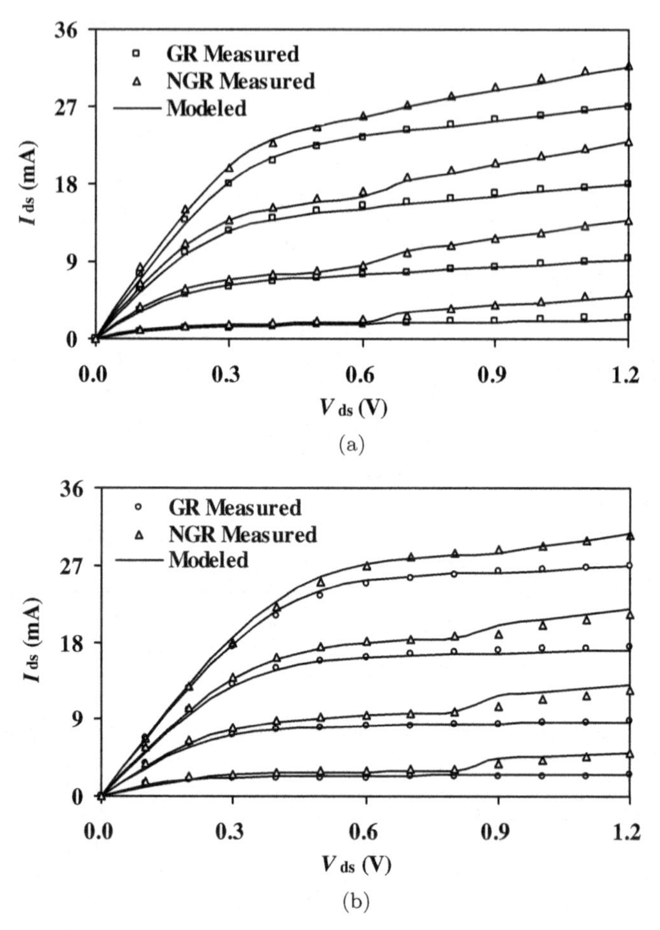

Figure 5.9. Comparison between the measured and modeled DC I–V characteristics of GR and NGR structures for 90 nm and 240 nm MOSFET devices: (a) 0.09 μm $\times$ 4 finger $\times 0.6\ \mu m$ $\times$ 18 cell; (b) 0.24 μm $\times$ 32 finger $\times$1.5 μm $\times$ 2 cell.

each device has two structures of GR and SGR. The comparison between the measured and modeled DC I–V characteristics of GR and SGR structures for 90 nm and 240 nm MOSFET devices is displayed in Figure 5.11. Compared with the NGR structure, the Kink effect current in MOSFET devices with SGR structure becomes very small, especially in deep submicron devices, this current becomes negligible. The comparison between the modeled and measured transconductance and output conductance for GR and SGR structures under

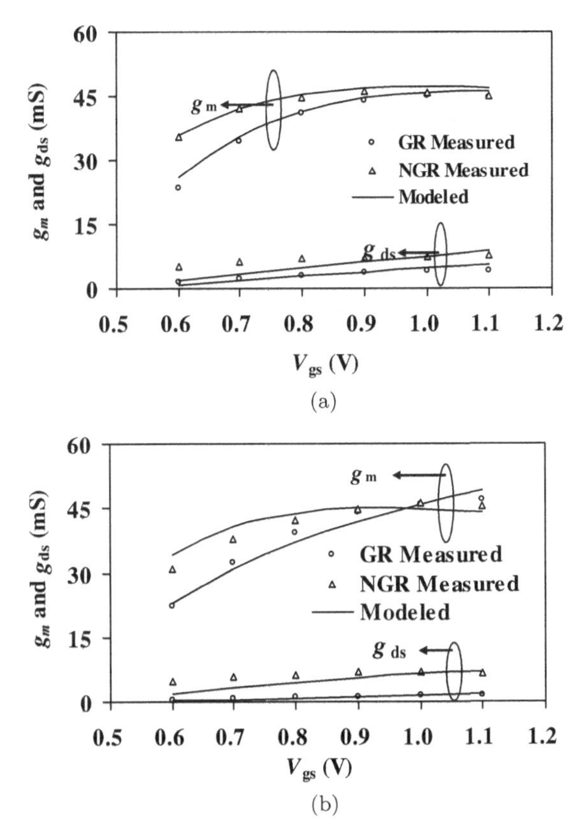

Figure 5.10. Transconductance and output conductance versus V_{gs} for 90 nm and 240 nm MOSFET devices (Bias: $V_{ds} = 1.2\,V$): (a) 0.09 μm$\times$4 finger $\times$0.6 μm$\times$ 18 cell; (b) 0.24 μm $\times$ 32 finger $\times$1.5 μm $\times$ 2 cell.

the bias condition of $V_{ds} = 1.2$ V is depicted in Figure 5.12. Apparently, for deep submicron (rather than submicron) MOSFET devices, most of the Kink effect can be eliminated by using the SGR structure.

5.3.3 *S parameters*

The influence of guard rings on the RF performance of the device is discussed in the following. It is worth noting that the topology of the equivalent circuit model for the device is the same regardless of whether it contains guard rings or not, however, the model parameters of the device with different GR structures are different.

Table 5.4. Comparison of DC model parameters between GR and SGR for MOSFET devices.

	0.09 μm × 8 finger × 1 μm × 6 cell		0.24 μm × 32 finger × 1.5 μm × 2 cell	
Parameter	GR	SGR	GR	SGR
$\beta(\times 10^{-2})$	1.75	1.76	4.98	3.15
α	5.4	5.0	5.8	7.0
V_{to}	0.42	0.42	0.4	0.4
λ	0.63	0.66	0.13	1.1
b	1.5	1.5	1.24	0.59
k_λ	0.5	0.5	0.3	0.55
k_{t}	0	0	0	−0.03
k_α	0	0.08	0	0.2

The intrinsic part of the small-signal equivalent circuit for single elementary cell shown in Figure 5.13 is characterized by the Y parameters:

$$Y_{11} = j\omega(C_{\text{gs}} + C_{\text{gd}}) \tag{5.24}$$

$$Y_{12} = -j\omega C_{\text{gd}} \tag{5.25}$$

$$Y_{21} = g_m e^{-j\omega\tau} - j\omega C_{\text{gd}} \tag{5.26}$$

$$Y_{22} = g_{\text{ds}} + j\omega(C_{\text{ds}} + C_{\text{gd}}) \tag{5.27}$$

The corresponding S parameters can be expressed as

$$S_{11} = \frac{(Y_{\text{o}} - Y_{11})(Y_{\text{o}} + Y_{22}) + Y_{12}Y_{21}}{(Y_{\text{o}} + Y_{11})(Y_{\text{o}} + Y_{22}) - Y_{12}Y_{21}} \tag{5.28}$$

$$S_{12} = \frac{-2Y_{12}Y_{\text{o}}}{(Y_{\text{o}} + Y_{11})(Y_{\text{o}} + Y_{22}) - Y_{12}Y_{21}} \tag{5.29}$$

$$S_{21} = \frac{2Y_{21}Y_{\text{o}}}{(Y_{\text{o}} + Y_{11})(Y_{\text{o}} + Y_{22}) - Y_{12}Y_{21}} \tag{5.30}$$

$$S_{22} = \frac{(Y_{\text{o}} + Y_{11})(Y_{\text{o}} - Y_{22}) + Y_{12}Y_{21}}{(Y_{\text{o}} + Y_{11})(Y_{\text{o}} + Y_{22}) - Y_{12}Y_{21}} \tag{5.31}$$

where Y_{o} is the characteristic admittance, and Y_{o} is equal to 0.02S for the standard 50 Ω system.

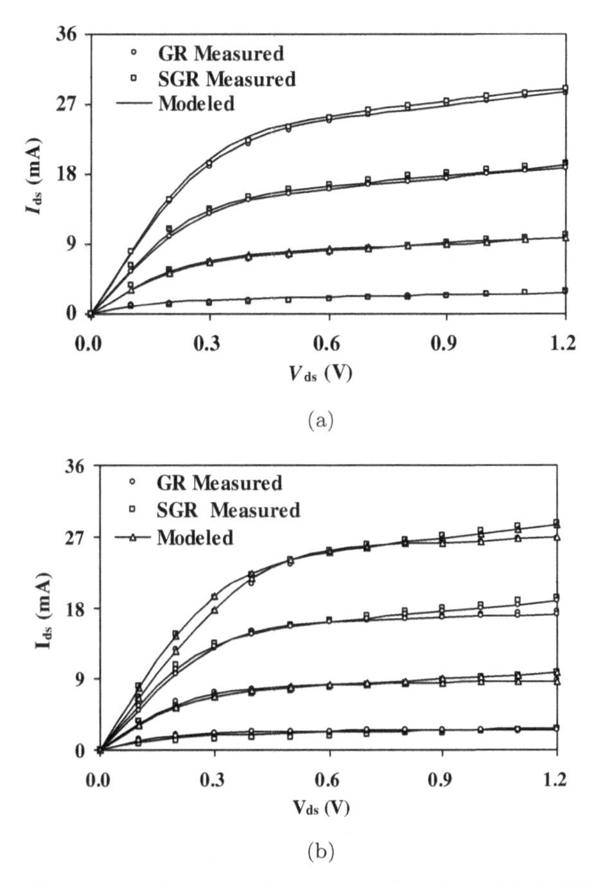

Figure 5.11. Comparison between the measured and modeled DC I–V characteristics of GR and SGR structures for 90 nm and 240 nm MOSFET devices: (a) 0.09 μm $\times$ 1 finger $\times$0.8 μm $\times$ 6 cell; (b) 0.24 μm $\times$ 32 finger $\times$1.5 μm $\times$ 2 cell.

From the physical structure perspective, the Kink effect only changes the transconductance and output conductance of the DC characteristic, while other intrinsic capacitance model parameters ($C_{\rm gs}$, $C_{\rm gd}$, and $C_{\rm ds}$) hold the line for different guard ring structures.

According to the above analysis, for the intrinsic capacitances of different structures, we have

$$C_{\rm gs}^{\rm GR} = C_{\rm gs}^{\rm NGR} = C_{\rm gs}^{\rm SGR} \tag{5.32}$$

$$C_{\rm gd}^{\rm GR} = C_{\rm gd}^{\rm NGR} = C_{\rm gd}^{\rm SGR} \tag{5.33}$$

$$C_{\rm ds}^{\rm GR} = C_{\rm ds}^{\rm NGR} = C_{\rm ds}^{\rm SGR} \tag{5.34}$$

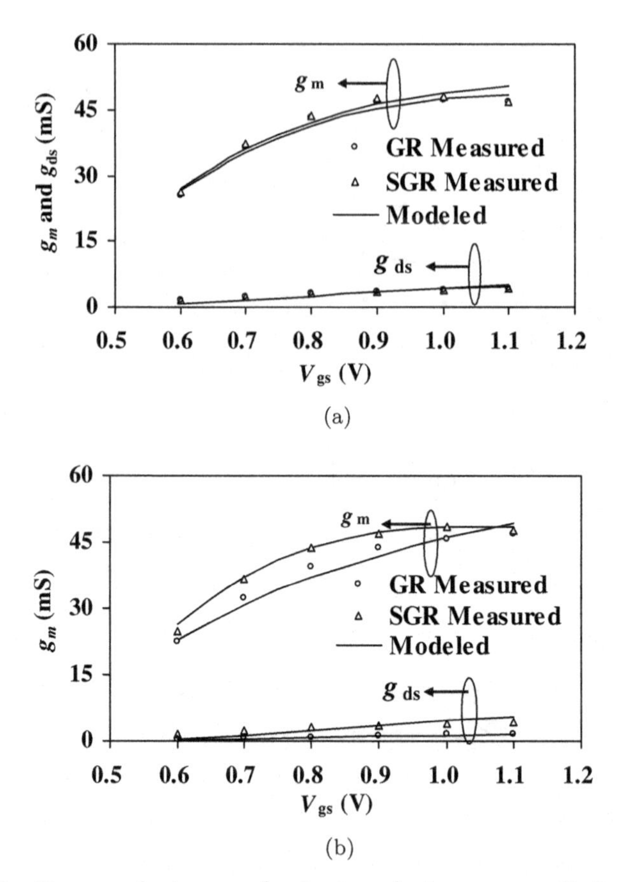

Figure 5.12. Transconductance and output conductance versus V_{gs} for 90 nm and 240 nm MOSFET devices (Bias: $V_{ds} = 1.2\,\text{V}$): (a) $0.09\,\mu\text{m} \times 1$ finger $\times 0.8\,\mu\text{m} \times$ 6 cell; (b) $0.24\,\mu\text{m} \times 32$ finger $\times 1.5\,\mu m \times 2$ cell.

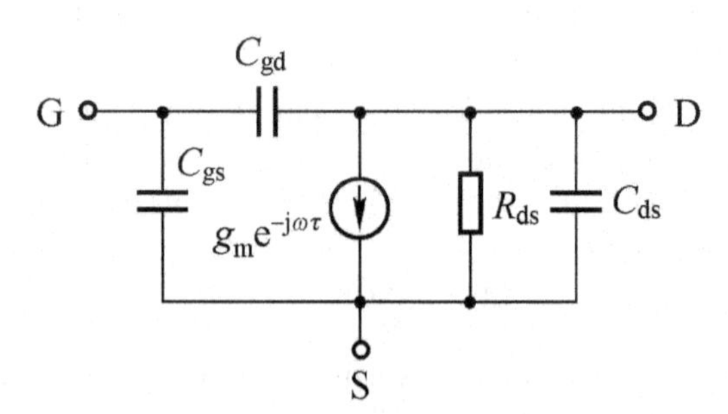

Figure 5.13. Intrinsic part of equivalent circuit model for single elementary cell.

For transconductance and output conductance, we have

$$g_m^{\text{NGR}} = (1 + k_{\text{NGR}}^{\text{g}})g_{\text{m}}^{\text{GR}} \tag{5.35}$$

$$g_{\text{ds}}^{\text{NGR}} = (1 + k_{\text{NGR}}^{\text{d}})g_{\text{ds}}^{\text{GR}} \tag{5.36}$$

$$g_m^{\text{SGR}} = (1 + k_{\text{SGR}}^{\text{g}})g_m^{\text{GR}} \tag{5.37}$$

$$g_{\text{ds}}^{SGR} = (1 + k_{\text{SGR}}^{\text{d}})g_{\text{ds}}^{\text{GR}} \tag{5.38}$$

where $k_{\text{NGR}}^{\text{g}}$ and $k_{\text{SGR}}^{\text{g}}$ are the kink factors of the transconductances for NGR and SGR structure device and $k_{\text{NGR}}^{\text{d}}$ and $k_{\text{SGR}}^{\text{d}}$ are the kink factors of the output conductances for NGR and SGR structure devices, respectively.

Based on the conversion between S parameters and Y parameters, we have

$$S_{11}^{\text{GR}} \approx S_{11}^{\text{NGR}} \approx S_{11}^{\text{SGR}} \tag{5.39}$$

$$S_{12}^{\text{NGR}} \approx \frac{Y_o + ng_{\text{ds}}^{\text{GR}}}{Y_o + ng_{\text{ds}}^{\text{NGR}}} S_{12}^{\text{GR}} \tag{5.40}$$

$$S_{12}^{\text{SGR}} \approx \frac{Y_o + ng_{\text{ds}}^{\text{GR}}}{Y_o + ng_{\text{ds}}^{\text{SGR}}} S_{12}^{\text{GR}} \tag{5.41}$$

$$S_{21}^{\text{NGR}} \approx \frac{g_m^{\text{NGR}}(Y_o + ng_{\text{ds}}^{\text{GR}})}{g_m^{\text{GR}}(Y_o + ng_{\text{ds}}^{\text{NGR}})} S_{21}^{\text{GR}} \tag{5.42}$$

$$S_{21}^{\text{SGR}} \approx \frac{g_m^{\text{SGR}}(Y_o + ng_{\text{ds}}^{\text{GR}})}{g_m^{\text{GR}}(Y_o + ng_{\text{ds}}^{\text{SGR}})} S_{21}^{\text{GR}} \tag{5.43}$$

$$S_{22}^{\text{NGR}} \approx S_{22}^{\text{GR}} = \frac{(Y_o - ng_{\text{ds}}^{\text{NGR}})(Y_o + ng_{\text{ds}}^{\text{GR}})}{(Y_o - ng_{\text{ds}}^{\text{GR}})(Y_o + ng_{\text{ds}}^{\text{NGR}})} S_{22}^{\text{GR}} \tag{5.44}$$

$$S_{22}^{\text{SGR}} \approx \frac{(Y_o - ng_{\text{ds}}^{\text{SGR}})(Y_o + ng_{\text{ds}}^{\text{GR}})}{(Y_o - ng_{\text{ds}}^{\text{GR}})(Y_o + ng_{\text{ds}}^{\text{SGR}})} S_{22}^{\text{GR}} \tag{5.45}$$

It can be found that the S_{11} remains invariant and S_{12} little changed for GR, NGR, and SGR structures, while S_{21} and S_{22} are proportional to the transconductance and the output conductance, respectively. From the expression above mentioned, the S parameters of devices with NGR and SGR structures can be obtained directly by utilizing a simple set of formulas from GR S parameters. All the relationships provide a bidirectional bridge for the transformation between NGR, SGR, and GR MOSFETs, respectively.

The small-signal model parameters of 0.09 μm $\times$ 4 finger $\times$ 0.6 μm $\times$ 18 cell GR-MOSFET are listed in Table 5.5. Under the bias condition of $V_{\text{gs}} = 1.0$ V and $V_{\text{ds}} = 1.0$ V, Figure 5.14 depicts the comparison between the measured and modeled S parameters for 0.09 μm $\times$ 4 finger $\times$ 0.6 μm $\times$ 18 cell GR-MOSFET devices. The comparison between the measured and modeled S parameters

Table 5.5. Small-signal model parameters. Device: 0.09 μm $\times$ 4 finger $\times$ 0.6 μm $\times$ 18 cell GR-MOSFET.

Parameter	Value	Parameter	Value
L_g(pH)	250	C_{jd}(fF)	0.8
L_d(pH)	150	C_{gs}(fF)	3.0
L_s(pH)	90	C_{gd}(fF)	1.55
$R_g(\Omega)$	20	C_{ds}(fF)	1.6
$R_d(\Omega)$	120	g_m(mS)	1.05
$R_s(\Omega)$	10	g_{ds}(mS)	0.077
$R_{\text{sub}}(\Omega)$	3000	$R_{\text{gs}}(\Omega)$	10

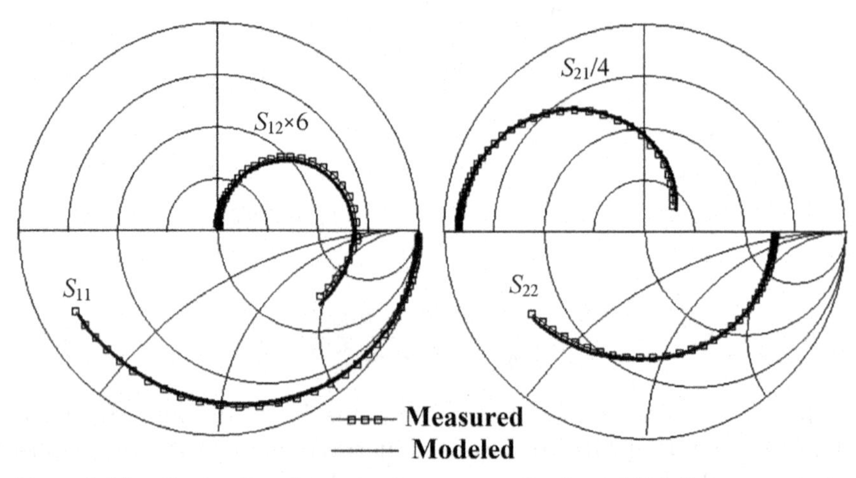

Figure 5.14. Comparison between the measured and modeled S parameters for 0.09 μm $\times$ 4 finger $\times$ 0.6 μm $\times$ 18 cell GR-MOSFET devices (Bias: $V_{\text{gs}} = 1.0$ V. $V_{\text{ds}} = 1.0$ V).

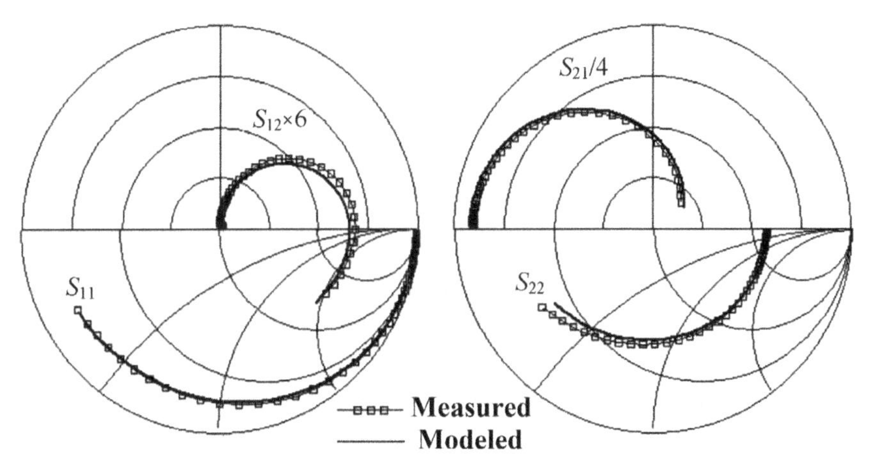

Figure 5.15. Comparison between the measured and the modeled S parameters for 0.09 μm $\times$ 4 finger $\times$0.6 μm $\times$ 18 cell NGR-MOSFET devices. $g_m^{\text{NGR}} = 2.8$ mS, $g_{\text{ds}}^{\text{NGR}} = 0.33$ mS. Bias: $V_{\text{gs}} = 1.0$ V, $V_{\text{ds}} = 1.0$ V.

Table 5.6. Small-signal model parameters. Device: 0.24 μm $\times$ 32 finger $\times$1.5 μm $\times$ 2 cell GR-MOSFET.

Parameter	Value	Parameter	Value
L_g(pH)	15	C_{jd}(fF)	5.4
L_d(pH)	15	C_{gs}(fF)	106
L_s(pH)	2	C_{gd}(fF)	28
R_g(Ω)	2.5	C_{ds}(fF)	22
R_d(Ω)	10	g_m(mS)	24.5
R_s(Ω)	1	g_{ds}(mS)	1
R_{sub}(Ω)	400	R_{gs}(Ω)	5

for 0.09 μm $\times$ 4 finger $\times$0.6 μm $\times$ 18 cell NGR-MOSFET devices is illustrated in Figure 5.15. Note that the modeled data are obtained by the bidirectional bridge for the transformation based on the GR-MOSFET. As seen, good agreements can be achieved to verify the validity of the approach.

Table 5.6 tabulates the small-signal model parameters of the 0.24 μm$\times$32 finger $\times$1.5 μm$\times$2 cell GR-MOSFET device. In addition, the comparison between the measured and modeled data for the bias condition of $V_{\text{gs}} = 1.0$ V and $V_{\text{ds}} = 1.0$ V is illustrated in Figure 5.16. For 0.24 μm $\times$ 32 finger $\times$1.5 μm $\times$ 2 cell NGR-MOSFET devices, the

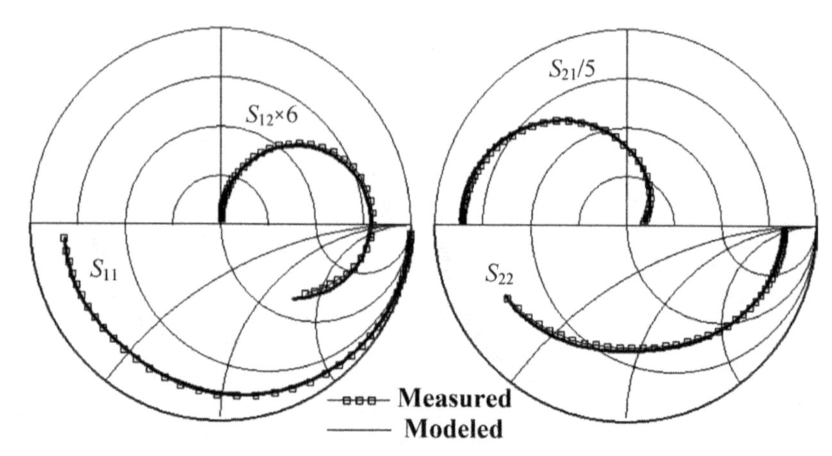

Figure 5.16. Comparison between the measured and the modeled S parameters for 0.24 μm × 32 finger ×1.5 μm × 2 cell GR-MOSFET devices. Bias: $V_{\mathrm{gs}} = 1.0$ V, $V_{\mathrm{ds}} = 1.0$ V.

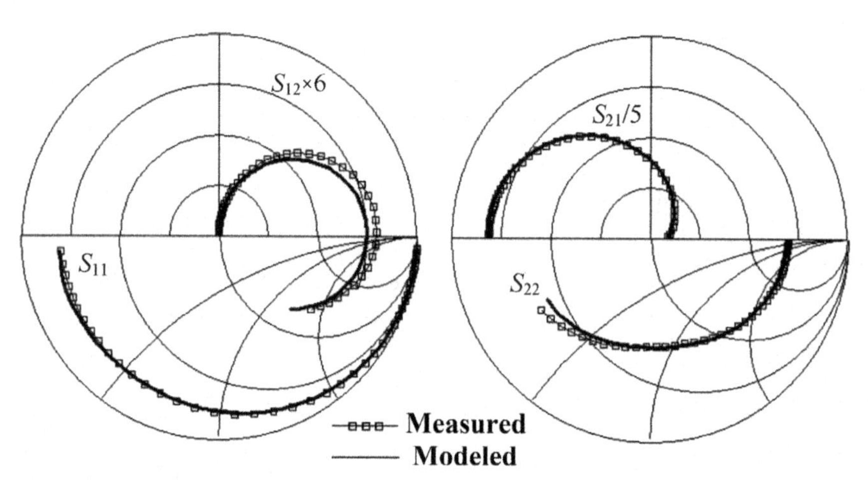

Figure 5.17. Comparison between the measured and the modeled S parameters for 0.24 μm × 32 finger ×1.5 μm × 2 cell NGR-MOSFET devices. $g_m^{\mathrm{NGR}} = 25.7$ mS, $g_{\mathrm{ds}}^{\mathrm{NGR}} = 1.92$ mS. Bias: $V_{\mathrm{gs}} = 1.0$ V, $V_{\mathrm{ds}} = 1.0$ V.

comparison of the measured and modeled S parameters calculated by the bidirectional bridge for the transformation according to the GR-MOSFET is plotted in Figure 5.17. As observed, the measured and modeled data fit well.

For GR-MOSFET devices with 0.09 μm × 8 finger ×1 μm × 6 cell, the small-signal model parameters are given in Table 5.7 and

Table 5.7. Small-signal model parameters. Device: 0.09 μm $\times$ 8 finger $\times 1\,\mu$m $\times$ 6 cell GR-MOSFET.

Parameter	Value	Parameter	Value
L_g(pH)	60	$C_{\rm jd}$(fF)	5.4
L_d(pH)	40	$C_{\rm gs}$(fF)	9
L_s(pH)	30	$C_{\rm gd}$(fF)	4.2
$R_g(\Omega)$	8	$C_{\rm ds}$(fF)	2
$R_d(\Omega)$	25	g_m(mS)	8.25
$R_s(\Omega)$	2	$g_{\rm ds}$(mS)	0.77
$R_{\rm sub}(\Omega)$	400	$R_{\rm gs}(\Omega)$	30

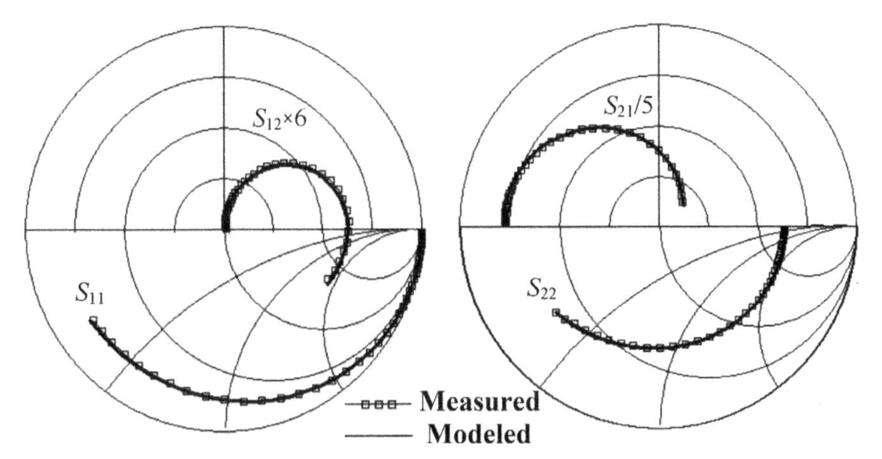

Figure 5.18. Comparison between the measured and the modeled S parameters for 0.09 μm $\times$ 8 finger $\times$ 1 μm $\times$ 6 cell GR-MOSFET devices. Bias: $V_{\rm gs} = 1.0$ V, $V_{\rm ds} = 1.0$ V.

the comparison between the measured and modeled data under bias conditions of $V_{\rm gs} = 1.0$ V and $V_{\rm ds} = 1.0$ V is sketched in Figure 5.18. Besides, the comparison of the measured and modeled S parameters for 0.09 μm $\times$ 8 finger $\times$ 1 μm $\times$ 6 cell SGR-MOSFET devices is illustrated in Figure 5.19. As explained earlier, the modeled data are achieved by the bidirectional bridge for the transformation based on the GR-MOSFET. As can be seen from Figure 5.19, good agreements can be obtained.

Table 5.8 presents the small-signal model parameters of the $0.24\,\mu$m $\times$ 32 finger $\times 1\,\mu$m $\times$ 2 cell GR-MOSFET device, and

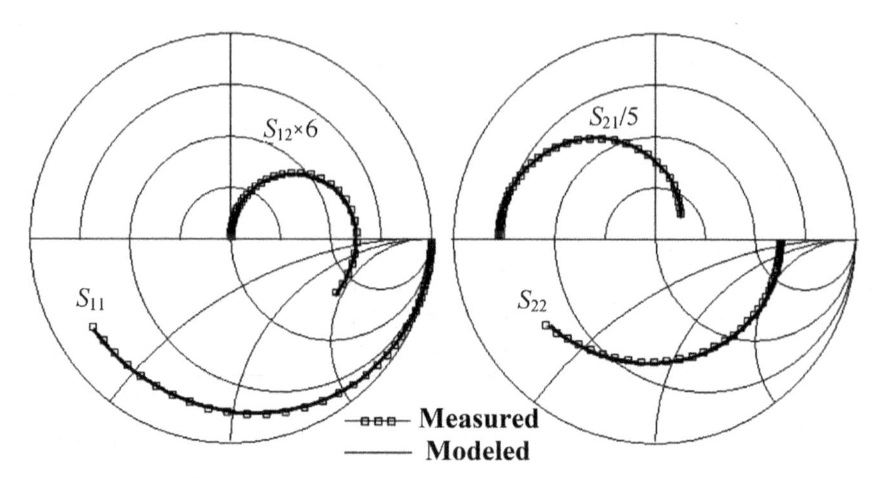

Figure 5.19. Comparison between the measured and the modeled S parameters for 0.09 μm × 8 finger × 1 μm × 6 cell SGR-MOSFET devices. $g_m^{\text{SGR}} = 8.35$ mS, $g_{\text{ds}}^{\text{SGR}} = 0.8$ mS. Bias: $V_{\text{gs}} = 1.0$ V, $V_{\text{ds}} = 1.0$ V.

Table 5.8. Small-signal model parameters. Device: 0.24 μm × 32 finger × 1 μm × 2 cell GR-MOSFET.

Parameter	Value	Parameter	Value
L_g(pH)	35	C_{jd}(fF)	4.4
L_d(pH)	40	C_{gs}(fF)	27
L_s(pH)	6	C_{gd}(fF)	14
R_g(Ω)	2	C_{ds}(fF)	20
R_d(Ω)	8	g_m(mS)	22.8
R_s(Ω)	1	g_{ds}(mS)	2.22
R_{sub}(Ω)	400	R_{gs}(Ω)	5

the corresponding comparison between the measured and modeled S parameters under bias condition of $V_{\text{gs}} = 1.0$ V and $V_{\text{ds}} = 1.0$ V is shown in Figure 5.20. Figure 5.21 illustrates the comparison of the measured and modeled S parameters for 0.24 μm × 32 finger ×1 μm × 2 cell SGR-MOSFET devices as well. Similarly, the modeled data are obtained by the bidirectional bridge for the transformation based on the GR-MOSFET and the measured and modeled data are in good agreement. It is obvious that the power gain S_{21} is almost the same for GR-, SGR-, and NGR-MOSFET devices of the same dimensions, the reason is that g_m and g_{ds} increase simultaneously.

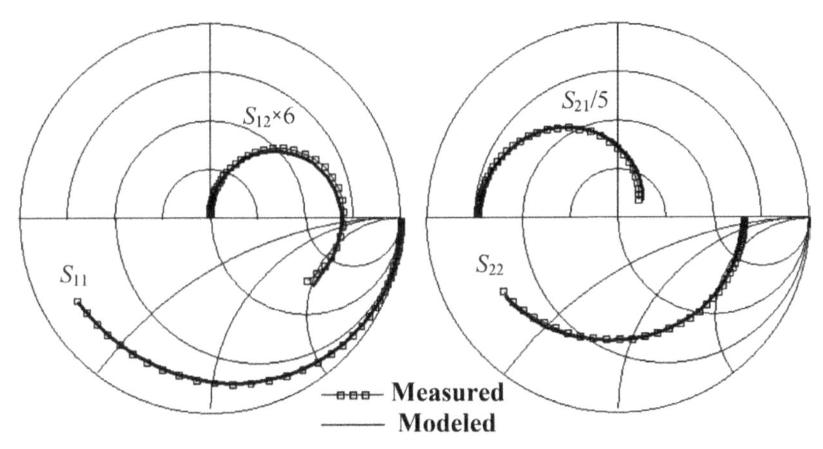

Figure 5.20. Comparison between the measured and the modeled S parameters for 0.24 μm $\times$ 32 finger $\times 1$ μm $\times$ 2 cell GR-MOSFET devices. Bias: $V_{\rm gs} = 1.0$ V, $V_{\rm ds} = 1.0$ V.

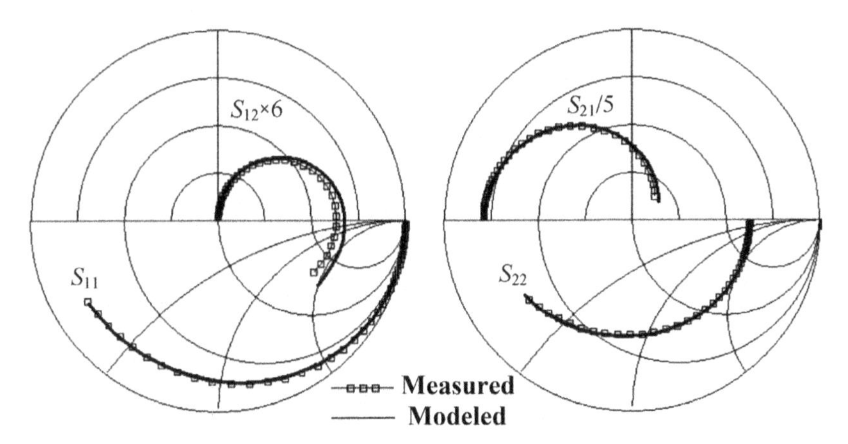

Figure 5.21. Comparison between the measured and the modeled S parameters for 0.24 μm $\times$ 32 finger $\times 1$ μm $\times$ 2 cell SGR-MOSFET devices. $g_m^{\rm SGR} = 25$ mS, $g_{\rm ds}^{\rm SGR} = 2.44$ mS. Bias: $V_{\rm gs} = 1.0$ V, $V_{\rm ds} = 1.0$ V.

5.4 DC/AC Dispersion

With the continuous reduction of process dimensions, higher requirements are put forward for the accurate characterization of device performance and the accuracy of large-signal modeling. In the modeling process, certain modifications need to be considered based on the conventional model, so as to adapt to the performance of the

MOSFET device. It is worth noting that the transconductance g_m and output conductance $g_{\rm ds}$ under DC and RF conditions are inconsistent; this phenomenon is similar to the high-frequency dispersion phenomenon of the III/V field effect device (FET) [14–17]. Hence, the DC/AC dispersion effect also needs to be considered in the modeling of Si-based MOSFET devices. This section mainly discusses the nonlinear model that can predict both the DC characteristics and DC/AC dispersion effect of the device [18–21].

5.4.1 *DC model*

The source-drain DC versus bias voltage is the heart of the matter in nonlinear equivalent circuit modeling. From the perspective of device physics, the output current of the device depends on the channel carrier mobility and carrier concentration, while the gate voltage of the device controls the carrier concentration, so as to achieve the purpose of controlling the current. In this section, the DC empirical model of GaAs-based device is utilized to model the DC characteristics of the MOSFET device. The DC model uses simple and clear mathematical expressions to describe and is compatible with the circuit computer-aided software. Usually, the choice of empirical model tends to use two independent functional expressions to describe the control of gate voltage and drain voltage on the drain output current which can be expressed as $I_{\rm ds} = f_1(V_{\rm gs})f_2(V_{\rm ds})$, and the advantage is that the model parameters can be easily extracted, and the control of each bias voltage on the current can also be readily understood. Note that the two independent parts are coupled. This section proposes the DC model which modifies the STATZ model to accurately describe the I–V characteristics of the MOSFET device, and the corresponding model parameter extraction method is also given.

Based on the STATZ model, the current model for MOSFET devices is proposed [21]:

$$I_{\rm ds} = \frac{\beta \{\ln [1 + \exp(B(V_{\rm gs} - V_{\rm t}))]\}^2}{1 + b(V_{\rm gs} - V_{\rm t})}(1 + \lambda V_{\rm ds}) \tan{\rm h}(\alpha V_{\rm ds}) \quad (5.46)$$

where

$$V_{\rm t} = V_{\rm to} - \gamma V_{\rm ds}$$

$$\beta = \frac{W \cdot M \cdot N}{W_{\rm e} \cdot M_{\rm e} \cdot N_{\rm e}} \mu_{\rm e}(V_{\rm gs}, V_{\rm ds})$$

where I_{ds} is the drain current under DC condition, V_{to} is the threshold voltage under zero bias condition, β is the transconductance parameter, λ is the channel length modulation effect parameter, b is the tailing factor, α is the saturation voltage parameter, γ indicates the parameter that the threshold voltage varies with the drain-source voltage V_{ds}, and B is the fitting parameter. Since the DUT consists of multiple cells in parallel, W, M, and N represent the gate width, the number of fingers, and the number of cells for DUT, respectively. While W_e, M_e, and N_e represent the single-finger gate width, the number of fingers, and the number of cells for elementary cell, respectively.

Each model parameter in the table can be regarded as the function of the bias voltage:

$$\lambda = \lambda_1 V_{gs}^2 + \lambda_2 V_{gs} + \lambda_3 \tag{5.47}$$

$$\alpha = \alpha_1 V_{gs}^2 + \alpha_2 V_{gs} + \alpha_3 \tag{5.48}$$

$$b = b_1 V_{ds} + b_2 \tag{5.49}$$

$$\mu_e(V_{gs}, V_{ds}) = (\mu_1 V_{gs}^2 + \mu_2 V_{gs} + \mu_3)(\mu_4 V_{ds} + \mu_5) \tag{5.50}$$

The MOSFET device with 90 nm gate length, 1 μm cell gate width, 12 cells, and four gate fingers was investigated to demonstrate the accuracy of the model. Table 5.9 tabulates the extracted DC model parameters. After optimization, the comparison between the modeled and measured data is shown in Figures 5.22 and 5.23. It can be seen from the figures that the modeled data are in good agreement with the measured data in both the linear region and the saturation region.

Table 5.9. Extracted parameters of DC model.

Parameter	Value	Parameter	Value
V_{to}	0.4	γ	0.0625
μ_1	−5.40e-5	μ_2	1.366e-4
μ_3	−2.93e-5	μ_4	−1.064
μ_5	4.816	b_1	1.081
b_2	2.722e-1	λ_1	−5.399
λ_2	1.074e1	λ_3	−2.283
α_1	−2.811	α_2	2.071
α_3	1.213e1	B	15

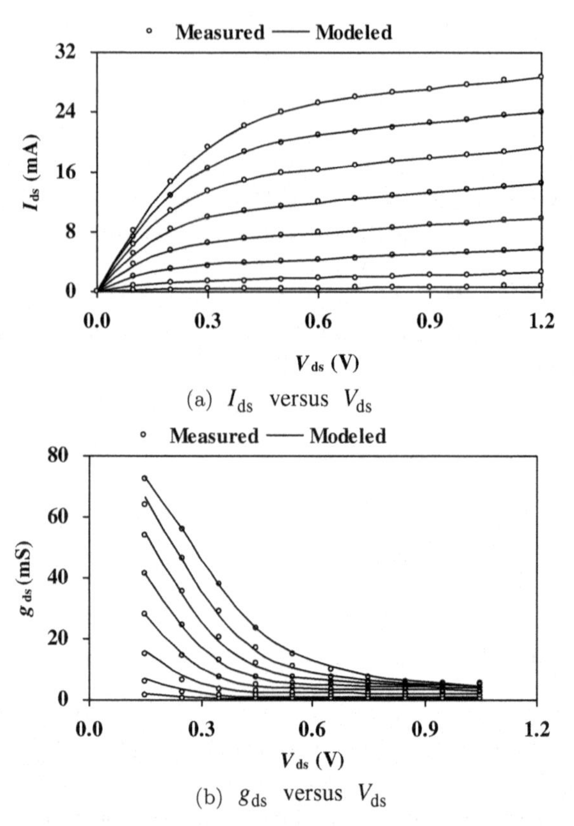

Figure 5.22. Comparison between the modeled and measured data for a 4 fingers $\times 1\,\mu$m $\times$ 12 cells MOSFET device. Bias voltage: $V_{gs} = 0.5$ V–1.2 V in 0.1 V steps: (a) I_{ds} versus V_{ds}; (b) g_{ds} versus V_{ds}.

To further estimate the error, Table 5.10 discusses the root mean square error (RMS) of the three models, and we can arrive at the RMS expression:

$$\varepsilon = \frac{1}{N}\sum_{i=0}^{i=N}\left(I_{\text{dmeas}} - I_{\text{dmod}}\right)^2 \tag{5.51}$$

where I_{dmeas} is the measured drain current and I_{dmol} is the modeled drain current.

The proposed DC model guarantees the scale properties of the model by applying simple scaling rules to the model parameters μ_e. To further verify the scalability of the model, for a MOSFET device

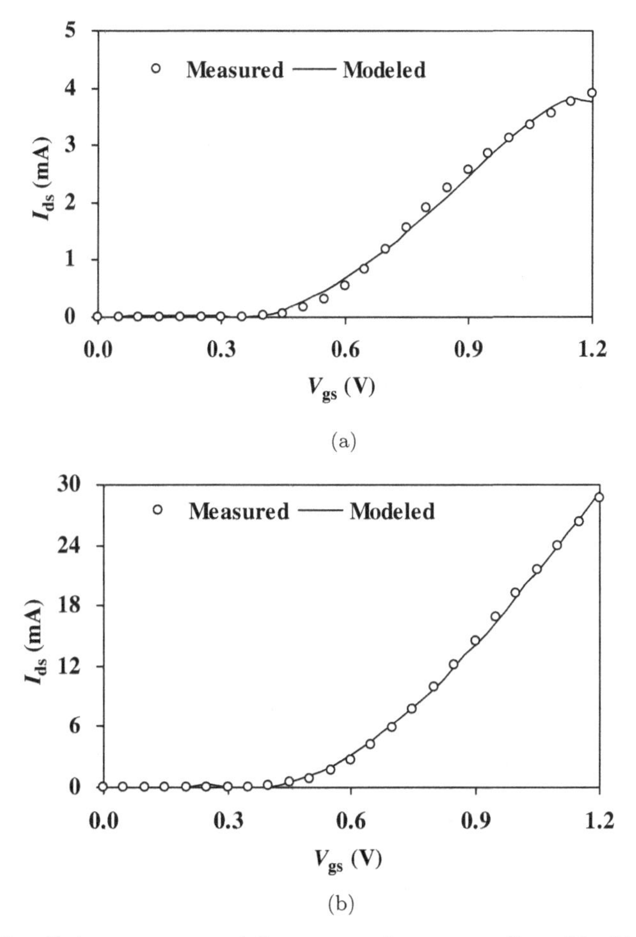

Figure 5.23. Gate-source current I_{ds} versus gate-source voltage V_{gs}. Device 4 fingers $\times 1\ \mu\mathrm{m} \times 12$ cells: (a) Bias voltage: $V_{\mathrm{ds}} = 0.05V$; (b) Bias voltage: $V_{\mathrm{ds}} = 1.2$ V.

Table 5.10. RMS error of DC model.

Number of model parameters		RMS error $V_{\mathrm{gs}} = 0.3 - 0.5$ V, $V_{\mathrm{ds}} = 0 - 1.2$ V	RMS error $V_{\mathrm{gs}} = 0.6 - 1.2$ V, $V_{\mathrm{ds}} = 0 - 1.2$ V
Proposed model	16	2.2293E-9	5.3831E-9
STATZ model	5	5.4584E-8	2.8931E-7
BSIM3v3 model	70+	1.0460E-8	8.9477E-2

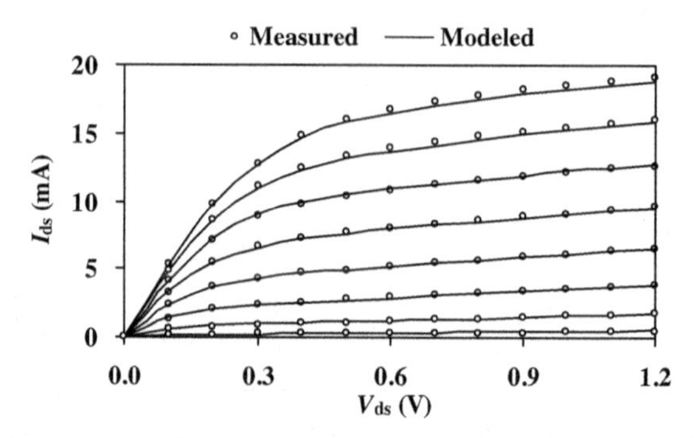

Figure 5.24. Comparison between the measured and modeled DC I–V characteristics scaled to 16 finger × 1 μm × 2 cell MOSFET device. Bias: $V_{gs} = 0.5 - 1.2$ V, step 0.1 V.

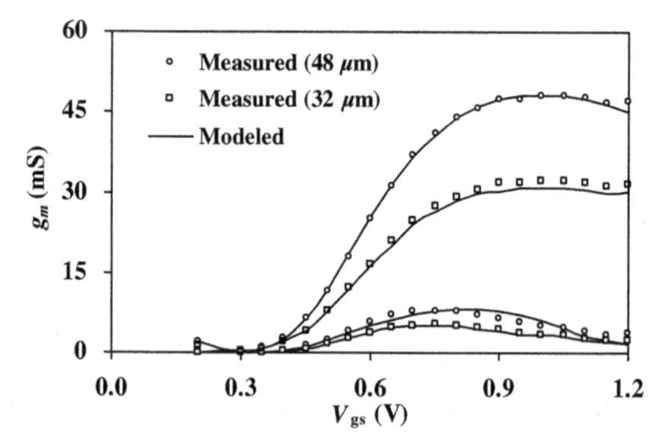

Figure 5.25. Measured and modeled transconductance g_m versus V_{gs} in case of 4 finger × 1 μm × 12 cell MOSFET device and scaled to 16 finger ×1 μm × 2 cell MOSFET device.

with 16 fingers ×1 μm × 2 cells, Figures 5.24–5.26 give the comparison between the measured and modeled data for MOSFET devices. Clearly, the measured and modeled data fit well, which verifies the scalability of the DC model.

5.4.2 *Modeling process*

As the operating frequency expands to the gigahertz ranges and the gate length shrinks to the deep submicron ranges, conventional methods have been unable to accurately describe the high-frequency

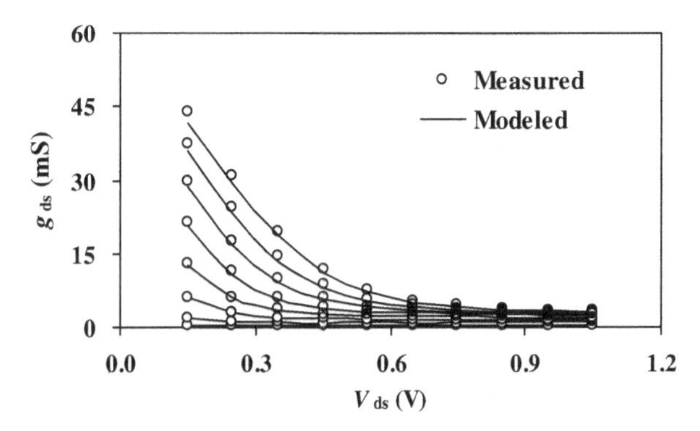

Figure 5.26. Measured and modeled output conductance $g_{\rm ds}$ scaled to 16 finger $\times$ 1 μm $\times$ 2 cell MOSFET device. Bias: $V_{\rm gs} = 0.5 - 1.2$ V in 0.1 V step.

behaviors of RF devices, especially for large-scale devices containing multiple cells. As mentioned, the dispersion effect causes the transconductance g_m and output conductance $g_{\rm ds}$ at high frequencies to be inconsistent with the g_m and output conductance $g_{\rm ds}$ measured under the DC condition. In order to model the DC/AC dispersion phenomena, a hybrid network including $I_{\rm db}$, $R_{\rm sh}$, and C_d is adapted. The shunt resistance $R_{\rm sh}$, placed in parallel with the current source, is used to enable the convergence of the simulator. The nonlinear equivalent circuit model of the MOSFET device considering the dispersion effect is sketched in Figure 5.27.

Two current sources are utilized in the model, where $I_{\rm ds,DC}$ are used to represent the transconductance and output conductance contributions in the DC state, and the RC branches ($R_{\rm sh}$, C_d) and $I_{\rm db}$ in the model topology are used to correct for deviations caused by transconductance and output conductance dispersion at high frequencies. Under DC conditions, due to the branch capacitance, the RC branch does not work, and the output current is $I_{\rm ds,DC}$. At high frequencies, the large capacitance is short-circuited, and the $I_{\rm db}$ current source begins to work. At this time, the output current is the sum of $I_{\rm ds,DC}$ and $I_{\rm db}$.

As shown in the dashed box in Figure 5.27, $I_{\rm db}$ represents the DC/AC dispersion current which can be expressed as

$$I_{\rm db}(V_{\rm gs}, V_{\rm ds}) = I_{\rm ds,RF} - I_{\rm ds,DC} \tag{5.52}$$

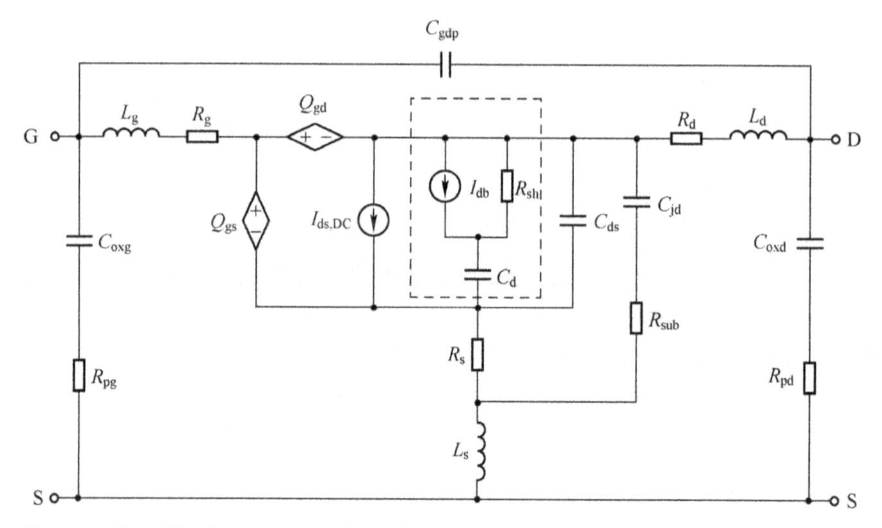

Figure 5.27. Nonlinear equivalent circuit model of MOSFET device considering dispersion effect.

where

$$I_{\mathrm{ds,RF}} = I_{\mathrm{ds}}(V_{\mathrm{gs}}, V_{\mathrm{ds}}, r_{\mathrm{RF}})$$
$$I_{\mathrm{ds,DC}} = I_{\mathrm{ds}}(V_{\mathrm{gs}}, V_{\mathrm{ds}}, r_{\mathrm{DC}})$$

The nonlinear equivalent circuit model diagram of the MOSFET device is presented in Figure 5.27. Compared with the small-signal equivalent circuit model, the elements in the intrinsic part are no longer the specific value, rather it is the nonlinear collection of the relationship between individual parameters and voltage. The flowchart of nonlinear model modeling is sketched in Figure 5.28. The modeling process is as follows:

(1) Source-drain DC current model: The output I–V characteristics in the linear and saturation region are accurately described, and the corresponding parameter extraction method is given.
(2) First, the extrinsic parameters' de-embedding is performed on the S-parameters under different bias voltages. Subsequently, the intrinsic parameters' transconductance $g_{m,\mathrm{RF}}$, output drain conductance $g_{\mathrm{ds,RF}}$, gate-source capacitance C_{gs}, and gate-drain capacitance C_{gd} with bias are obtained.

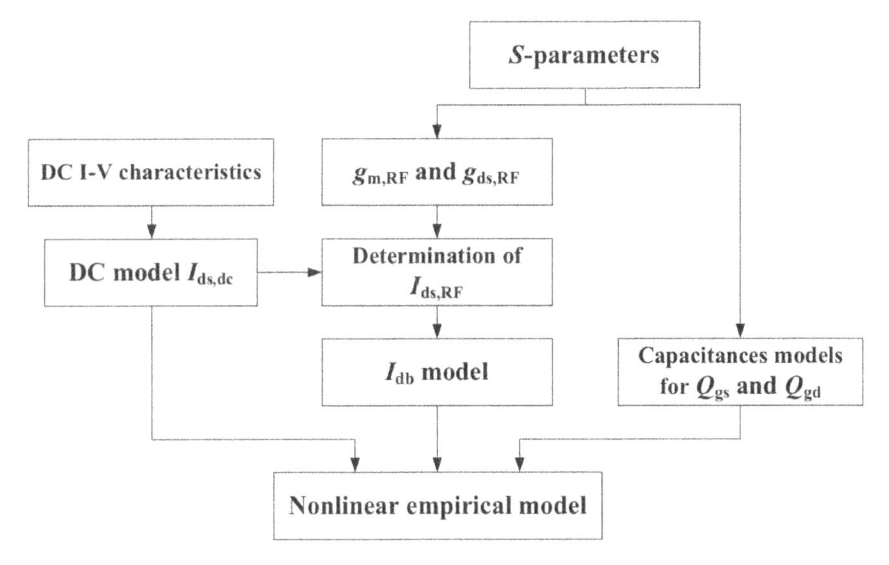

Figure 5.28. Flowchart of MOSFET nonlinear modeling.

(3) Calculate the RF source-drain current. First, we arrive at the RF transconductance and output conductance of the model.

$$g_{m,\mathrm{RF}}(V_{\mathrm{gs}}, V_{\mathrm{ds}}) = \frac{\partial}{\partial V_{\mathrm{gs}}} I_{\mathrm{ds,RF}}(V_{\mathrm{gs}}, V_{\mathrm{ds}}) \tag{5.53}$$

$$g_{\mathrm{ds,RF}}(V_{\mathrm{gs}}, V_{\mathrm{ds}}) = \frac{\partial}{\partial V_{\mathrm{ds}}} I_{\mathrm{ds,RF}}(V_{\mathrm{gs}}, V_{\mathrm{ds}}) \tag{5.54}$$

Based on the above equations, the RF source-drain current can be calculated by integrating the transconductance and output conductance:

$$I_{\mathrm{ds,RF}} = \int_0^{V_{\mathrm{gs}}} g_m \mathrm{d}V_{\mathrm{gs}} = \int_0^{V_{\mathrm{ds}}} g_m \mathrm{d}V_{\mathrm{ds}} \tag{5.55}$$

(4) Calculate the dispersion current I_{db} based on the DC source-drain current $I_{\mathrm{ds,DC}}$ and the RF source-drain current $I_{\mathrm{ds,RF}}$.
(5) The gate-source charge Q_{gs} and gate-drain charge Q_{gd} can be achieved by integrating the gate-source capacitance C_{gs} and gate-drain capacitance C_{gd} where the relationship between the

capacitances and charges can be expressed as [18–20]

$$C_{\text{gd}} = \frac{\partial Q_{\text{gs}}}{\partial V_{\text{ds}}} + \frac{\partial Q_{\text{gd}}}{\partial V_{\text{ds}}} \tag{5.56}$$

$$C_{\text{gs}} = \frac{\partial Q_{\text{gs}}}{\partial V_{\text{gs}}} + \frac{\partial Q_{\text{gd}}}{\partial V_{\text{gs}}} \tag{5.57}$$

In order to ensure the law of charge conservation, so as not to cause nonlinear simulation errors or nonconvergence of the simulation, it is necessary to satisfy

$$\frac{\partial C_{\text{gd}}}{\partial V_{\text{gs}}} = \frac{\partial C_{\text{gs}}}{\partial V_{\text{ds}}} \tag{5.58}$$

5.4.3 *Model parameter extraction*

In a bid to analyze the influence of the DC/AC dispersion effect, the MOSFET devices with 4 fingers $\times 1\,\mu$m $\times$ 12 cells and 16 fingers $\times$ $1\,\mu$m $\times$ 12 cell (number of gate finger $\times$ gate width of single finger $\times$ number of cell) fabricated by 90 nm process were investigated. Simultaneously, the S parameters under different bias conditions ($V_{\text{gs}} = 0.4$ V–1.2 V, step size is 0.2 V and $V_{\text{ds}} = 0.4$ V–1.2 V, step size is 0.2 V) are utilized to obtain the RF intrinsic elements.

The extracted transconductance and conductance versus bias voltage are illustrated in Figures 5.29 and 5.30, respectively. It is significant to observe that the transconductance has an obvious dispersion effect under high gate voltage bias conditions (data are inconsistent), and the RF transconductance is significantly higher than DC transconductance. The dispersion effect of transconductance increases with the increase in V_{gs}, while the dispersion effect of output conductance decreases with the increase in V_{ds}. This phenomenon is speculated to be caused by electron traps. The extracted DC model parameters are listed in Table 5.11. The comparison between the modeled and extracted data of the 4 finger $\times$ 1 μm $\times$ 12 cell MOSFET device is illustrated in Figure 5.31. As exhibited, the extracted data of $g_{m,\text{RF}}$ and $g_{\text{ds,RF}}$ are in good agreement with the modeled data. The RF current $I_{\text{ds,RF}}$ can be directly obtained by integrating the RF transconductance $g_{m,\text{RF}}$ and RF output conductance $g_{\text{ds,RF}}$.

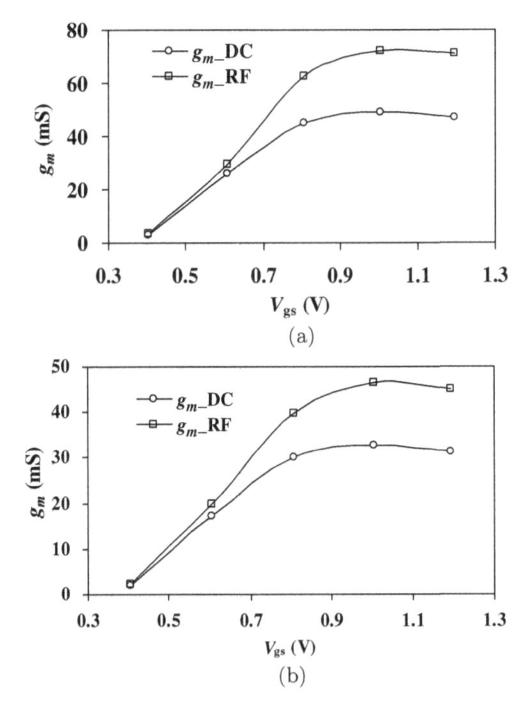

Figure 5.29. Extracted transconductance versus bias voltage $V_{\rm gs}$. Bias voltage: $V_{\rm ds} = 1.2$ V: (a) 4 finger $\times$ 1 μm $\times$ 12 cell; (b) 16 finger $\times$ 1 μm $\times$ 2 cell.

Figure 5.32 depicts the comparison between the measured DC I–V characteristic and the RF output current calculated from RF measurement of 4 finger $\times$ 1 μm $\times$ 12 cell MOSFET devices. As seen, the frequency dispersion effect causes the current to be inconsistent under high frequency and DC conditions, and the RF current is significantly higher than the DC.

The DC I–V characteristics of the device are different from the I–V characteristics at high frequencies, that is, the DC/AC dispersion effect occurs. To illustrate the efficiency of the extraction methodology, two of the most important bias-dependent small-signal elements of equivalent circuit models, whose values are extracted directly from a 4 finger $\times$ 1 μm $\times$ 12 cell MOSFET device and scaled from a 16 finger $\times$ 1 μm $\times$ 2 cell MOSFET device, are present in Figure 5.33. From the figures, it can be seen that the scaling properties are well good overall bias conditions. As a general rule, the RF

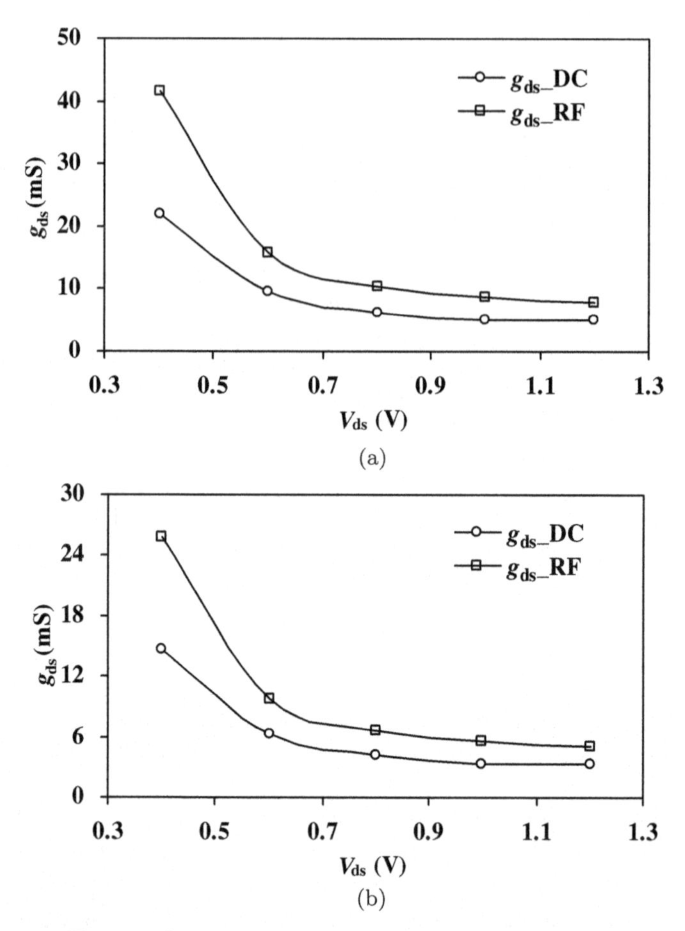

Figure 5.30. Extracted conductance versus bias voltage $V_{\rm ds}$. Bias voltage: $V_{\rm gs}$ = 1.2 V: (a) 4 finger × 1 μm × 12 cell; (b) 16 finger × 1 μm × 2 cell.

Table 5.11. DC model parameters.

Parameter	Value	Parameter	Value
$\gamma_{\rm RF}$	0.08	$\mu_{\rm 1RF}$	$-2.80E$-4
$\mu_{\rm 2RF}$	$5.33E$-4	$\mu_{\rm 3RF}$	$-1.69E$-4
$\mu_{\rm 4RF}$	-0.19	$\mu_{\rm 5RF}$	2.57
$\lambda_{\rm 1RF}$	-0.31	$\lambda_{\rm 2RF}$	0.99
$\lambda_{\rm 3RF}$	0.82		

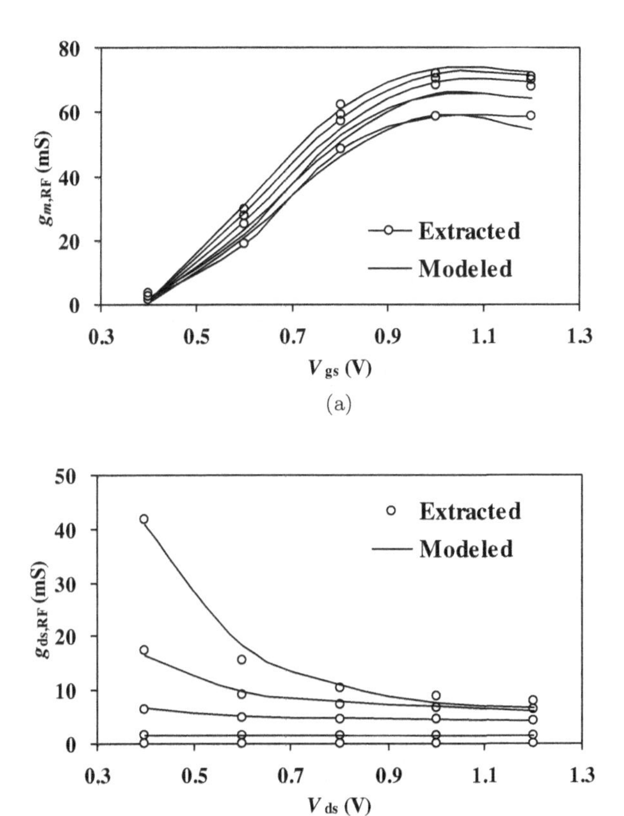

Figure 5.31. Comparison between the measured and the modeled data of 4 finger × 1 μm × 12 cell MOSFET devices. Bias voltage: $V_{\mathrm{gs}} = 0.4$ V–1.2 V, step: 0.2 V: (a) RF transconductance $g_{m,\mathrm{RF}}$ versus bias voltage V_{gs}; (b) Output conductance $g_{\mathrm{ds,RF}}$ versus bias voltage V_{ds}.

transconductance and output conductance follow a straightforward scaling pattern. To better emphasize this phenomenon, the relative errors are adopted to evaluate the discrepancies of RF transconductance and output conductance with scaled ones. The relative error formula can be expressed as follows:

$$g_{m\_\mathrm{err}} = \left| \frac{g_m - g_{m\_\mathrm{sc}}}{g_{m\_\mathrm{sc}}} \right| \cdot 100\% \tag{5.59}$$

$$C_{\mathrm{gs\_err}} = \left| \frac{C_{\mathrm{gs}} - C_{\mathrm{gs\_sc}}}{C_{\mathrm{gs\_sc}}} \right| \cdot 100\% \tag{5.60}$$

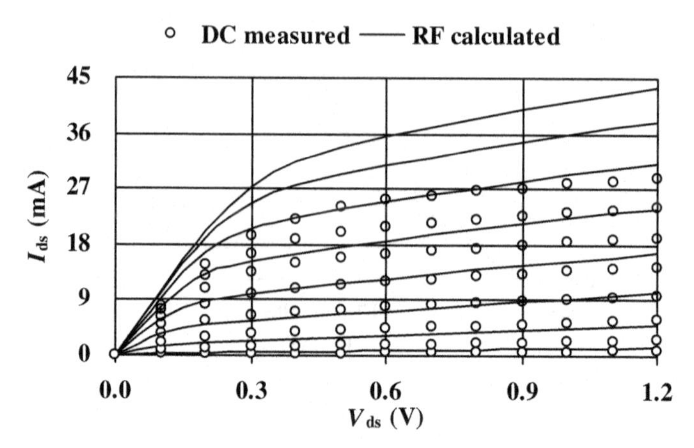

Figure 5.32. Comparison between the measured DC I–V characteristic and RF output current calculated from RF measurement of 4 finger $\times$ 1 μm $\times$ 12 cell MOSFET devices. Bias voltage: V_{gs} = 0.5–1.2V, step: 0.1 V.

where C_{gs} and g_m represent the extracted gate-source capacitance and transconductance, respectively. C_{gs_sc} and g_{m_sc} represent the scaled gate-source capacitance and transconductance, respectively. The relative error of two intrinsic parameters is exhibited in Figure 5.34. It can be seen that the relative errors are all below 5%, which proves that the gate-source capacitance and transconductance have good scalability.

The nonlinear capacitance model describes the gate-source capacitance C_{gs} and the gate-drain capacitance C_{gd} versus the bias voltage (V_{gs} and V_{ds}). By extracting the parameters of the MOSFET device under different bias conditions, C_{gs} and C_{gd} versus V_{gs} or V_{ds} can be achieved. It is noticeable that the drain-source capacitance C_{ds} is mainly related to the channel charge and has small variation with bias voltage. In nonlinear analysis, the capacitance is usually regarded as constant. Hence, the following work mainly focuses on the gate-source capacitance C_{gs} and the gate-drain capacitance C_{gd} modeling. In the meantime, in the process of establishing the C–V model, it is necessary to ensure the conservation of charge to avoid the nonconvergence of the model.

The hyperbolic tangent function is the most commonly used in the description of nonlinear behaviors. The good choice to characterize

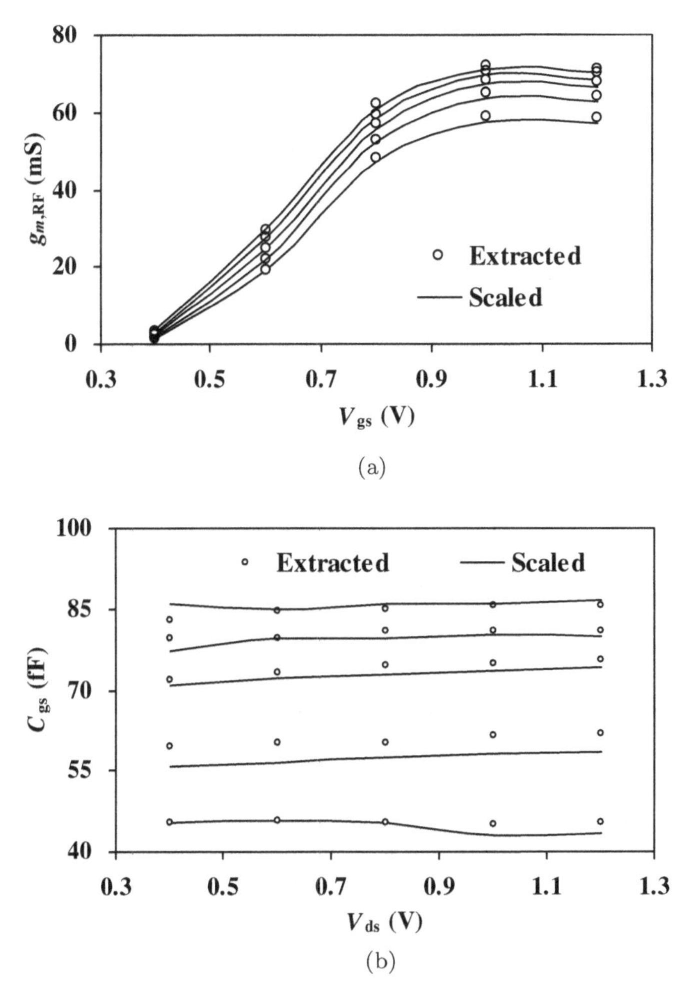

Figure 5.33. Extracted transconductance g_m and gate-source capacitance C_{gs} versus bias voltage of 4 finger $\times$ 1 μm $\times$ 12 cell MOSFET device. Bias voltage: V_{gs} = 0.4–1.2V, step: 0.2 V: (a) Transconductance g_m; (b) Gate-source capacitance C_{gs}.

gate-source capacitance C_{gs} and gate-drain capacitance C_{gd} versus bias voltage is to combine the hyperbolic tangent function with the exponential function. In general, there exist partial derivatives of the two charge terms, Q_{gs} and Q_{gd} with regard to V_{gs} and V_{ds}, respectively. Here, the voltage-dependent gate source capacitance

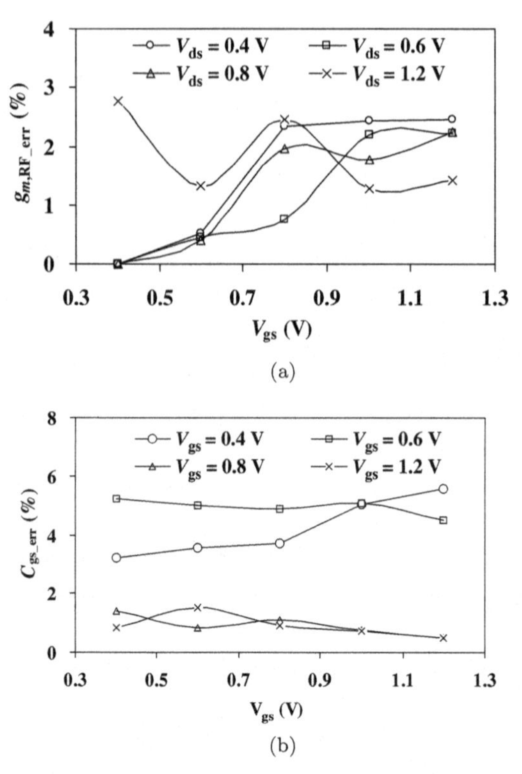

Figure 5.34. Relative error of transconductance and gate-source capacitance: (a) Transconductance g_m; (b) Gate-source capacitance C_{gs}.

$C_{gs}(V_{gs}, V_{ds})$ and gate-drain capacitance $C_{gd}(V_{gs}, V_{ds})$ are empirically modeled based on exponential and hyperbolic tangent functions [21]:

$$C_{gs} = \frac{\partial Q_{gs}}{\partial V_{gs}} + \frac{\partial Q_{gd}}{\partial V_{gs}} = C_{gsp} + \frac{C_{gs0}}{1 + e^{-a_1 V_{gs} + a_2}} \times \tan\mathrm{h}(a_3 V_{ds}) \quad (5.61)$$

$$C_{gd} = \frac{\partial Q_{gs}}{\partial V_{ds}} + \frac{\partial Q_{gd}}{\partial V_{ds}} = C_{gdp} + \frac{C_{gd0}}{1 + e^{-c_1 V_{dg} + c_2}} \times \tan\mathrm{h}(c_3 V_{gs}) \quad (5.62)$$

where Q_{gs} and Q_{gd} represent the gate-source charge and gate-drain charge, respectively. C_{gsp} and C_{gdp} represent the capacitance under zero bias condition. C_{gs0}, C_{gd0}, a_1, a_2, a_3, c_1, c_2, and c_3 are fitting parameters.

The total charge of the network is $Q_g = Q_{\mathrm{gs}} + Q_{\mathrm{gd}}$, where Q_{gs} and Q_{gd} can be obtained by calculus the capacitance:

$$Q_{\mathrm{gs}} = C_{\mathrm{gsp}} V_{\mathrm{gs}} + \frac{C_{\mathrm{gs0}}}{a_1} \{\ln[1 + \exp(-a_1 V_{\mathrm{gs}} + a_2)] + (a_1 V_{\mathrm{gs}} - a_2)\} \cdot \tan\mathrm{h}(a_3 V_{\mathrm{ds}}) \tag{5.63}$$

$$Q_{\mathrm{gd}} = C_{\mathrm{gdp}} V_{\mathrm{ds}} + \frac{C_{\mathrm{gd0}}}{c_1} \{\ln[1 + \exp(-c_1 V_{\mathrm{dg}} + c_2)] + (c_1 V_{\mathrm{dg}} - c_2)\} \cdot \tan\mathrm{h}(c_3 V_{\mathrm{gs}}) \tag{5.64}$$

The MOSFET device with 4 fingers $\times$ 1 μm $\times$ 12 cells is utilized to verify the validity of the model. The comparison between the measured and modeled data for gate-source capacitance C_{gs} and gate-drain capacitance C_{gd} is presented in Figure 5.35. As seen, the measured and modeled data are in good agreement. Besides, the extracted capacitance model parameters are tabulated in Table 5.12.

5.4.4 *Model implementation and verification*

The purpose of device modeling is to use the model to predict the performance of circuit characteristics. Significantly, the model can be incorporated with circuit simulation software. Therefore, in order to verify the circuit model, it is first necessary to implement the proposed model in the circuit simulation software. Currently, the common approach is to utilize the symbolic defined devices (SDD) method to incorporate the custom device model and to establish the device model by defining the number of ports, port voltages, port currents, and formula derivatives. The topology and node numbers of the nonlinear equivalent circuit model for a MOSFET device are illustrated in Figure 5.36. As seen, the model is implemented in circuit simulation software using resistances, capacitances, and nonlinear controlled sources. Moreover, four node voltages are also annotated.

The MOSFET device with 4 finger $\times$ 1 μm $\times$ 12 cell was investigated to demonstrate the accuracy of the model. In the interest of better observing the agreement, the Smith chart is replaced by the real part and imaginary part of Y parameters. In the frequency range of 1–50 GHz, the comparison between the measured and modeled

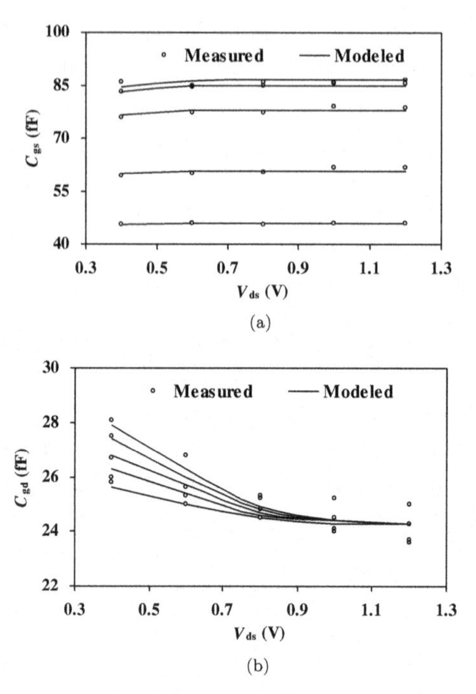

Figure 5.35. Comparison between the measured and modeled data. Bias: V_{gs} = 0.4 V–1.2 V in 0.2 V step: (a) Gate-source capacitance C_{gs}; (b) Gate-drain capacitance C_{gd}.

Table 5.12. Extracted capacitance model parameters.

Parameter	Value	Parameter	Value
C_{gsp}(fF)	39.5	C_{gs0}(fF)	47.4
a_1	8.387	a_2	5.241
a_3	4.826	C_{gdp}(fF)	24.3
C_{gd0}(fF)	8.4	c_1	−7.845
c_2	−4.437	c_3	0.515

Y parameters under the bias condition of $V_{gs} = V_{ds} = 1.2$ V is illustrated in Figure 5.37. As seen, good agreement is obtained between the measured and modeled data.

To better verify the scalability of the model, Figure 5.38 plots the comparison of measured and scaled Y parameters for the 16 fingers × 1 μm × 2 cell MOSFET device. Obviously, good agreements are obtained over the measured frequency range of 1–40 GHz. However, as frequency increases (above 40 GHz), the modeled data deviate from the measured data.

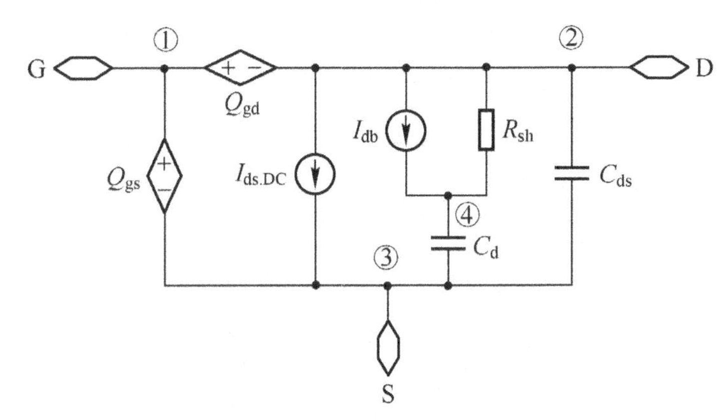

Figure 5.36. Topology and node numbers of the nonlinear equivalent circuit model for MOSFET device.

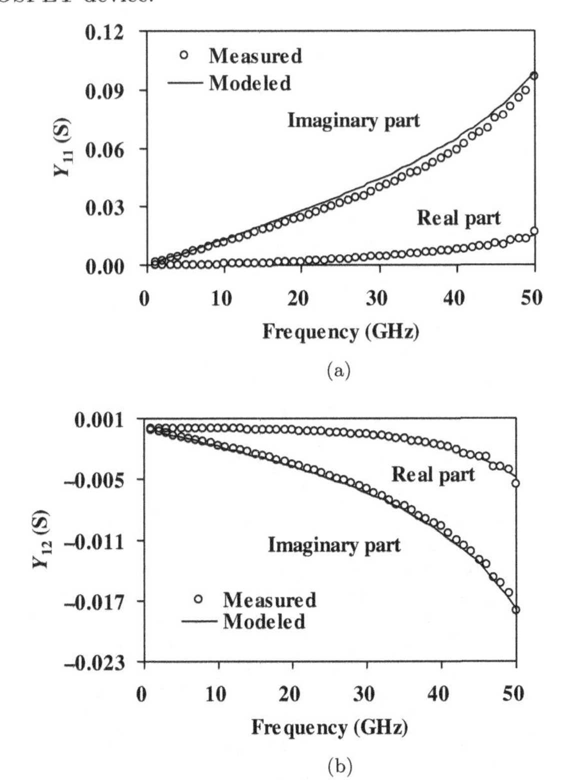

Figure 5.37. Comparison between the measured and modeled Y parameters for 4 fingers $\times\, 1\,\mu$m $\times$ 2 cell MOSFET device. Bias: $V_{gs} = V_{ds} = 1.2$ V. (a) Y_{11}; (b) Y_{12}; (c) Y_{21}; (d) Y_{22}.

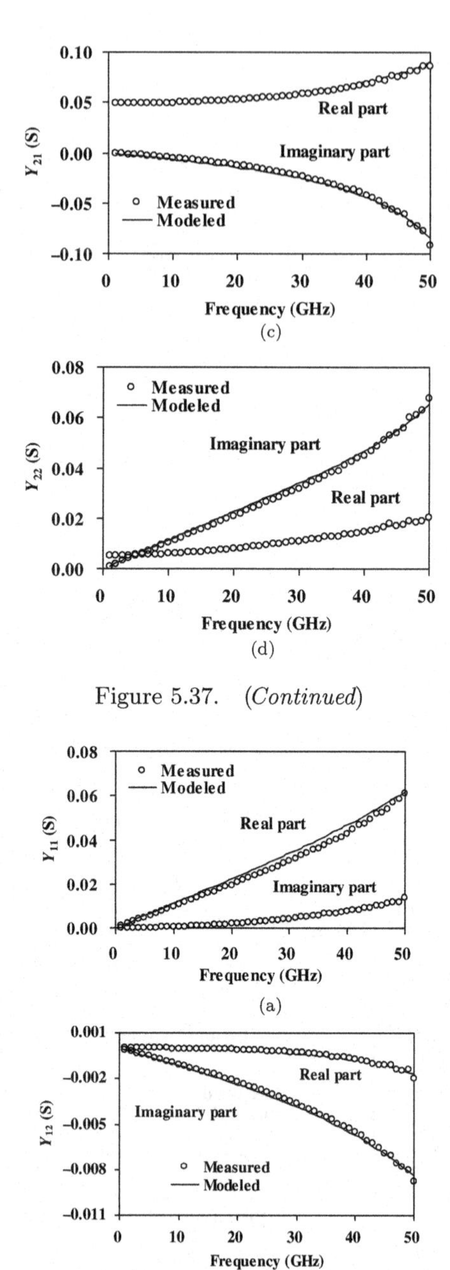

Figure 5.37. *(Continued)*

Figure 5.38. Comparison between the measured and scaled Y parameters for 16 fingers $\times\, 1\ \mu$m $\times$ 2 cell MOSFET device. Bias: $V_{\mathrm{gs}} = V_{\mathrm{ds}} = 1.2$ V. (a) Y_{11}; (b) Y_{12}; (c) Y_{21}; (d) Y_{22}.

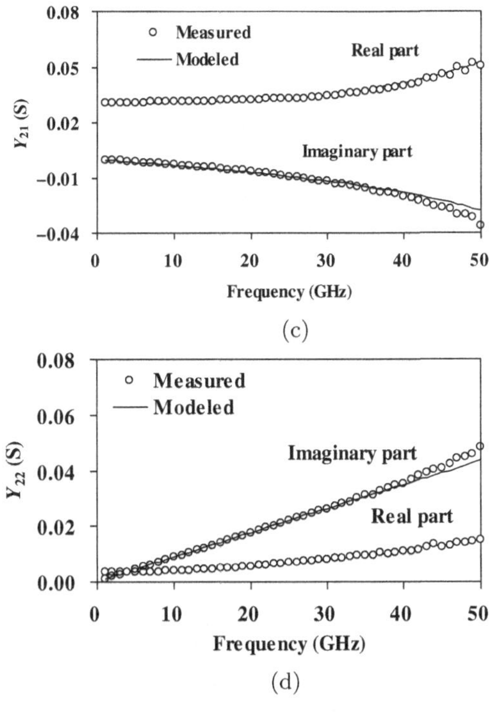

Figure 5.38. (*Continued*)

5.5 Summary

This chapter first introduces the commonly used nonlinear models of MOSFET devices and then discusses the effect of guard rings on the DC and high-frequency performance of deep-submicron MOSFET devices and gives a detailed comparison of the performance of MOSFET devices with four different guard ring structures. A set of simple and effective closed-form expressions are derived to provide bidirectional calculation expressions for the S parameter conversion of devices with different guard ring structures. Finally, the large-signal behaviors and modeling of MOSFET devices considering high-frequency dispersion effects are deliberated. The corresponding model parameter extraction method is given and the complete large-signal equivalent circuit model including the dispersion effect is established to verify the accuracy of the model.

References

[1] E. Ngoya, C. Quindroit, and J. M. Nebus, "On the continuous-time model for nonlinear-memory modeling of RF power amplifiers," *IEEE Transactions on Microwave Theory and Techniques*, 57(12): 3278–3292, 2009.

[2] Y. P. Tsividis and K. Suyama, "MOSFET modeling for analog circuit CAD: Problems and prospects," *IEEE Journal of Solid State Circuits*, 29(3): 210–216, 1994.

[3] M. Rudolph, C. Fager, and D. E. Root, *Nonlinear Transistor Model Parameter Extraction Techniques.* Sweden: Cambridge University Press, 2011, pp. 200–220.

[4] Y. S. Chauhan, S. Venugopalan, M. A. Chalkiadaki, M. A. U. Karim, and H. Agarwal, "BSIM6: Analog and RF compact model for bulk MOSFET," *IEEE Transactions on Electron Devices*, 61(2): 234–244, 2014.

[5] Y. Cheng and C. Hu, *MOSFET Modeling & BSIM3 User's Guide.* Berlin: Springer, 2002.

[6] J. Cheng, "Parameter extraction and modeling technology of RF microwave MOS device," Ph.D Dissertation, East China Normal University, 2012.

[7] I. Angelov, N. Rorsman, J. Stenarson, *et al.,* "An empirical table-based FET model," *IEEE Transactions on Microwave Theory and Techniques*, 47(12): 2350–2357, 1999.

[8] L. Sun, J. Gao, and A. Werthof, "Effect of guard-ring on the DC and high-frequency performance of deep-submicrometer metal oxide semiconductor field effect transistor," *International Journal of RF and Microwave Computer-Aided Engineering*, 24(2): 259–267, 2014.

[9] R. Howes and W. R. White, "A small-signal model for the frequency-dependent drain admittance in floating-substrate MOSFETs," *IEEE Journal Solid-State Circuits*, 27(8): 1186–1193, 1992.

[10] D. Suh and J. G. Fossum, "A physical charge-based model for non-fully depleted SOI MOSFET's and its use in assessing floating-body effects in SOI CMOS circuits," *IEEE Transactions on Electron Devices*, 42(4): 728–737, 1995.

[11] K. Young and J. Burns, "Avalanche-induced drain-source breakdown in silicon-on-insulator n-MOSFET's," *IEEE Transaction on Electron Devices*, 35(4): 426–431, 1988.

[12] A. Siligaris, G. Dambrine, D. Schreurs, and F. Danneville, "130-nm partially depleted SOI MOSFET nonlinear model including the kink effect for linearity properties investigation," *IEEE Transaction on Electron Devices*, 52(12): 2809–2812, 2005.

[13] H. Statz, P. Newman, I. W. Smith, R. A. Pucel, and H. Haus, "GaAs FET device and circuit simulation in SPICE," *IEEE Transactions on Electron Devices*, 34(2): 160–169, 1987.

[14] P. H. Ladbrooke and S. R. Blight, "Low-field low-frequency dispersion of transconductance in GaAs MESFETs with implications for other rate dependent anomalies," *IEEE Transactions on Electron Devices*, 35(3): 257–267, 1988.

[15] Y. Hasumi, T. Oshima, N. Matsunaga, and H. Kodera, "Analysis of the frequency dispersion of transconductance and drain conductance in GaAs MESFETs," *Electronics and Communications in Japan*, 89(4): 20–28, 2006.

[16] K. I. Jeon, Y. S. Kwon, and S. C. Hong, "A frequency dispersion model of GaAs MESFET for large-signal applications," *IEEE Microwave and Guided Wave Letters*, 7(3): 78–80, 2002.

[17] C. Camacho-Peñalosa and C. S. Aitchison, "Modelling frequency dependence of output impedance of a microwave MESFET at low frequencies," *Electronics Letters*, 21(12): 528–529, 1985.

[18] M. A. Cirit, "The meyer model revisited: Why is charge not conserved?" *IEEE Transactions on Computer-Aided Design*, 8(10): 1033–1037, 1989.

[19] B. J. Sheu, W. J. Hsu, and P. K. Ko, "An MOS transistor charge model for VLSI design," *IEEE Transactions on Computer-Aided Design*, 7(4): 520–527, 1988.

[20] A. D. Snider, "Charge conservation and the transcapacitance element: An exposition," *IEEE Transactions on Education*, 38(4): 376–379, 1995.

[21] P. Yu, "Microwave modeling and parameter extraction for 90-nm gate-length MOSFET devices," Ph.D Dissertation, East China Normal University, 2018.

Chapter 6

MOSFET Noise Model

The research on noise is important because it represents a lower limit to the size of electrical signal that can be amplified by a circuit without significant deterioration in signal quality, and the level of noise determines the sensitivity and dynamic range of the circuit. The noise phenomena considered here are caused by the small current and voltage fluctuations that are generated within the devices themselves.

In order to accurately predict and describe the noise performance of semiconductor devices, establishing an equivalent circuit model that accurately reflects the noise characteristics of semiconductor devices is very important for the design of low-noise amplifiers and oscillators. It is worth noting that the noise equivalent circuit model of the device is based on the small-signal equivalent circuit model.

As the channel length of the MOSFET device is made smaller, the cutoff frequency increases significantly, enabling higher operating frequencies and better noise performance. The complete characterization of these devices in terms of noise and scattering parameters is necessary for computer-aided design (CAD) of monolithic microwave integrated circuits (MMICs) or optoelectronic integrated circuits (OEICs). The typical flowchart for the MOSFET device noise modeling procedure is sketched in Figure 6.1 [1,2].

The full noise characterization of the MOSFET device requires the determination of four noise parameters: minimum noise figure $F_{\min}$, noise resistance R_n, optimum source conductance G_{opt}, and optimum source susceptance B_{opt}. From the circuit point of view, the

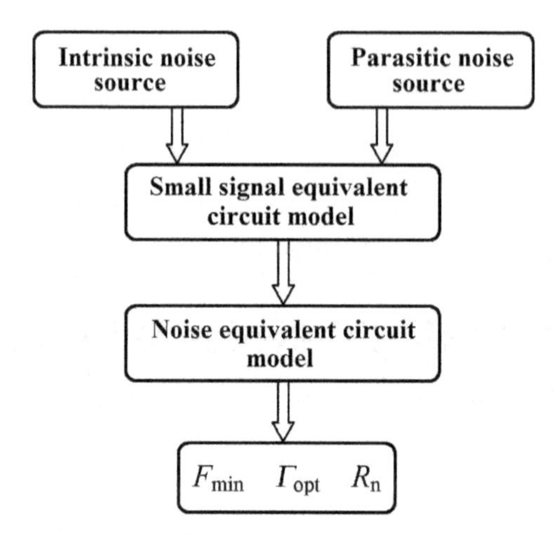

Figure 6.1. Flowchart for MOSFET device noise modeling.

MOSFET device can be treated as a black box of a noisy two-port. To the best of our knowledge, the noise behavior of a linear noisy two-port network can be characterized by the four noise parameters, $F_{\min}$, R_n, G_{opt}, and B_{opt}, with

$$F = F_{\min} + \frac{R_n}{G_s}\left[(G_s - G_{\mathrm{opt}})^2 + (B_s - B_{\mathrm{opt}})^2\right] \tag{6.1}$$

where F is the noise factor, $Y_s = G_s + jB_s$ is the source admittance, and $Y_{\mathrm{opt}} = G_{\mathrm{opt}} + jB_{\mathrm{opt}}$ is the optimum source admittance.

6.1 Noise Modeling

The noise equivalent circuit model of semiconductor devices usually includes intrinsic noise source, extrinsic resistance thermal noise source, and small-signal equivalent circuit model. The intrinsic noise sources are the gate-induced noise, drain channel noise, and the related noise between them. The parasitic noise source refers to the thermal noise generated by the parasitic resistance.

The complete MOSFET small-signal and noise equivalent circuit model is shown in Figure 6.2 [3–6]. Figures 6.2(a) and 6.2(b) illustrate the intrinsic and extrinsic networks, respectively. The circuit

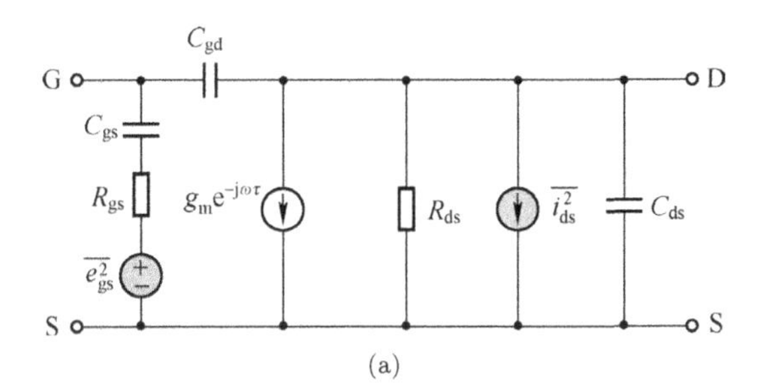

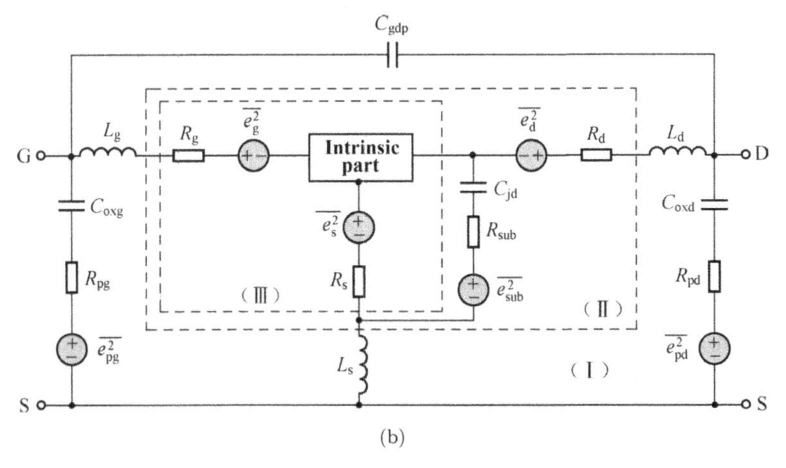

Figure 6.2. MOSFET small-signal and noise equivalent circuit model: (a) intrinsic part; (b) extrinsic part.

model comprises the well-known small-signal equivalent circuit and the following eight noise sources:

$\overline{e_g^2}$: thermal noise of gate extrinsic resistance

$\overline{e_d^2}$: thermal noise of drain extrinsic resistance

$\overline{e_s^2}$: thermal noise of source extrinsic resistance

$\overline{e_{sub}^2}$: thermal noise of substrate resistance between the drain and lossy substrate

$\overline{e_{pg}^2}$: thermal noise of gate pad substrate loss

$\overline{e_{pd}^2}$: thermal noise of drain pad substrate loss

$\overline{e_{gs}^2}$: thermal noise of gate intrinsic resistance

$\overline{i_{ds}^2}$: thermal noise of drain output admittance

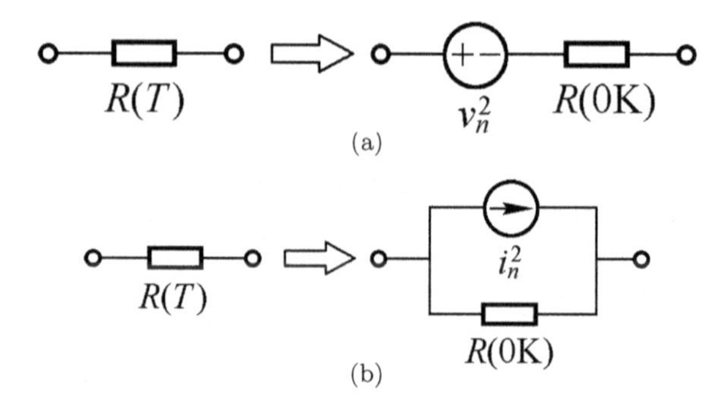

Figure 6.3. Equivalent circuit model of a noisy resistor. (a) Series; (b) parallel.

6.1.1 *Thermal noise caused by extrinsic resistances*

Thermal noise is also called Johnson or Nyquist noise and is generated by thermal energy causing random electron motion. Thermal noise can be found in most of the passive and active devices.

A noisy resistor at a temperature can be modeled by an ideal noiseless resistor at 0°K in series with a noise voltage source (or modeled by an ideal noiseless resistor at 0°K in parallel with a noise current source), as shown in Figure 6.3.

The root-mean-square thermal noise voltage and current in resistor over a frequency range Δf can be expressed as follows:

$$\overline{e_i^2} = 4kT_oR_i\Delta f(i = \text{pg, pd, g, d, s, sub}) \tag{6.2}$$

where T_o is the Kelvin temperature of the resistor (in Kelvin) and normally regarded as 290 K. k represents Boltzmann's constant $(= 1.38 \times 10^{-23}\ \text{V} \cdot \text{C/K})$. Δf represents bandwidth (Hz). R_i represents the extrinsic resistor (Ω).

6.1.2 *Intrinsic noise sources*

When the transistor is under normal bias condition, the carrier moves from source to drain, thence the channel thermal noise is generated by random thermal motion. The channel can be regarded as a parallel plate capacitor. Due to the thermal movement of carriers in the channel, the thermal noise is coupled to the gate through the gate capacitance to produce induced gate noise, which is proportional to the frequency.

The "PRC" model has emerged as one of the most accurate and convenient ways to obtain the noise model parameters for FET devices in microwave simulators. Following these pioneering works, Pospieszalski [7,8] proposes an alternative high-frequency noise model. The aim of this model consists in dissociating the noise on the gate from the noise on the drain. Both the above-mentioned models are well suited to the case of MOSFET devices seeing that the high-frequency noise mechanisms are similar.

In Figure 6.2, the two uncorrelated current noise sources $\overline{e_{\rm gs}^2}$ and $\overline{i_{\rm ds}^2}$ represent the intrinsic noise sources of the MOSFET device. The two noise sources are characterized by their mean quadratic value in a bandwidth Δf centered on the frequency f and can be given by the following expressions:

$$\overline{e_{\rm gs}^2} = 4kT_{\rm g}R_{\rm gs}\Delta f \tag{6.3}$$

$$\overline{i_{\rm ds}^2} = 4kT_{\rm d}g_{\rm ds}\Delta f \tag{6.4}$$

Seeing that the two noise sources $\overline{e_{\rm gs}^2}$ and $\overline{i_{\rm ds}^2}$ are independent, we arrive at

$$\overline{e_{\rm gs}^* i_{\rm ds}} = 0 \tag{6.5}$$

where T_g and T_d are the equivalent noise temperature of the intrinsic resistance $R_{\rm gs}$ and output conductance $g_{\rm ds}$, respectively.

The noise temperatures associated with the six extrinsic resistances have been selected to be equal to the room temperature.

6.2 Noise De-embedding Method

In order to obtain the noise parameters of the device, it is necessary to eliminate the influence of parasitic elements by utilizing the de-embedding method; the open-circuit equivalent circuit model and short-circuit equivalent circuit model are used to eliminate the influence of pads and feedlines.

6.2.1 *Noise correlation matrix*

The circuit theory of linear noisy networks shows that any noisy two-port can be replaced by a noise equivalent circuit consisting of

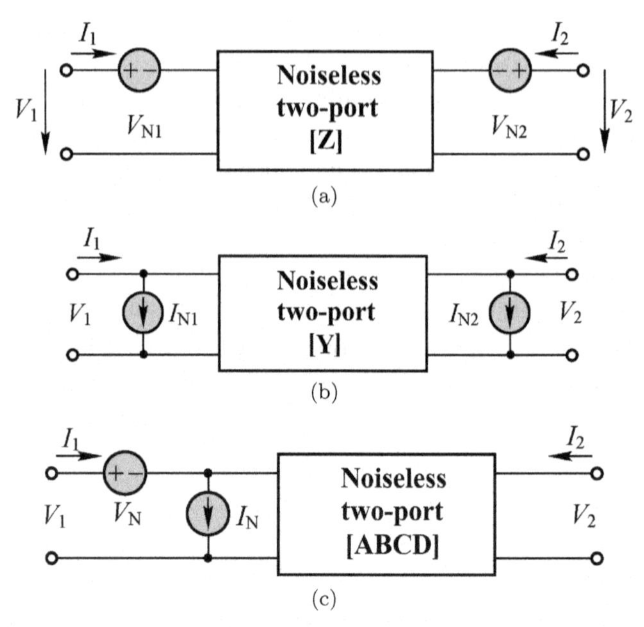

Figure 6.4. Noise representation of noisy two-port network: (a) impedance; (b) admittance; (c) chain.

the original two-port (now assumed to be noiseless) and two additional correlated noise sources. Two noise sources can be completely described by their noise power spectral densities and correlated spectral density.

The two-port noise network can usually be characterized by impedance noise correlation matrix, admittance noise correlation matrix, cascade correlation noise matrix, etc., as shown in Figure 6.4 [9].

The impedance noise representation of the noisy two-port network can be described as follows:

$$\begin{bmatrix} V_1 \\ V_2 \end{bmatrix} = \begin{bmatrix} Z_{11} & Z_{12} \\ Z_{21} & Z_{22} \end{bmatrix} \begin{bmatrix} I_1 \\ I_2 \end{bmatrix} + \begin{bmatrix} V_{\mathrm{N1}} \\ V_{\mathrm{N2}} \end{bmatrix} \tag{6.6}$$

The impedance noise representation normalization with respect to the thermal noise available power $kT\Delta f$ is used:

$$C_Z = \frac{1}{2\Delta f} \begin{bmatrix} \langle V_{N1} \cdot V_{N1}^* \rangle & \langle V_{N1} \cdot V_{N2}^* \rangle \\ \langle V_{N2} \cdot V_{N1}^* \rangle & \langle V_{N2} \cdot V_{N2}^* \rangle \end{bmatrix} \tag{6.7}$$

where V_{N1} and V_{N2} are the noise voltage sources of the input and output ports, respectively, and also known as the open-circuit noise voltage sources.

The admittance noise representation of the noisy two-port network can be written as follows:

$$\begin{bmatrix} I_1 \\ I_2 \end{bmatrix} = \begin{bmatrix} Y_{11} & Y_{12} \\ Y_{21} & Y_{22} \end{bmatrix} \begin{bmatrix} V_1 \\ V_2 \end{bmatrix} + \begin{bmatrix} I_{N1} \\ I_{N2} \end{bmatrix} \tag{6.8}$$

The admittance noise representation normalization with respect to the thermal noise available power $kT\Delta f$ is utilized:

$$C_Y = \frac{1}{2\Delta f} \begin{bmatrix} \langle I_{N1} \cdot I_{N1}^* \rangle & \langle I_{N1} \cdot I_{N2}^* \rangle \\ \langle I_{N2} \cdot I_{N1}^* \rangle & \langle I_{N2} \cdot I_{N2}^* \rangle \end{bmatrix} \tag{6.9}$$

where I_{N1} and I_{N1} are the noise current sources of the input and output ports, respectively, and also known as the short-circuit noise current sources.

The chain noise representation of the noisy two-port network can be described by

$$\begin{bmatrix} V_1 \\ I_1 \end{bmatrix} = \begin{bmatrix} A & B \\ C & D \end{bmatrix} \begin{bmatrix} V_2 \\ -I_2 \end{bmatrix} + \begin{bmatrix} V_N \\ I_N \end{bmatrix} \tag{6.10}$$

The admittance noise representation normalization with respect to the thermal noise available power $kT\Delta f$ is used:

$$C_A = \frac{1}{2\Delta f} \begin{bmatrix} \langle V_N \cdot V_N^* \rangle & \langle V_N \cdot I_N^* \rangle \\ \langle I_N \cdot V_N^* \rangle & \langle I_N \cdot I_N^* \rangle \end{bmatrix} \tag{6.11}$$

where V_N and I_N are the noise voltage source and current source of the input ports, respectively.

The above three noise correlation matrices can be converted to each other, and the conversion relationship is shown in Table 6.1 [9], where C and C' denote the correlation matrix of the original and resulting representations, respectively. T is the transformation matrix.

There are three interconnections of a two-port network: series, parallel, and chain. As shown in Figure 6.5, it is assumed that there are two noiseless networks N_1 and N_2. C_{Z1}, C_{Y1}, and C_{A1} are the impedance noise correlation matrix, admittance noise correlation

Table 6.1. Conversion between different network noise representations C_Y, C_Z, and C_A.

Resulting matrix C'	Original matrix C		
	C_Y	C_Z	C_A
C'_Y	$\begin{bmatrix} 1 & 0 \\ 0 & 1 \end{bmatrix}$	$\begin{bmatrix} Y_{11} & Y_{12} \\ Y_{21} & Y_{22} \end{bmatrix}$	$\begin{bmatrix} -Y_{11} & 1 \\ -Y_{21} & 0 \end{bmatrix}$
C'_Z	$\begin{bmatrix} Z_{11} & Z_{12} \\ Z_{21} & Z_{22} \end{bmatrix}$	$\begin{bmatrix} 1 & 0 \\ 0 & 1 \end{bmatrix}$	$\begin{bmatrix} 1 & -Z_{11} \\ 0 & -Z_{21} \end{bmatrix}$
C'_A	$\begin{bmatrix} 0 & A_{12} \\ 1 & A_{22} \end{bmatrix}$	$\begin{bmatrix} 1 & -A_{11} \\ 0 & -A_{21} \end{bmatrix}$	$\begin{bmatrix} 1 & 0 \\ 0 & 1 \end{bmatrix}$

matrix, and chain noise correlation matrix of N_1, respectively. C_{Z2}, C_{Y2}, and C_{A2} are the impedance noise correlation matrix, admittance noise correlation matrix, and chain noise correlation matrix of N_2, respectively.

The resulting impedance noise matrix in a series of networks is the sum of the respective noise matrix in impedance of the original two-port:

$$C_Z = C_{Z1} + C_{Z2} \tag{6.12}$$

The resulting impedance noise matrix in a parallel of networks is the sum of the respective noise matrix in impedance of the original two-port:

$$C_Y = C_{Y1} + C_{Y2} \tag{6.13}$$

The resulting impedance noise matrix in a cascade of networks can be expressed as follows:

$$C_A = C_{A1} + A_1 C_{A2} A_1^{+} \tag{6.14}$$

It is noticeable that a more complicated relationship is obtained in a cascade of networks, which additionally contains the electrical matrix A_1 of the first two-port.

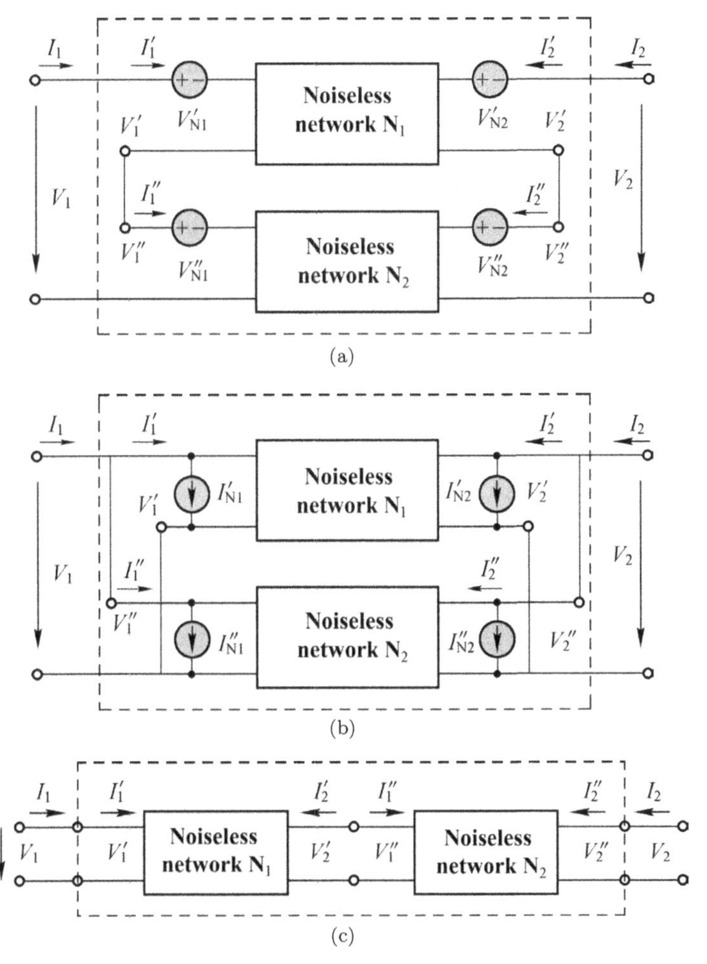

Figure 6.5. Interconnections of two-port networks: (a) series; (b) parallel; (c) cascade.

6.2.2 *Noise parameters extraction*

(1) Calculation of the chain noise correlation matrix for MOSFET device

$$C_A^{\text{meas}} = 4kT \begin{bmatrix} R_n^{\text{meas}} & \frac{F_{\min}^{\text{meas}}-1}{2} - R_n^{\text{meas}} Y_{\text{opt}}^{\text{meas}} \\ \frac{F_{\min}^{\text{meas}}-1}{2} + R_n^{\text{meas}} \left(Y_{\text{opt}}^{\text{meas}}\right)^* & R_n^{\text{meas}} \left|Y_{\text{opt}}^{\text{meas}}\right|^2 \end{bmatrix} \tag{6.15}$$

(2) Transformation of the chain noise correlation matrix into the admittance noise correlation matrix and subtraction of pad parasitics

$$C_Y = C_Y^{\text{meas}} - C_Y^{\text{open}} \tag{6.16}$$

where C_Y^{open} is the chain noise correlation matrix of the open-test structure:

$$C_Y^{\text{open}} = 4kT\,\text{Re}([Y_{\text{open}}]) = 4kT \begin{bmatrix} \frac{\omega^2 C_{\text{oxg}}^2 R_{\text{pg}}}{1+(\omega C_{\text{oxg}} R_{\text{pg}})^2} & 0 \\ 0 & \frac{\omega^2 C_{\text{oxd}}^2 R_{\text{pd}}}{1+(\omega C_{\text{oxd}} R_{\text{pd}})^2} \end{bmatrix} \tag{6.17}$$

(3) Transformation of the admittance noise correlation matrix into the impedance noise correlation matrix and subtraction of external feedline inductances. Due to the external feedline inductances network being a noiseless network, the impedance noise matrix retains invariant.

$$C_Z^{\text{II}} = C_Z - C_Z^{\text{s}} \tag{6.18}$$

where C_Z^{II} is the impedance noise correlation matrix of network II in Figure 6.2 and C_Z^{s} is the impedance noise correlation matrix of the feedline network. If the feedline network is lossless, C_Z^{s} is the zero matrix.

A typical de-embedding result of noise parameters is illustrated in Figure 6.6. It can be clearly seen that the data after de-embedding become smaller [10].

6.3 Noise Parameter Expressions

A set of analytical expressions for the minimum noise temperature, noise resistance, and the optimum source impedance have been derived based on the simple equivalent circuit only considering the intrinsic gate-to-source capacitance and resistance, transconductance, and output conductance. Unfortunately, these expressions are based on a simple noise model that neglects the substrate

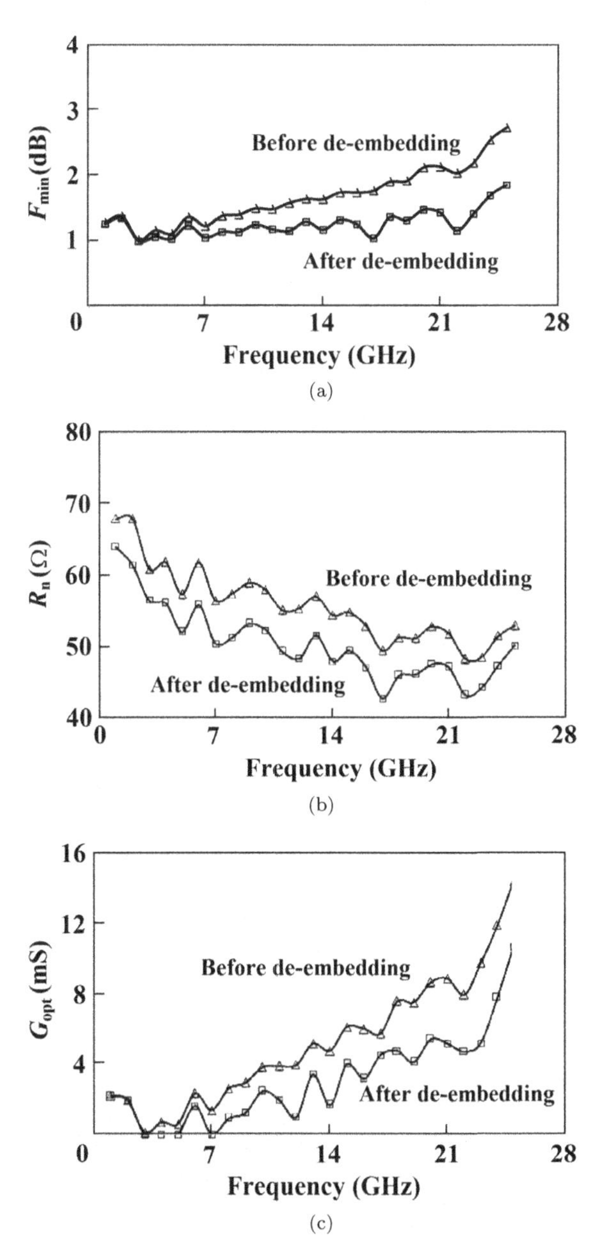

Figure 6.6. (*Continued*)

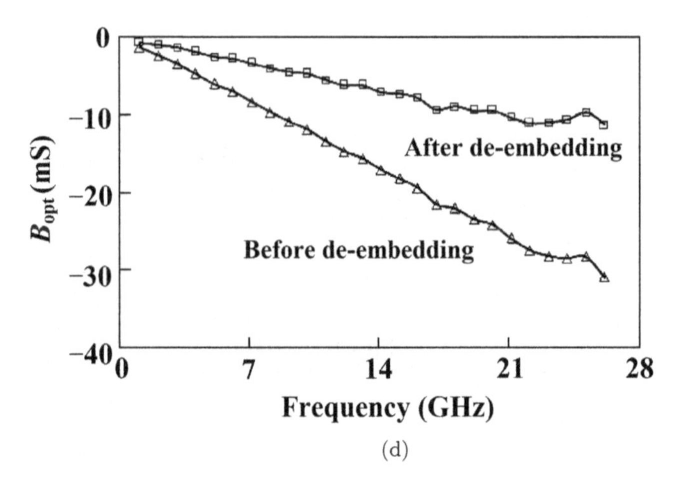

(d)

Figure 6.6. De-embedding of noise parameters for MOSFET device: (a) minimum noise figure; (b) noise resistance; (c) optimum source conductance; (d) optimum source susceptance.

losses and the extrinsic resistances, inductances, and capacitances. Actually, these expressions are not suitable for calculating high-frequency noise parameters of MOSFET devices and therefore need to be improved [11]. A set of new expressions for the four noise parameters of silicon-based MOSFET devices are derived from an accurate noise equivalent circuit model based on the Pospieszalski model without any assumptions and approximations. The effects which become important at higher frequencies such as the substrate parasitics and series inductances need to be taken into account.

Based on the noise correlation matrix technique, the extraction of the four unknown noise parameters can be carried out as follows:

(1) Calculation of the chain noise correlation matrix of a MOSFET device (dashed box I in Figure 6.2) as follows:

$$C_A^{\mathrm{III}} = C_A + \begin{bmatrix} R_g + R_s & 0 \\ 0 & 0 \end{bmatrix} \tag{6.19}$$

where C_A is the chain noise correlation matrix of the intrinsic part.
(2) Calculation of the chain noise correlation matrix of the network I and substrate parasitics network with R_d in cascade (see dashed box II in Figure 6.2(b)).

For the chain noise correlation matrix of two cascaded two ports, we have

$$C_{\mathrm{A}}^{\mathrm{II}} = C_{\mathrm{A}}^{\mathrm{III}} + A_{\mathrm{III}} C_{\mathrm{A}}^{\mathrm{SA}} A_{\mathrm{III}}^{+} \tag{6.20}$$

where A_{III} is the chain matrix of the network III and the plus sign is used to denote the Hermitian transpose. $C_{\mathrm{A}}^{\mathrm{SA}}$ is the chain matrix substrate parasitics network with R_{d} in cascade.

The noise parameters of the network II can be expressed as follows:

$$R_n^{\mathrm{II}} = C_{\mathrm{A11}}^{\mathrm{II}} \tag{6.21}$$

$$G_{\mathrm{opt}}^{\mathrm{II}} = \sqrt{\frac{C_{\mathrm{A22}}^{\mathrm{II}}}{C_{\mathrm{A11}}^{\mathrm{II}}} - \left[\frac{\mathrm{Im}(C_{\mathrm{A12}}^{\mathrm{II}})}{C_{\mathrm{A11}}^{\mathrm{II}}}\right]^2} \tag{6.22}$$

$$B_{\mathrm{opt}}^{\mathrm{II}} = \frac{\mathrm{Im}(C_{\mathrm{A12}}^{\mathrm{II}})}{C_{\mathrm{A11}}^{\mathrm{II}}} \tag{6.23}$$

$$F_{\mathrm{min}}^{\mathrm{II}} = 1 + 2[\mathrm{Re}(C_{\mathrm{A12}}^{\mathrm{II}}) + G_{\mathrm{opt}}^{\mathrm{II}} C_{\mathrm{A11}}^{\mathrm{II}}] \tag{6.24}$$

For low frequencies (normally less than 6 GHz), the influence of the extrinsic inductances and substrate high-frequency effect can be neglected. By neglecting the small high-order terms of angular frequency ω, the noise parameters can be simplified to

$$R_n^L = N + R_s + R_g \tag{6.25}$$

$$B_{\mathrm{opt}}^L = -\omega\left[\frac{(C_{\mathrm{gs}} + C_{\mathrm{gd}})N - C_{\mathrm{gs}} T_g R_{\mathrm{gs}}/T_o}{N + R_s + R_g} + C_{\mathrm{pg}}\right] \tag{6.26}$$

$$G_{\mathrm{opt}}^L = \frac{\omega C_{\mathrm{gs}}}{g_m(N + R_s + R_g)}\sqrt{\frac{T_{\mathrm{d}} g_{\mathrm{ds}}}{T_{\mathrm{o}}}\left[R_s + R_g + \frac{T_g R_{\mathrm{gs}}}{T_o}\right]} \tag{6.27}$$

$$F_{\mathrm{min}}^L = 1 + 2G_{\mathrm{opt}}^L R_n^L \tag{6.28}$$

where superscript L denotes the noise parameters in the low-frequency range and N is a constant defined as follows:

$$N = \frac{T_g R_{\mathrm{gs}}}{T_o} + \frac{T_d g_{\mathrm{ds}}}{T_o g_m^2} \tag{6.29}$$

It is noted that for the determination of the noise parameters in the low-frequency range, only the four resistances (R_g, R_s, R_{gs}, and g_{ds}), three capacitances (C_{pg}, C_{gs}, and C_{gd}), and one transconductance (g_m) are necessary. It also can be found that the equivalent noise resistance R_n is frequency independent, and optimum source conductance G_{opt} and optimum source susceptance B_{opt} are proportional to angular frequency ω. The optimum noise figure $F_{\min}$ is the linear function of angular frequency ω.
(3) The corresponding noise model parameters can be extracted from the admittance noise correlation matrix of the intrinsic part for a MOSFET device:

$$T_g = \frac{C_{Y11}(1+\omega^2 C_{\text{gs}}^2 R_{\text{gs}}^2)}{4k\Delta f \omega^2 C_{\text{gs}}^2 R_{\text{gs}}} \tag{6.30}$$

$$T_d = \frac{C_{Y22} - C_{Y11} g_m^2/(\omega C_{\text{gs}})^2}{4k\Delta f g_{\text{ds}}} \tag{6.31}$$

where C_{Y11} and C_{Y22} are the admittance noise correlation matrix for the intrinsic network.

In order to verify the equations for the four noise parameters, silicon-based MOSFET devices with 16 fingers, 5 μm channel width per finger, and 0.35 μm channel length have been characterized. The extracted values of the bias-independent small-signal elements are summarized in Table 6.2, and the intrinsic parameters C_{gs}, C_{gd}, C_{ds}, g_m, τ, R_i, and g_{ds} at the constant drain–source voltage $V_{\text{ds}} = 2.0$ V

Table 6.2. Extrinsic parameters of MOSFET device (0.35 μm gate length, 16 finger $\times$ 5 μm gate width).

Parameters	Values	Parameters	Values
L_g (pH)	35	$R_{\text{pd}}(\Omega)$	12
L_d (pH)	35	$R_g(\Omega)$	1.5
L_s (pH)	3.5	$R_d(\Omega)$	3
$R_{\text{pg}}(\Omega)$	12	$R_s(\Omega)$	2.5
C_{sub} (fF)	25	$R_{\text{sub}}(\Omega)$	200

Table 6.3. Intrinsic parameters of MOSFET device (0.35 μm gate length, 16 finger $\times$ 5 μm gate width).

Parameters	$V_{gs} = 0.6$ V	$V_{gs} = 0.8$ V	$V_{gs} = 1.0$ V
I_{ds} (mA)	2.90	8.24	14.86
C_{gs} (fF)	224	260	270
C_{gd} (fF)	40	41	41
C_{ds} (fF)	14	18	22
g_m (pH)	25.4	38	44.2
τ (pS)	1.9	1.3	1.2
g_{ds} (mS)	0.4	0.7	1.0
R_i (Ω)	5	3	3
T_g (K)	290	670	870
T_d (K)	10270	14000	15000

and $V_{gs} = \{0.6\text{ V}, 0.8\text{ V}, 1.0\text{ V}\}$ are tabulated in Table 6.3. Moreover, the corresponding noise temperature model parameters under different bias conditions are summarized in Table 6.3.

In the frequency range of 0.1 GHz–40 GHz, Figures 6.7(a)–6.7(c) depict the comparison between the measured and modeled S parameters of a MOSFET device under three different bias conditions (V_{gs} = 0.6 V, 0.8 V, 1.0 V, V_{ds} = 2 V). An excellent agreement over the whole frequency range is obtained. In Figures 6.8–6.10, the noise parameters versus the frequency of a MOSFET device under the same bias conditions (V_{gs} = 0.6 V, 0.8 V, 1.0 V, V_{ds} = 2 V) are shown, and the comparison between the measured and calculated data is also illustrated. As seen, good agreements can be observed from the figures to verify the approach.

The comparison between measured and calculated noise parameters versus gate-source voltage V_{gs} is depicted in Figure 6.11. The operating frequency is 10 GHz. An excellent agreement can be observed. It is noted that with the increase in V_{gs}, $F_{\min}$ and R_n increase. Moreover, with the increase in V_{gs}, the magnitude of Γ_{opt} decreases, and with increased V_{gs}, the phase of Γ_{opt} increases. In Figure 6.12, the calculated noise parameters in the low-frequency range are compared with measured data versus frequency for the MOSFET device under the bias condition of V_{gs} = 1.4 V and V_{ds} = 2.0 V.

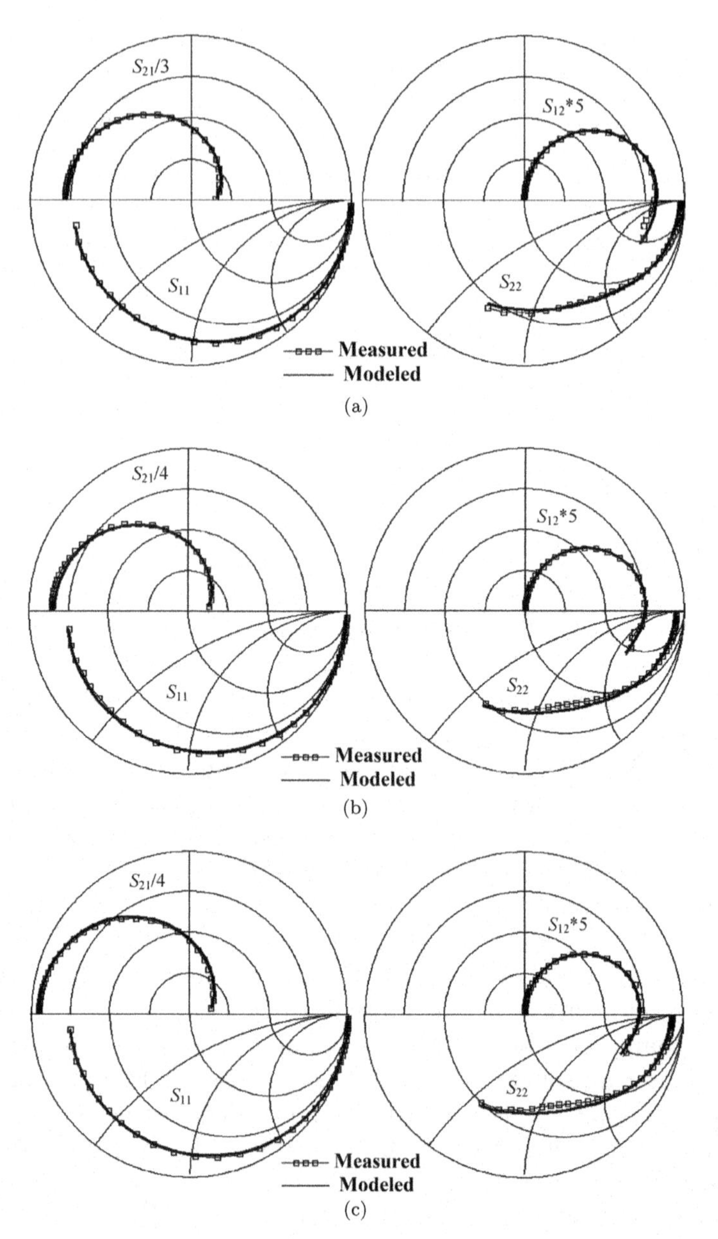

Figure 6.7. Comparison between modeled and measured S parameter of MOSFET device. Bias condition: (a) $V_{\mathrm{gs}} = 0.6$ V, $V_{\mathrm{ds}} = 2$ V; (b) $V_{\mathrm{gs}} = 0.8$ V, $V_{\mathrm{ds}} = 2$ V; (c) $V_{\mathrm{gs}} = 1.0$ V, $V_{\mathrm{ds}} = 2$ V.

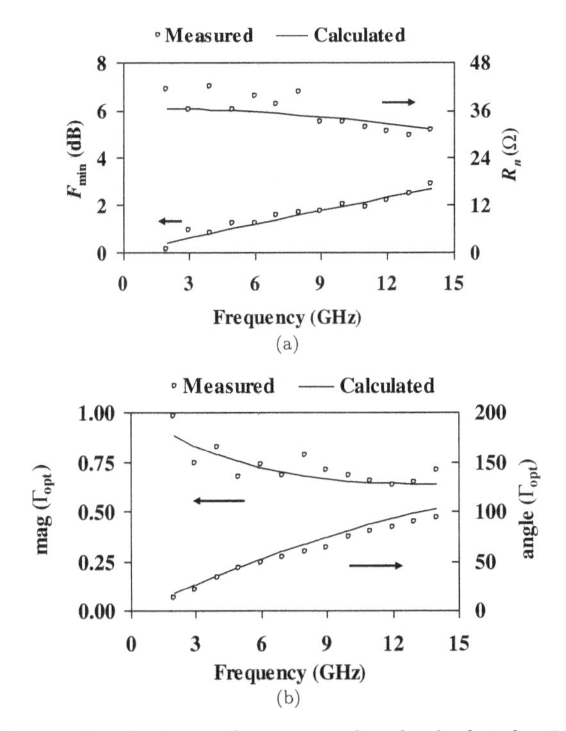

Figure 6.8. Comparison between the measured and calculated noise parameters. Bias condition: $V_{gs} = 0.6$ V, $V_{ds} = 2$ V. (a) $F_{\min}$ and R_n; (b) Γ_{opt}.

6.4 Noise Parameters Extraction Methods

The commonly used methods for the determination of noise parameters for MOSFET devices are the tuner-based method [12] and 50 Ω noise figure measurement system-based method [13].

6.4.1 *Tuner-based extraction method*

The determination of the noise parameters is typically performed by analyzing the variation of the measured noise figure as a function of the source impedance. A minimum of four independent measurements is required. Nevertheless, for increasing the accuracy, more than four measurements are performed usually and curve-fitting techniques are used then to determine the noise parameters. The tuner-based noise parameters measurement system is illustrated in Figure 6.13.

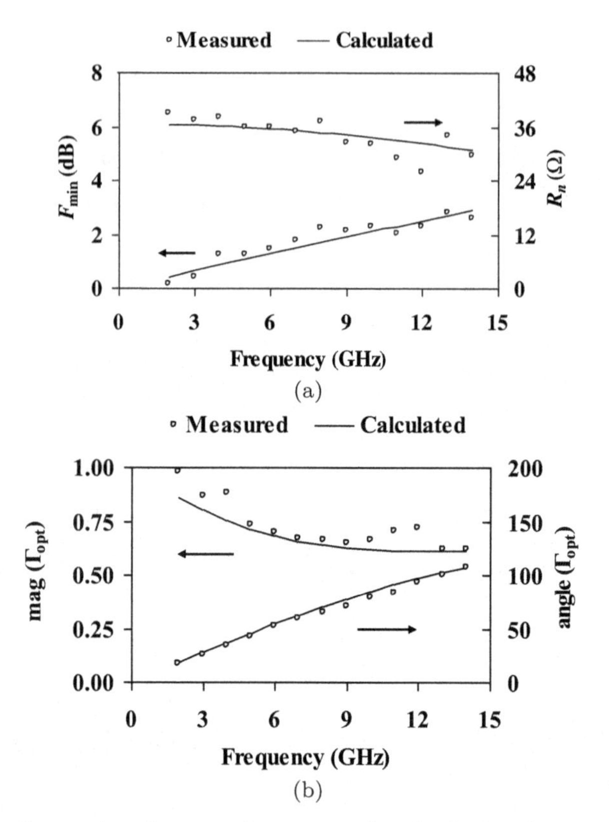

Figure 6.9. Comparison between the measured and calculated noise parameters. Bias condition: $V_{\rm gs} = 0.8$ V, $V_{\rm ds} = 2$ V. (a) $F_{\min}$ and R_n; (b) $\Gamma_{\rm opt}$.

Assuming

$$A = F_{\min} - 2R_n G_{\rm opt} \tag{6.32}$$

$$B = R_n \tag{6.33}$$

$$C = R_n(G_{\rm opt}^2 + B_{\rm opt}^2) \tag{6.34}$$

$$D = -2R_n B_{\rm opt} \tag{6.35}$$

The noise figure can be regarded as a nonlinear function of the source admittance [12]:

$$F = A + BG_s + \frac{C + BB_s^2 + DB_s}{G_s} \tag{6.36}$$

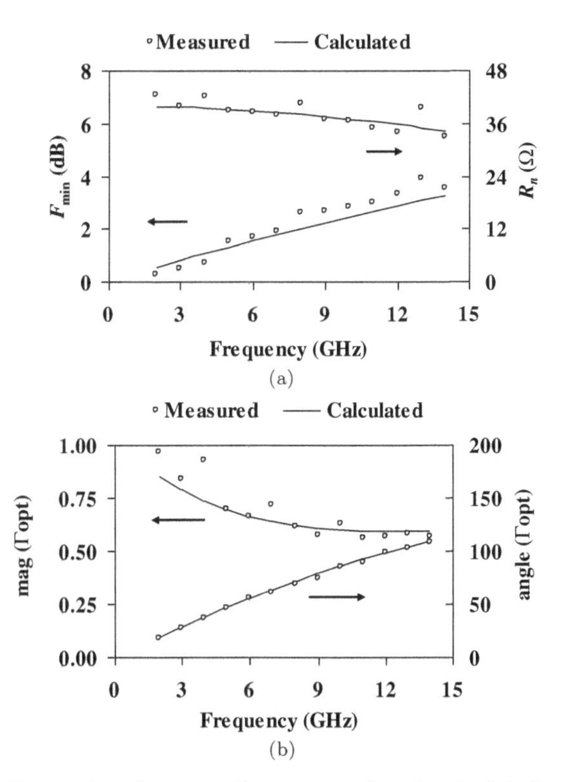

Figure 6.10. Comparison between the measured and calculated noise parameters. Bias condition: $V_{gs} = 1.0$ V, $V_{ds} = 2$ V. (a) F_{min} and R_n; (b) Γ_{opt}.

The error criterion ε is

$$\varepsilon = \frac{1}{2}\sum_{i=1}^{n}\left[A + B\left(G_i + \frac{B_i^2}{G_i}\right) + \frac{C}{G_i} + \frac{DB_i}{G_i} - F_i\right]^2 \tag{6.37}$$

where G_i and B_i $(i = 1, 2, \ldots, n)$ are the source conductances and susceptances, respectively, and F_i is the corresponding noise figures.

The typical experimental setup for the determination of the noise parameters is illustrated in Figure 6.14 [14]. It is composed of a wafer-probe station, an automatic network analyzer HP8510C up to 40 GHz, the noise measurement system (NMS) up to 26.5 GHz, and an electronic broadband noise source HP346C up to 50 GHz.

Most algorithms rely on the source-pull measurement technique to extract the noise parameters from a large set of parameters at a single frequency, for example, by employing the correlation matrix method to de-embed the parasitics and to determine the intrinsic

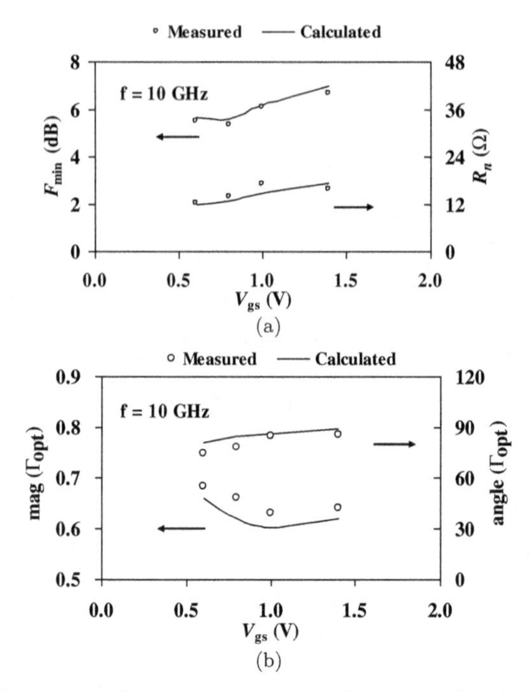

Figure 6.11. Comparison between the measured and calculated noise parameters versus V_{gs} at 10 GHz. (a) F_{min} and R_n; (b) Γ_{opt}.

noise sources. A minimum of four independent measurements are required, however frequently more measurements are performed to achieve higher accuracy. Curve-fitting techniques are then used to determine the noise parameters. Although this method gives accurate results, it is time consuming and requires an expensive automatic broadband microwave tuner that involves complex calibration procedures.

6.4.2 *Noise parameters based on noise figure measurement*

A full analytical method to determine the four noise parameters of the MOSFET device is proposed without any optimization; this method has the following advantages:

(1) A set of new expressions for the four noise parameters of a silicon-based MOSFET device is derived from an accurate noise

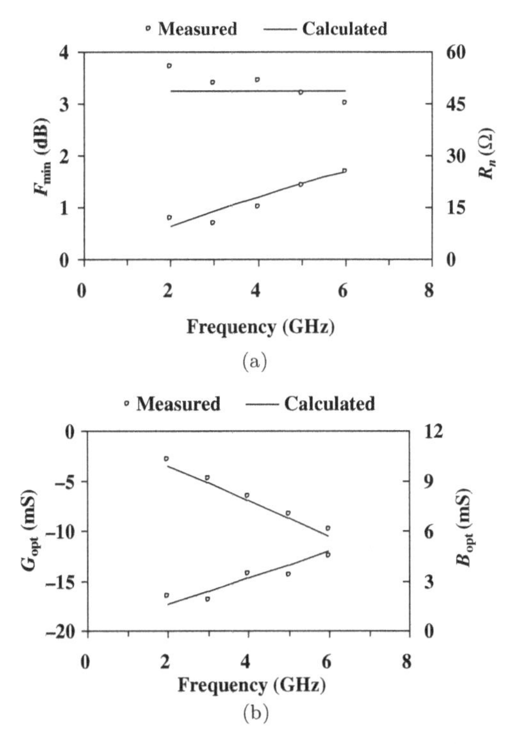

Figure 6.12. Comparison between the measured and calculated noise parameters versus frequency. Bias condition: V_{gs} = 1.4 V, V_{ds} = 2.0 V. (a) F_{min} and R_n; (b) G_{opt} and B_{opt}.

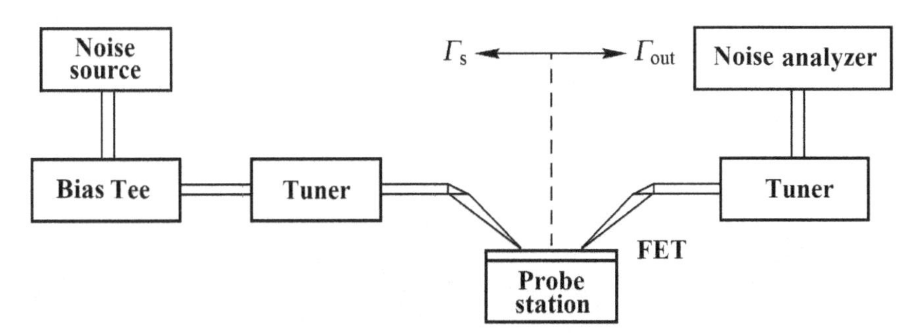

Figure 6.13. Tuner-based noise parameters measurement system.

equivalent circuit model that is based on the Pospieszalski model without any assumptions and approximations. The effects which become important at higher frequencies such as the substrate

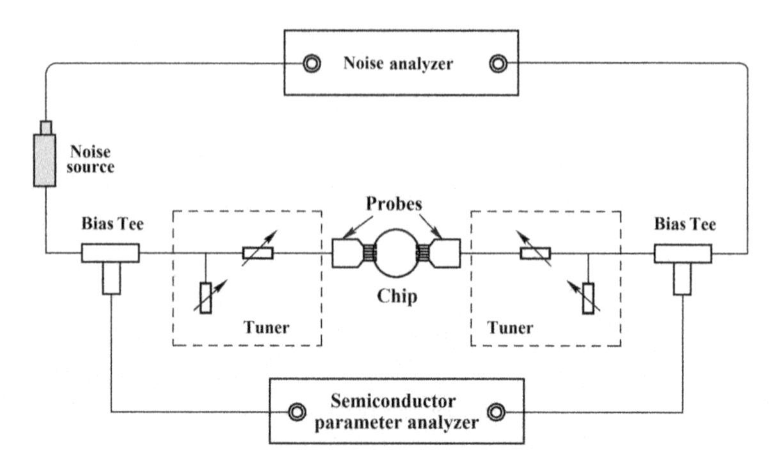

Figure 6.14. Commercial noise parameters measurement setup.

parasitics, the pad capacitances, and series inductances are taken into account.

(2) Further simplified expressions of noise parameters for the MOSFET device in the low-frequency range are derived.

(3) All four noise parameters can be determined directly from the noise figure on wafer measurement at low frequencies based on the accurate analytical expressions of noise parameters.

Such a method would be beneficial in acceptance testing and faster than full testing with a source tuner and more likely to be done with a test setup not requiring a tuner. Nevertheless, an accurate small-signal model parameters extraction is need.

Based on the noise correlation matrix technique [15], the expressions of the four noise parameters can be carried out as follows:

(1) Calculation of the admittance noise correlation matrix of the intrinsic part for the MOSFET device as follows:

$$C_{Y11}^{INT} = 4kT_g \Delta f R_{\mathrm{gs}} \left| \frac{j\omega C_{\mathrm{gs}}}{1 + j\omega C_{\mathrm{gs}} R_{\mathrm{gs}}} \right|^2 \tag{6.38}$$

$$C_{Y22}^{INT} = 4k\Delta f \left(T_d g_{\mathrm{ds}} + T_g R_{\mathrm{gs}} \left| \frac{g_m}{1 + j\omega C_{\mathrm{gs}} R_{\mathrm{gs}}} \right|^2 \right) \tag{6.39}$$

$$C_{Y12}^{INT} = 4kT_g \Delta f \frac{g_m^* \omega C_{\mathrm{gs}} R_{\mathrm{gs}}}{|1 + j\omega C_{\mathrm{gs}} R_{\mathrm{gs}}|^2} \tag{6.40}$$

The corresponding noise parameters of the intrinsic network can be expressed as follows:

$$R_n^{INT} = \frac{T_g R_{\text{gs}}}{T_o k_1} + \frac{T_d g_{\text{ds}}(1 + \omega^2 C_{\text{gs}}^2 R_{\text{gs}}^2)}{T_o k_1 g_m^2} \tag{6.41}$$

$$B_{\text{opt}}^{INT} = -\omega \left(C_{\text{gs}} + C_{\text{gd}} - C_{\text{gs}} \frac{T_g R_{\text{gs}}}{T_o R_n} \right) \tag{6.42}$$

$$G_{\text{opt}}^{INT} = \omega C_{\text{gs}} \frac{\sqrt{k_1 k_3 (k_2 - k_3)}}{k_2} \tag{6.43}$$

$$F_{\min}^{INT} = 1 + 2k_4 + 2G_{\text{opt}} R_n \tag{6.44}$$

with

$$k_1 = 1 + \frac{2\omega^2 C_{\text{gd}}(\tau + R_{\text{gs}} C_{\text{gs}})}{g_{\text{m}}}$$

$$k_2 = \frac{T_d g_{\text{ds}}}{T_o g_m} + k_3$$

$$k_3 = \frac{T_g g_m R_{\text{gs}}}{T_o(1 + \omega^2 C_{\text{gs}}^2 R_{\text{gs}}^2)}$$

$$k_4 = \frac{\omega^2 C_{\text{gs}}(g_m \tau + C_{gd}) k_3}{g_m^2}$$

It can be seen that the equivalent noise resistance R_n can be regarded as frequency independent. In addition, G_{opt} and B_{opt} can be regarded as proportional to angular frequency ω.

(2) Transformation of the admittance noise correlation matrix to the chain noise correlation matrix by readily adding of the extrinsic resistances R_g and R_s (dashed box III in Figure 6.2(b)). We arrive at the noise parameters of the network III:

$$R_n^{\text{III}} = R_n^{INT} + R_g + R_s \tag{6.45}$$

$$B_{\text{opt}}^{\text{III}} = \frac{R_n^{INT} B_{\text{opt}}^{INT}}{R_n^I} \tag{6.46}$$

$$G_{\text{opt}}^{\text{III}} = \sqrt{(G_{\text{opt}}^{INT})^2 + \frac{k_5 g_m^2 (R_g + R_s)}{k_2^2} \cdot \frac{R_n^{INT}}{R_n^{\text{I}}}} \tag{6.47}$$

$$F_{\min}^{\text{III}} = 1 + 2k_4 + 2k_5(R_g + R_s) + 2G_{\text{opt}}^{\text{I}} R_n^{\text{I}} \tag{6.48}$$

with

$$k_5 = \frac{\omega^2}{g_m}\left[k_2(C_{\text{gs}} + C_{\text{gd}})^2 - k_3 C_{\text{gs}}(C_{\text{gs}} + 2C_{\text{gd}})\right]$$

Seeing that the parasitic resistances R_g and R_s only affect the constant terms, the relationship between the noise parameters and frequency remains invariant roughly.

(3) Calculation of the chain noise correlation matrix of the network III and substrate parasitics network with R_d in cascade (see dashed box II in Figure 6.2(b)).

$$C_{\text{A}}^{\text{II}} = C_{\text{A}}^{\text{III}} + A_{\text{III}} C_{\text{A}}^{S} A_{\text{III}}^{+} \tag{6.49}$$

with

$$C_{\text{A}}^{S} = R_d \begin{bmatrix} 1 & \dfrac{\omega^2 R_{\text{sub}} C_{\text{jd}}^2 - j\omega C_{\text{jd}}}{1} + \omega^2 R_{\text{sub}}^2 C_{\text{jd}}^2 \\ \dfrac{\omega^2 R_{\text{sub}} C_{\text{jd}}^2 + j\omega C_{\text{jd}}}{1+\omega^2 R_{\text{sub}}^2 C_{\text{jd}}^2} & \dfrac{\omega^2 (1+R_{\text{sub}}/R_d) C_j^2}{1+\omega^2 R_{\text{sub}}^2 C_{\text{jd}}^2} \end{bmatrix}$$

A_{III} is the chain matrix of the network III, and the plus sign is used to denote the Hermitian conjugation.

The noise parameters of the network II can be expressed as follows:

$$B_{\text{opt}}^{\text{II}} = B_{\text{opt}}^{\text{I}} \tag{6.50}$$

$$G_{\text{opt}}^{\text{II}} = G_{\text{opt}}^{\text{I}} \tag{6.51}$$

$$R_n^{\text{II}} = R_n^{\text{I}} + R_{\text{d}} \frac{g_{\text{ds}}^2}{g_m^2} + \left(\frac{1}{g_m} + R_{\text{s}}\right)^2 \frac{\omega^2 (R_{\text{sub}} + R_d) C_{\text{jd}}^2}{1 + \omega^2 R_{\text{sub}}^2 C_{\text{jd}}^2} \tag{6.52}$$

$$F_{\min}^{\text{II}} = F_{\min}^{\text{I}} + 2G_{\text{opt}}^{\text{I}} R_d \frac{g_{\text{ds}}^2}{g_m^2} \tag{6.53}$$

It can be found that G_{opt} and B_{opt} remain invariant and R_n and $F_{\min}$ change a little bit, that means the substrate parasitics network with R_{d} has a weak influence on the noise parameters.

(4) Addition of the inductances L_g and L_s.

Since $F_{\min}$ is invariant with respect to lossless transformation at the input and output ports of the two port network, $R_n G_{\text{opt}}$ is also invariant with respect to lossless transformation, and Y_{opt} can be calculated by

$$Y_{\text{opt}}^{\text{I}} = \frac{1}{\frac{1}{Y_{\text{opt}}^{\text{II}}} - j\omega(L_g + L_s)} \tag{6.54}$$

In terms of the above-mentioned transformation of the noise parameters along a lossless network, the reference plane can be moved readily to the input port, and L_g and L_s are included:

$$F_{\min}^{\text{I}} = F_{\min}^{\text{II}} \tag{6.55}$$

$$G_{\text{opt}}^{\text{I}} = \frac{G_{\text{opt}}^{\text{II}}}{1 + \omega^2(L_g + L_s)^2|Y_{\text{opt}}^{\text{II}}|^2 + 2\omega B_{\text{opt}}^{\text{II}}(L_g + L_s)} \tag{6.56}$$

$$B_{\text{opt}}^{\text{I}} = \frac{B_{\text{opt}}^{\text{II}} + \omega(L_g + L_s)|Y_{\text{opt}}^{\text{II}}|^2}{1 + \omega^2(L_g + L_s)^2|Y_{\text{opt}}^{\text{II}}|^2 + 2\omega B_{\text{opt}}^{\text{II}}(L_g + L_s)} \tag{6.57}$$

$$R_n^{\text{I}} = \frac{R_n^{\text{II}} G_{\text{opt}}^{\text{II}}}{G_{\text{opt}}^{\text{I}}} \tag{6.58}$$

(5) Addition of the pad network.
Since the substrate is resistive, the pad network for the MOSFET device is more complicated compared to III–V compound semiconductor devices. After adding the effect of the pad network, we arrive at the noise parameters of the whole device:

$$B_{\text{opt}}^{T} = B_{\text{opt}}^{\text{III}} - \omega C_{\text{pg}} \tag{6.59}$$

$$R_n^{T} = R_n^{\text{III}} + \frac{\omega^2 R_{\text{pd}} C_{\text{pd}}^2}{g_m^2} \tag{6.60}$$

$$G_{\text{opt}}^{T} = \sqrt{(G_{\text{opt}}^{\text{III}})^2 + \frac{F_{\min}^{\text{III}} \omega^2 C_{\text{pg}}^2 R_{\text{pg}}}{R_n^{\text{III}}}} - \omega^2 C_{\text{pg}}^2 R_{\text{pg}} \tag{6.61}$$

$$F_{\min}^{T} = F_{\min}^{\text{III}} + \omega^2 R_{\text{pg}} C_{\text{pg}}^2 R_n^{\text{III}} \tag{6.62}$$

At low frequencies, the influence of the extrinsic inductances and substrate high-frequency effect can be neglected. By neglecting the small high-order ω-terms, we have

$$R_n = N + R_s + R_g \tag{6.63}$$

$$B_{\text{opt}} = -\omega \left[\frac{(C_{\text{gs}} + C_{\text{gd}})N - C_{\text{gs}} T_g R_{\text{gs}} / T_o}{N + R_s + R_g} + C_{\text{pg}} \right] \tag{6.64}$$

$$G_{\text{opt}} = \frac{\omega C_{\text{gs}}}{g_m (N + R_s + R_g)} \sqrt{\frac{T_d g_{\text{ds}}}{T_o} \left[R_s + R_g + \frac{T_g R_{\text{gs}}}{T_o} \right]} \tag{6.65}$$

$$F_{\min} = 1 + 2 G_{\text{opt}} R_n \tag{6.66}$$

with

$$N = \frac{T_g R_{\text{gs}}}{T_{\text{o}}} + \frac{T_{\text{d}} g_{\text{ds}}}{T_{\text{o}} g_m^2}$$

It is noted that in order to determine the noise parameters in the low-frequency range, only the four resistances (R_g, R_s, R_{gs}, and g_{ds}), three capacitances (C_{pg}, C_{gs}, and C_{gd}), and one transconductance (g_m) are required. It also can be found that the equivalent noise resistance R_n is frequency independent, optimum source conductance G_{opt} and optimum source susceptance B_{opt} are proportional to angular frequency ω, and optimum noise figure $F_{\min}$ is a linear function of ω.

In the case of the 50 Ω generator impedance, the noise figure can be rewritten as

$$F_{50} = 1 + R_n G_0 + \frac{R_n}{G_0} |Y_{\text{opt}}|^2 \tag{6.67}$$

The 50 Ω noise figure measurement system is illustrated in Figure 6.15, where $Y_s = G_0 = 20$ mS.

R_n is nearly frequency independent, while $|Y_{\text{opt}}|$ varies proportional to ω^2, the plot of F_{50} versus ω^2 is linear and the value at $\omega = 0$ is equal to $(1 + R_n G_0)$. Thus, R_n can be readily deduced from the F_{50} extrapolation for $\omega = 0$:

$$R_n = \frac{(F_{50}^{\omega=0} - 1)}{G_0} \tag{6.68}$$

The slope of F_{50} versus ω^2 provides the magnitude of the optimum generator admittance $|Y_{\text{opt}}|$.

$$|Y_{\text{opt}}|^2 = \frac{dF_{50}}{d\omega^2} \cdot \frac{\omega^2 G_0}{R_n} \tag{6.69}$$

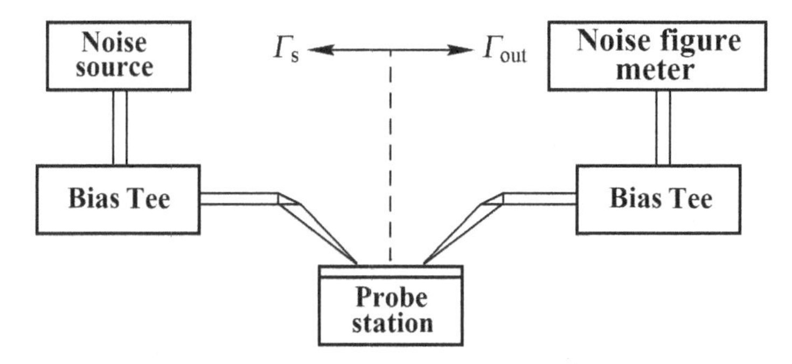

Figure 6.15. 50 Ω noise figure measurement system.

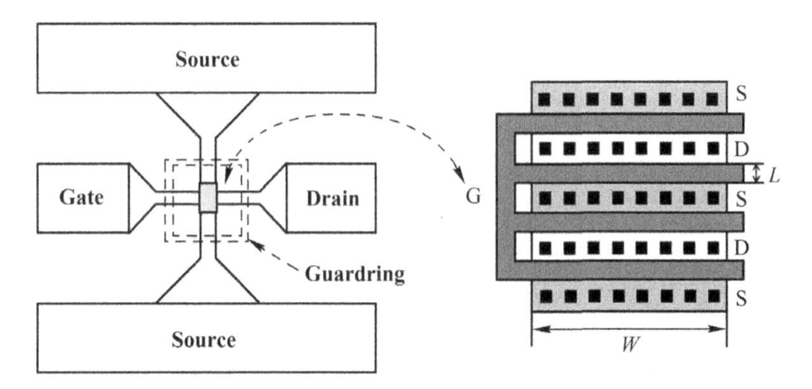

Figure 6.16. Layout of MOSFET devices.

The noise model parameter can be obtained as follows:

$$T_d = \frac{\left[G_o \frac{dF_{50}}{d\omega^2} - R_n \left(C_{\text{pg}} + \frac{NC_{\text{gd}}}{R_n}\right)^2\right] g_m^2 T_o}{C_{\text{gs}} \left(C_{\text{gs}} + 2C_{\text{pg}} + \frac{2NC_{\text{gd}}}{R_n}\right) g_{\text{ds}}} \tag{6.70}$$

$$T_g = \frac{T_{\text{o}}}{R_{\text{gs}}}(R_n - R_s - R_g) - \frac{T_d g_{\text{ds}}}{R_{\text{gs}} g_m^2} \tag{6.71}$$

Three different gate-length silicon-based MOSFET devices (0.5 μm × 5 μm × 16 finger, 0.35 μm × 5 μm × 16 finger, and 0.18 μm × 5 μm × 16 finger) with the same pad structure have been characterized. The S parameter measurements for model extraction and verification were made up to 40 GHz. Microwave noise parameter measurements are carried out on a wafer over the frequency

Table 6.4. PAD parasitics.

Parameters	Values	Parameters	Values
C_{pg} (fF)	42	L_g (pH)	35
C_{pd} (fF)	42	L_d (pH)	35
C_{pgd} (fF)	1.5	L_s (pH)	3.5
$R_{\mathrm{pg}}(\Omega)$	12	R_{pd} (Ω)	12

Table 6.5. Small-signal model parameters.

Parameters	$L = 0.5$ μm	$L = 0.35$ μm	$L = 0.18$ μm
$R_g(\Omega)$	2	1.5	2.0
$R_d(\Omega)$	5	3	7
$R_s(\Omega)$	1.5	1.5	0.8
$R_{\mathrm{sub}}(\Omega)$	300	300	300
C_{sub} (fF)	30	40	50
C_{gs} (fF)	370	265	125
C_{gd} (fF)	47	44	41
C_{ds} (fF)	20	24	30
g_m (mS)	34	42	57
τ (pS)	2.0	1.2	0.4
g_{ds} (mS)	0.85	1.5	3.12
R_i (Ω)	3	3	2

range 2–14 GHz. The MOSFET devices layout structure considered for our studies is depicted in Figure 6.16. It is worth noting that all the devices have the same pad profile.

The extracted pad parasitics which include pad capacitances, substrate losses, and feedline inductances are summarized in Table 6.4. Once the values of the parasitic elements are known, all bias-dependent elements can be readily calculated utilizing the de-embedding technique. Table 6.5 tabulates the intrinsic parameters with extrinsic resistances and substrate parasitics for 0.5 μm $\times$ 5 μm $\times$ 16 finger, 0.35 μm $\times$ 5 μm $\times$ 16 finger, and 0.18 μm $\times$ 5 μm $\times$ 16 finger MOSFET devices.

In the frequency range of 0.1–40 GHz, the comparison between the measured and modeled S parameters for the 0.5 μm $\times$ 5 μm $\times$ 16 finger, 0.35 μm $\times$ 5 μm $\times$ 16 finger, and 0.18 μm $\times$ 5 μm$\times$ 16 finger MOSFET devices under the bias condition of $V_{\text{gs}} = 1.2$ V and $V_{\text{ds}} = 2$ V is exhibited in Figure 6.17. An excellent agreement over the whole frequency range is achieved. Furthermore, the corresponding cutoff frequencies f_t are 12 GHz, 20 GHz, and 42 GHz respectively under the bias condition mentioned above.

Figure 6.18 illustrates the experimental evaluations of F_{50} versus the square of the frequency for two different MOSFET devices (0.5 μm $\times$ 5 μm $\times$ 16 finger and 0.35 μm $\times$ 5 μm $\times$ 16 finger). It can be observed that with the decrease in the gate length, the noise figure decreases.

Figure 6.19 illustrates the extracted noise model parameters as a function of the gate length. It can be found that the noise temperatures (T_d and T_g) for different size devices are almost independent of the gate length and also independent of the gate width certainly.

In Figures 6.20–6.22, the comparisons of the measured and modeled noise parameters versus frequency are given for 0.5 μm $\times$ 5 μm $\times$ 16 finger, 0.35 μm $\times$ 5 μm $\times$ 16 finger, and 0.18 μm $\times$ 5 μm $\times$ 16 finger MOSFET devices under the same bias conditions ($V_{\text{gs}} = 1.2$ V, $V_{\text{ds}} = 2$ V). Good agreement between measured and modeled results can be indicated and the validity of the method is confirmed.

Figure 6.23 depicts the variation of the four noise parameters as a function of the gate length, the operating frequency being 4 GHz under the same bias conditions of $V_{\text{gs}} = 1.2$ V and $V_{\text{ds}} = 2$ V. It can be observed that the optimum noise figure F_{min} decreases with the decrease in the gate length; the reason is that with the decrease in gate length, the cutoff frequency increases. Additionally, the noise resistance R_n increases with the decrease in the gate length, owing to the variation of the factor g_{ds}/g_m^2. With the decrease in the gate length, the magnitude of the optimum source reflection Γ_{opt} increases and the phase of the optimum source reflection Γ_{opt} decreases, respectively. The main reason is that with decreased gate length, the intrinsic capacitance C_{gs} decreases.

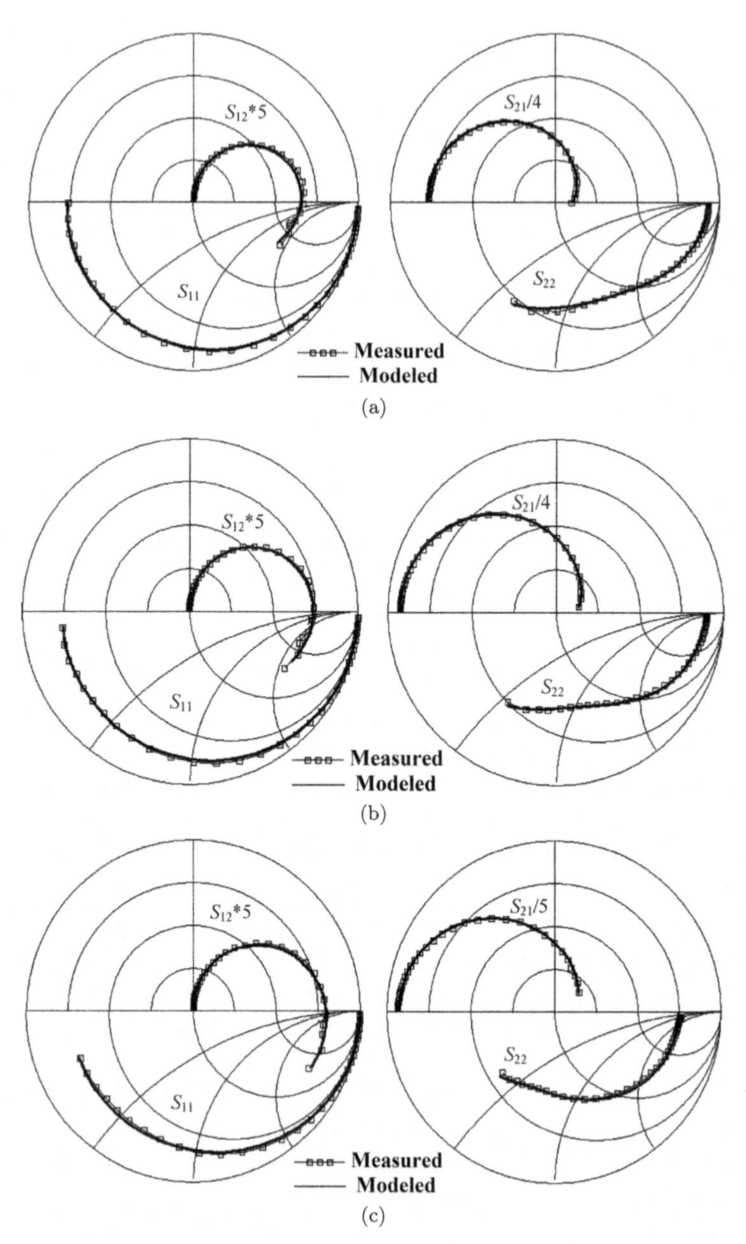

Figure 6.17. Comparison between the modeled and measured S parameter for the MOSFET. Bias condition: $V_{\text{gs}} = 1.2$ V, $V_{\text{ds}} = 2$ V: (a) 0.5 μm × 5 μm× 16 finger; (b) 0.35 μm × 5 μm × 16 finger; (c) 0.18 μm × 5 μm × 16 finger.

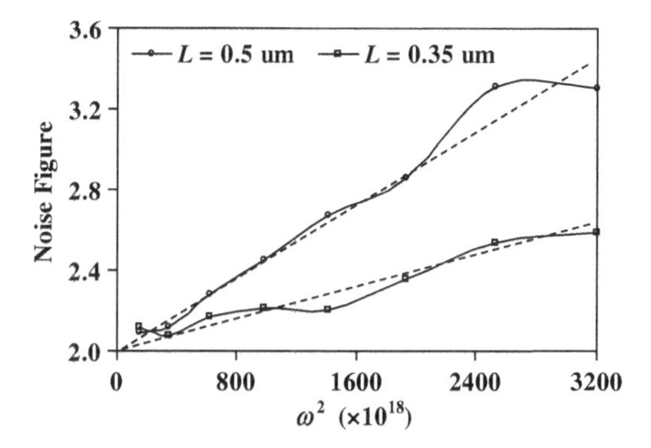

Figure 6.18. Evaluation of the noise figures F_{50} for MOSFET devices versus the square of the frequency ω^2. Bias condition: $V_{\rm gs} = 1.2$ V, $V_{\rm ds} = 2$ V.

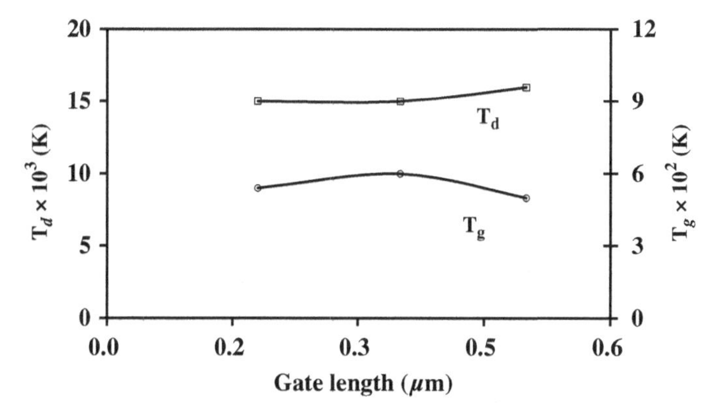

Figure 6.19. Noise model parameters versus gate length of MOSFET device.

6.5 Scalable Noise Model

The conventional small-signal models of MOSFET devices are based on the device structures which consist of only one elementary cell and are not suitable for the large-size device which consists of multiple elementary cells actually. The main reason is that the effect of the interconnection between the elementary cells has not been taken into account in the conventional models of the MOSFET device. Hence, these models have to be improved. In this section, an improved model that is suitable for both a single-elementary cell device and multiple elementary cells device is developed [15].

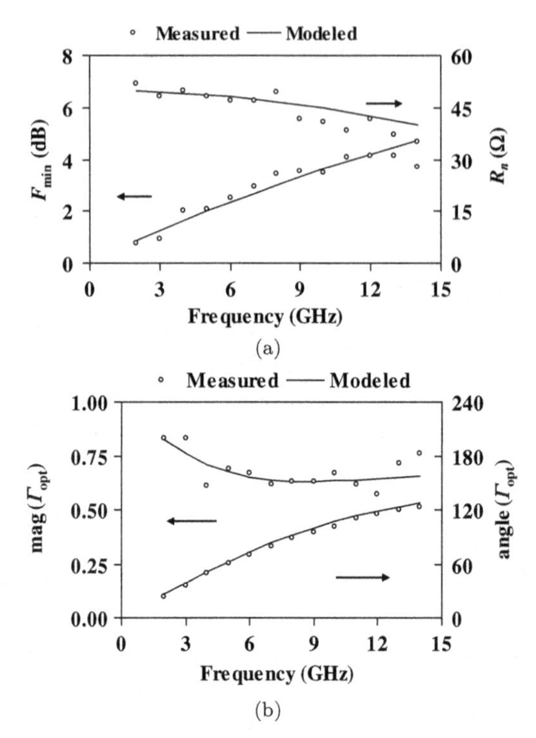

Figure 6.20. Comparison between the modeled and measured noise parameters of the $0.5 \times 5 \times 16$ μm MOSFET device. Bias condition: $V_{gs} = 1.2$ V, $V_{ds} = 2$ V. (a) F_{min} and R_n; (b) Γ_{opt}.

6.5.1 *Multi-cell noise model*

Figure 6.24 illustrates the schematic layout of the large-size MOSFET device which consists of multiple elementary cells; the bulk series resistance is minimized by surrounding each elementary cell with a guard ring, which is connected to the top bulk contact.

Figure 6.25 displays the proposed noise and small-signal model for a large-size MOSFET device which consists of multiple elementary cells. It can be observed that the effect of the interconnections between the elementary cells has been taken into account. This equivalent circuit model can be divided into two parts: the outer part which contains just pad parasitics and external feedline inductances and the inner part which contains n elementary cells.

Three inductances L_{gx}, L_{dx}, and L_{sx} represent the feedlines between pads and part of elementary cells, and another three

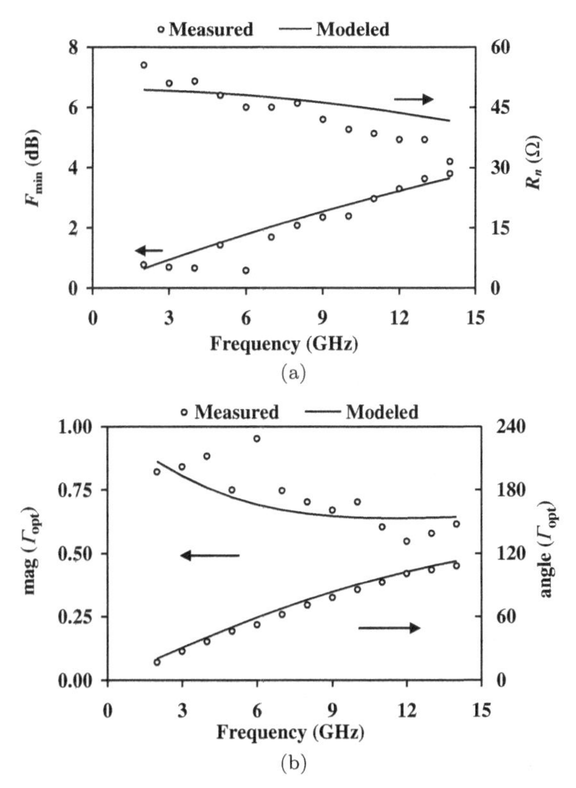

Figure 6.21. Comparison between the modeled and measured noise parameters of 0.35 × 5 × 16 μm MOSFET device. Bias condition: $V_{gs} = 1.2$ V, $V_{ds} = 2$ V. (a) F_{min} and R_n; (b) Γ_{opt}.

inductances L_g, L_d, and L_s represent the inductances of the gate, drain, and source interconnections between elementary cells, respectively.

In Pospieszalski's model, the corresponding admittance noise matrix of the elementary cell can be expressed as follows [7]:

$$C_{Y11}^{INT} = \overline{e_{gs}^2}\left|\frac{j\omega C_{gs}}{1 + j\omega C_{gs}R_{gs}}\right|^2 \tag{6.72}$$

$$C_{Y22}^{INT} = \overline{i_{ds}^2} + \overline{e_{gs}^2}\left|\frac{g_m}{1 + j\omega C_{gs}R_{gs}}\right|^2 \tag{6.73}$$

$$C_{Y12}^{INT} = \overline{e_{gs}^2}\frac{g_m^*\omega C_{gs}}{|1 + j\omega C_{gs}R_{gs}|^2} \tag{6.74}$$

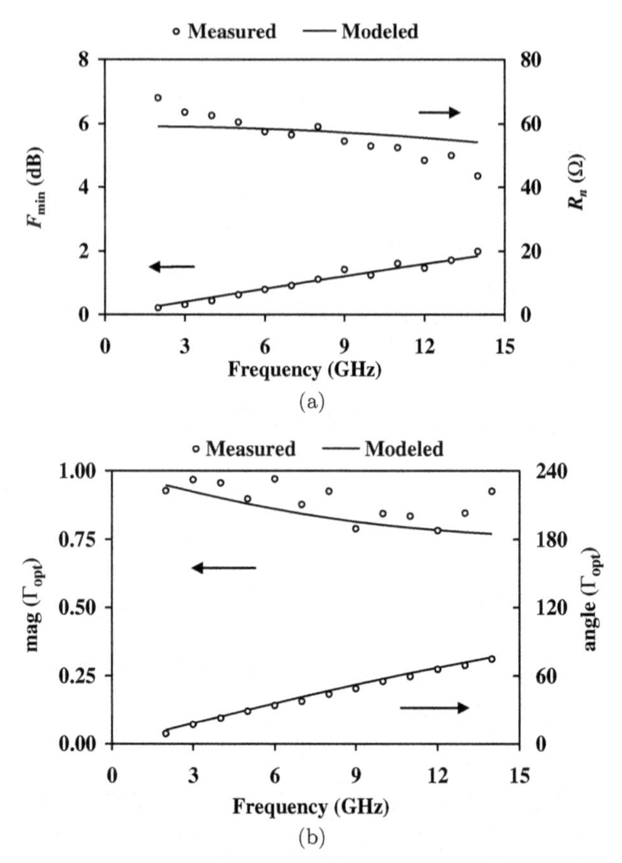

Figure 6.22. Comparison between the modeled and measured noise parameters of the 0.18 μm $\times$ 5 μm $\times$ 16 finger MOSFET device. Bias condition: $V_{gs} = 1.2$ V, $V_{ds} = 2$ V. (a) F_{min} and R_n; (b) Γ_{opt}.

In PRC's model [16–18], the short-circuit noise currents at the drain and gate can be expressed as follows:

$$\overline{i_g^2} = 4kT_o \frac{(\omega C_{gs})^2 R}{g_m} \Delta f \tag{6.75}$$

$$\overline{i_d^2} = 4kT_o g_m P \Delta f \tag{6.76}$$

with

$$\overline{i_d^* i_g} = 4kT_o \omega C_{gs} C \sqrt{PR} \Delta f \tag{6.77}$$

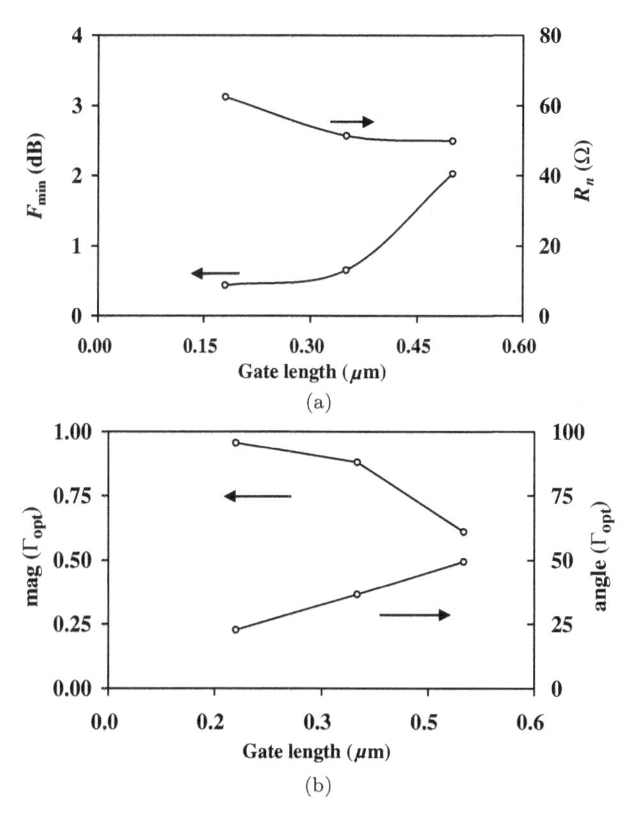

Figure 6.23. Noise parameters versus gate length at 4 GHz. (a) $F_{\min}$ and R_n; (b) Γ_{opt}.

where R and P are the gate and drain noise model parameters, respectively, and C is the correlation coefficient.

For the condition $\omega C_{\text{gs}} \ll 1$, the coefficients P, R, and C can be written in terms of the gate and drain temperatures for the Pospieszalski model [19]:

$$R = g_m R_{\text{gs}} \frac{T_g}{T_o} \tag{6.78}$$

$$P = R + \frac{T_d}{g_m R_{\text{ds}} T_o} \tag{6.79}$$

$$\text{Im}(C) = -\sqrt{\frac{R}{P}} \tag{6.80}$$

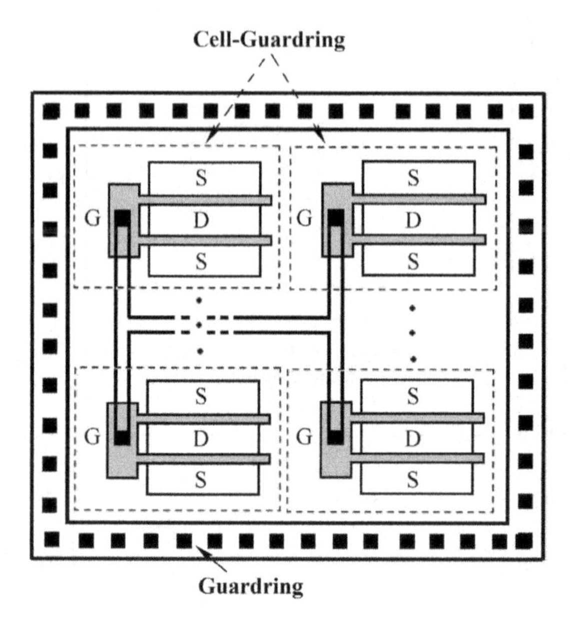

Figure 6.24. Schematic layout of large-size MOSFET device which consists of multiple elementary cells.

6.5.2 *Noise parameters extraction*

Once the small-signal elements are extracted from the S parameter measurements, the extraction of the noise model parameters can be carried out using the procedure based on the noise correlation matrix technique as follows:

(1) Calculation of the chain noise correlation matrix for the MOSFET device.

$$C_A = 4kT \begin{bmatrix} R_n & \frac{F_{\min}-1}{2} - R_n Y_{\text{opt}} \\ \frac{F_{\min}-1}{2} - R_n Y_{\text{opt}}^* & R_n \left|Y_{\text{opt}}\right|^2 \end{bmatrix} \tag{6.81}$$

(2) Transformation of the chain noise correlation matrix into the admittance noise correlation matrix and subtraction of pad parasitics (C_{oxg}, C_{oxd}, R_{pg}, and R_{pd}).

(3) Transformation of the admittance noise correlation matrix into the impedance noise correlation matrix and subtraction of external feedline inductances (L_{gx}, L_{dx}, and L_{sx}). Since the external feedline inductances network is noiseless, the impedance noise matrix remains invariant.

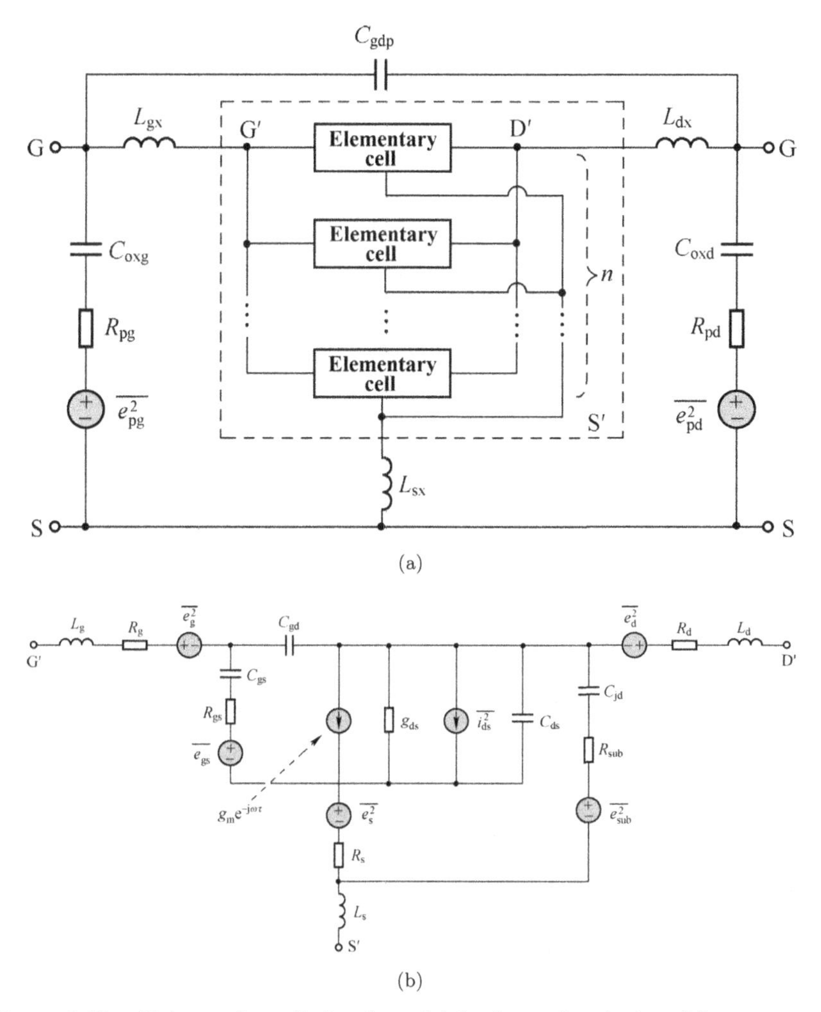

Figure 6.25. Noise and small-signal model for large-size device: (a) outer part; (b) elementary cell.

(4) Transformation of the impedance noise correlation matrix into the admittance noise correlation matrix and calculation of the admittance noise correlation matrix of the elementary cell [15]:

$$Y = nY^c \tag{6.82}$$

$$C_Y = nC_Y^c \tag{6.83}$$

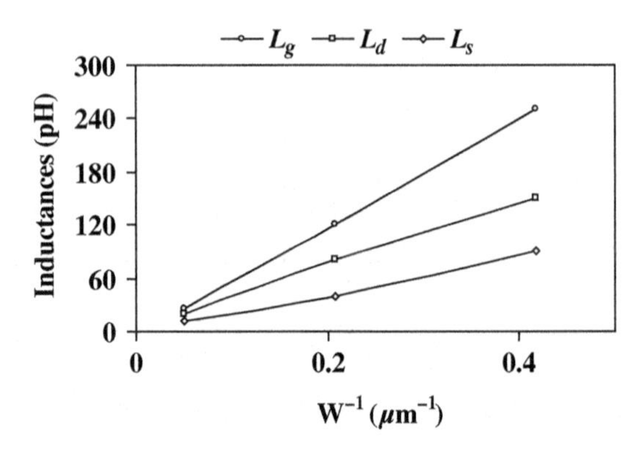

Figure 6.26. Extrinsic inductances versus gate width for elementary cell.

where Y and C_Y are the Y and noise matrices of the inner part of the MOSFET device and Y^c and C_Y^c are the Y and noise matrices of the elementary cell.

(5) Transformation of the admittance noise correlation matrix into the impedance noise correlation matrix and subtraction of interconnection feedline inductances (L_g, L_d, and L_s) and extrinsic resistances (R_g, R_d, and R_s).

(6) Transformation of the impedance noise correlation matrix into the admittance noise correlation matrix, and subtraction of substrate parasitics (C_{jd} and R_{sub}).

6.5.3 *Noise scaling rules*

Figures 6.26 and 6.27 illustrate the extracted extrinsic inductances and resistances of elementary cell versus device gate width. It can be found that the extrinsic inductances and resistances are inversely proportional to the gate width of the elementary cell. The substrate parasitics versus gate width for the elementary cell is depicted in Figure 6.28. As seen, the drain-to-bulk capacitance C_{jd} is proportional to the gate width of the elementary cell, and the series bulk resistance R_{sub} is inversely proportional to the gate width.

From Figures 6.26–6.28, it is obvious that the scaling formulas are determined to be as follows for extrinsic elements of the

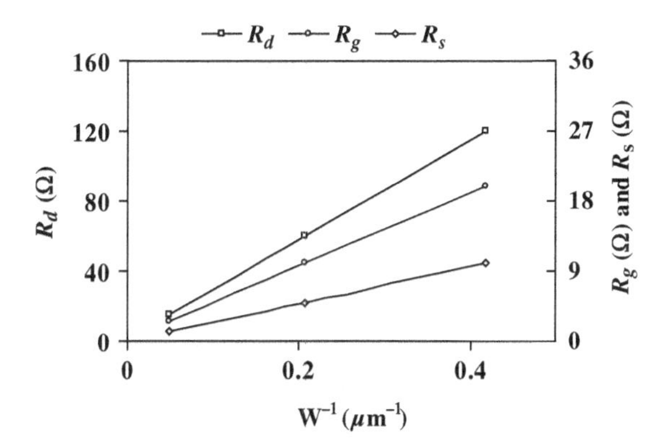

Figure 6.27. Extrinsic resistances versus gate width for elementary cell.

elementary cell:

$$\begin{bmatrix} L_{\mathrm{g}} \\ L_{\mathrm{d}} \\ L_s \\ R_g \\ R_d \\ R_s \\ C_{\mathrm{jd}} \\ R_{\mathrm{sub}} \end{bmatrix} = \begin{bmatrix} W^{-1} & 0 & 0 & 0 & 0 & 0 & 0 & 0 \\ 0 & W^{-1} & 0 & 0 & 0 & 0 & 0 & 0 \\ 0 & 0 & W^{-1} & 0 & 0 & 0 & 0 & 0 \\ 0 & 0 & 0 & W^{-1} & 0 & 0 & 0 & 0 \\ 0 & 0 & 0 & 0 & W^{-1} & 0 & 0 & 0 \\ 0 & 0 & 0 & 0 & 0 & W^{-1} & 0 & 0 \\ 0 & 0 & 0 & 0 & 0 & 0 & W & 0 \\ 0 & 0 & 0 & 0 & 0 & 0 & 0 & W^{-1} \end{bmatrix} \times \begin{bmatrix} L_g^{\mathrm{c}} \\ L_d^{\mathrm{c}} \\ L_s^{\mathrm{c}} \\ R_g^{\mathrm{c}} \\ R_d^{\mathrm{c}} \\ R_s^{\mathrm{c}} \\ C_{\mathrm{jd}}^{\mathrm{c}} \\ R_{\mathrm{sub}}^{\mathrm{c}} \end{bmatrix} \quad (6.84)$$

where W is the gate width of the elementary cell and superscript c denotes the elementary cell.

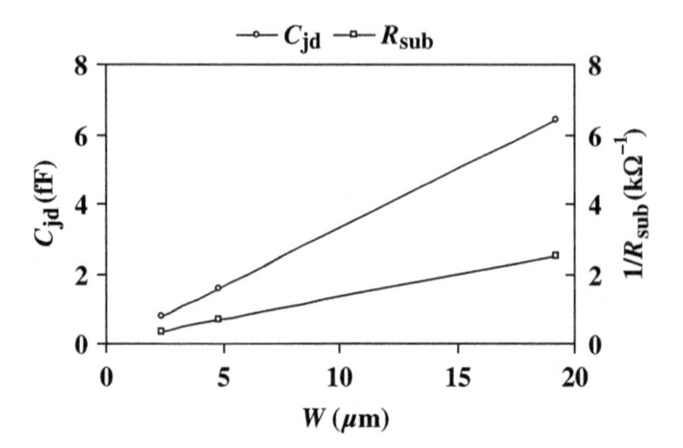

Figure 6.28. Substrate parasitics versus gate width for elementary cell.

The extracted intrinsic inductances and resistances of the elementary cell versus device gate width are exhibited in Figure 6.29. The capacitances are proportional to the gate width, the resistances are inversely proportional to the gate width, and the transconductance and drain conductance are proportional to the gate width.

$$\begin{bmatrix} C_{gs} \\ C_{gd} \\ C_{ds} \\ g_m \\ g_{ds} \\ R_{gs} \end{bmatrix} = \begin{bmatrix} W & 0 & 0 & 0 & 0 & 0 \\ 0 & W & 0 & 0 & 0 & 0 \\ 0 & 0 & W & 0 & 0 & 0 \\ 0 & 0 & 0 & W & 0 & 0 \\ 0 & 0 & 0 & 0 & W & 0 \\ 0 & 0 & 0 & 0 & 0 & W^{-1} \end{bmatrix} \begin{bmatrix} C_{gs}^{c} \\ C_{gd}^{c} \\ C_{ds}^{c} \\ g_m^{c} \\ g_{ds}^{c} \\ R_{gs}^{c} \end{bmatrix} \tag{6.85}$$

Based on the scalable rules, Z parameters of the elementary cell are inversely proportional to the gate width:

$$Z_{ij}^{c}(i = 1, 2; j = 1, 2) \propto W^{-1} \tag{6.86}$$

The impedance noise correlation matrix for the elementary cell is proportional to the gate width:

$$C_Z^C \propto \frac{1}{W} \tag{6.87}$$

The impedance noise correlation matrix for the intrinsic part of the elementary cell can be obtained by de-embedding the extrinsic

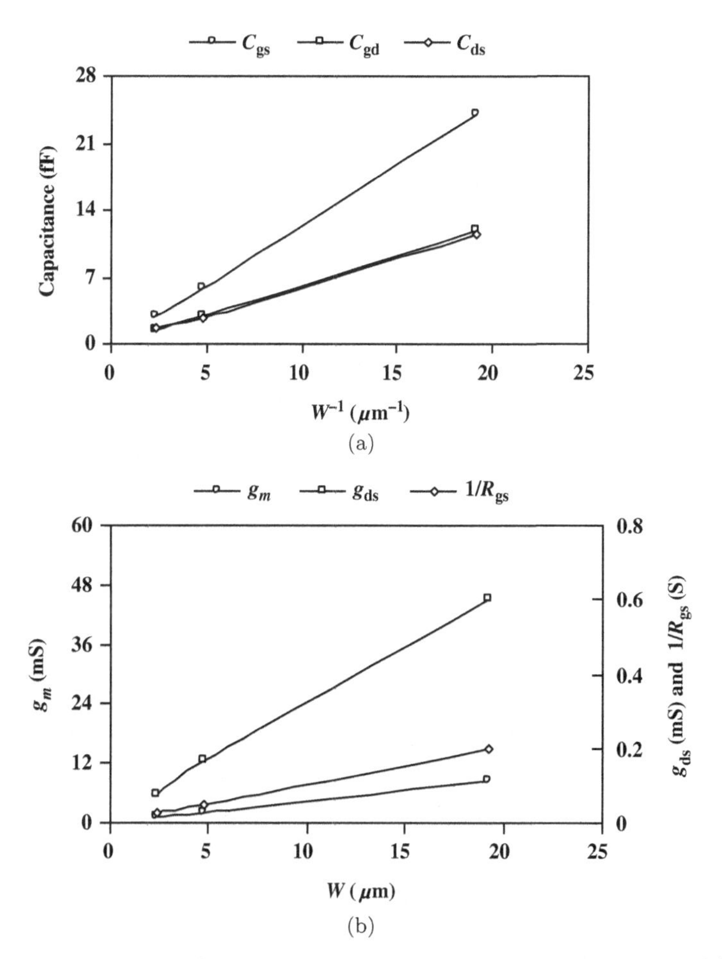

Figure 6.29. Intrinsic elements versus gate width for elementary cell: (a) C_{gs}, C_{gd}, and C_{ds}; (b) g_m, g_{ds}, and R_i.

resistances and inductances:

$$C_Z^{INT} = C_Z^C - 4kT_o \begin{bmatrix} R_g + R_s & R_s \\ R_s & R_d + R_s \end{bmatrix} \tag{6.88}$$

Thence,

$$C_Z^{INT} \propto \frac{1}{W} \tag{6.89}$$

The corresponding admittance noise matrix is obtained by translating the impedance noise correlation matrix:

$$C_Y^{INT} = Y^{INT} C_Z^{INT} (Y^{INT})^+ \propto W \tag{6.90}$$

The noise model parameters T_g and T_d are independent of the device size and remain invariant for scalable MOSFET devices:

$$\begin{bmatrix} T_g \\ T_d \\ P \\ R \\ C \end{bmatrix} = \begin{bmatrix} 1 & 0 & 0 & 0 & 0 \\ 0 & 1 & 0 & 0 & 0 \\ 0 & 0 & 1 & 0 & 0 \\ 0 & 0 & 0 & 1 & 0 \\ 0 & 0 & 0 & 0 & 1 \end{bmatrix} \begin{bmatrix} T_g^c \\ T_d^c \\ P^c \\ R^c \\ C^c \end{bmatrix} \tag{6.91}$$

6.5.4 *Model verification*

A set of noise and small-signal model parameters are determined for 90 nm NMOSFET transistors with 4 finger $\times$ 0.6 μm $\times$ 18 cell (number of gate fingers $\times$ unit gate width of single cell $\times$ number of cells) and then scaled for larger-size devices with 8 finger $\times$ 0.6 μm $\times$ 12 cell and 32 finger $\times$ 0.6 μm $\times$ 2 cell gate widths using the scalable rules described above. The MOSFET devices at a constant drain–source voltage V_{ds} = 0.6 V and I_{ds} = 1.76 mA are summarized in Table 6.6. In the frequency range of 50 MHz–40 GHz, the comparison between the measured and modeled S parameters for the 4 finger $\times$ 0.6 μm $\times$ 18 cell MOSFET devices under the bias condition of V_{gs} = 0.6 V and V_{ds} = 0.6 V is depicted in Figure 6.30. The modeled S parameters agree very well with the measured ones to validate the accuracy of the proposed model. The proposed model is also compared with the conventional model, and the comparison of accuracy between both models is illustrated in Figure 6.31. Figure 6.32 shows good agreement in S parameters between the measured and the scaled models determined by utilizing the scalable rules for 8 finger $\times$ 0.6 μm $\times$ 12 cell and 32 finger $\times$ 0.6 μm $\times$ 2 cell MOSFET devices.

Figure 6.33 illustrates the comparison between the measured and modeled noise parameters for a 4 finger $\times$ 0.6 μm $\times$ 18 cell MOSFET device under the bias condition of V_{gs} = 0.6 V and V_{ds} = 0.6 V. The excellent agreement over the whole frequency range is obtained.

Table 6.6. Model parameters of elementary cell (4 finger × 0.6 μm × 18 cell).

Parameter	Value	Unit	Parameter	Value	Unit
C_{oxg}	115	fF	R_g	20	Ω
C_{oxd}	110	fF	R_d	120	Ω
C_{pgd}	1.2	fF	R_s	10	Ω
R_{pg}	9	Ω	R_{sub}	3000	Ω
R_{pd}	8	Ω	C_{jd}	0.8	fF
L_{gx}	40	pH	C_{gs}	3.0	fF
L_{dx}	35	pH	C_{gd}	1.55	fF
L_{sx}	5	pH	C_{ds}	1.6	fF
L_g	250	pH	g_m	1.05	mS
L_d	150	pH	g_{ds}	0.077	mS
L_s	90	pH	R_{gs}	40	Ω

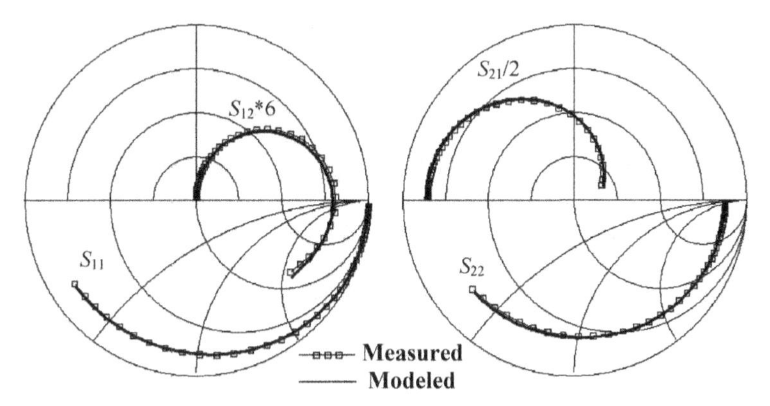

Figure 6.30. Comparison between the modeled and measured S parameters for the 4 finger × 0.6 μm × 18 cell MOSFET device. Bias condition: $V_{\text{gs}} = 0.6$ V, $V_{\text{ds}} = 0.6$ V.

The corresponding noise model parameters are given by the following:

$$\text{For Pospieszalski model:} \quad T_g = 300 \text{ K} \quad T_d = 4200 \text{ K}$$

$$\text{For PRC model:} \quad P = 1.12 \quad R = 0.042 \quad C = 0.195$$

Figure 6.34 shows good agreement in noise parameters between the measured and the scaled models determined by using the scalable rules for 8 finger × 0.6 μm × 12 cell MOSFET devices.

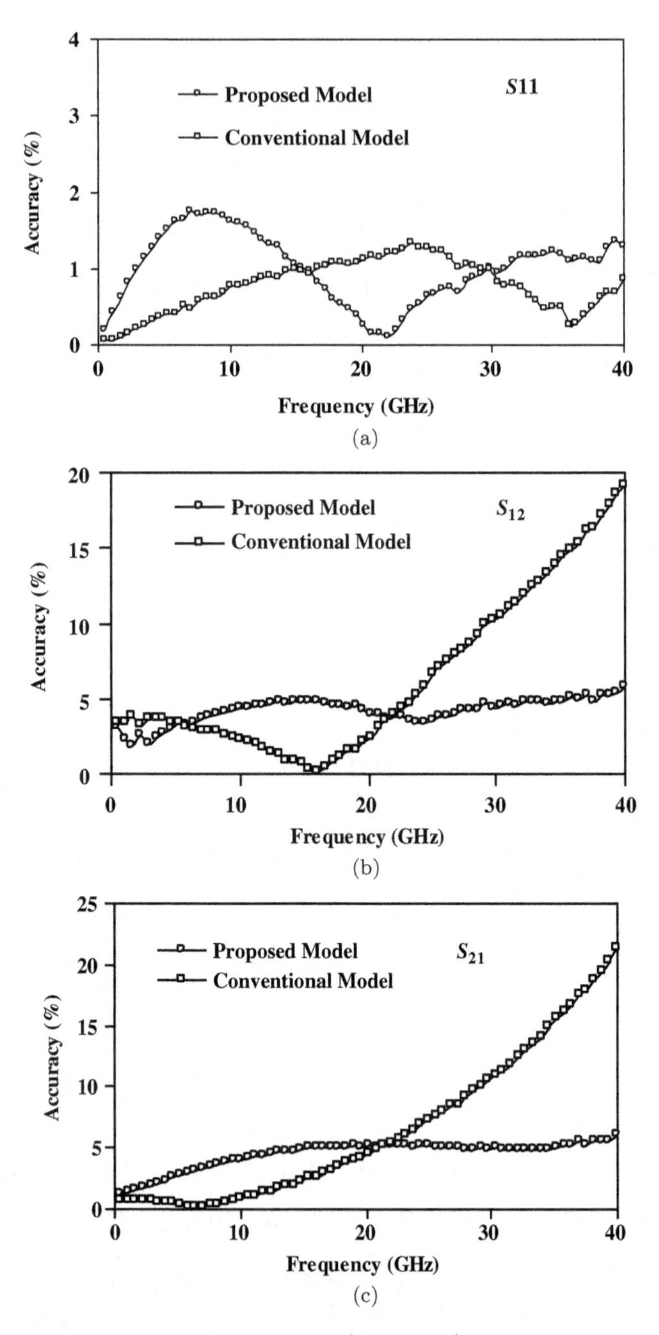

Figure 6.31. *(Continued)*

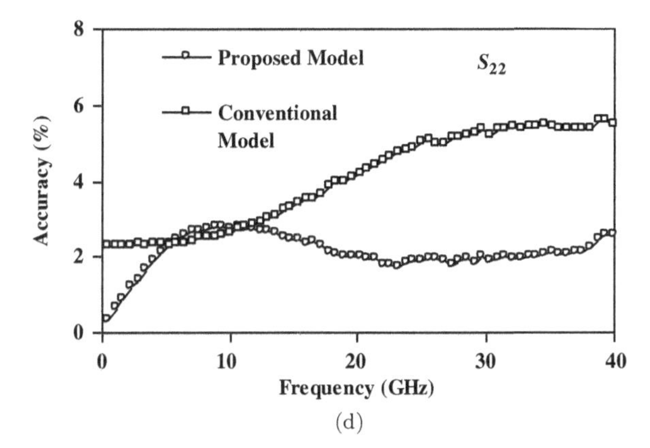

(d)

Figure 6.31. Comparison of accuracy between the proposed and conventional models. (a) S_{11}; (b) S_{12}; (c) S_{21}; (d) S_{22}.

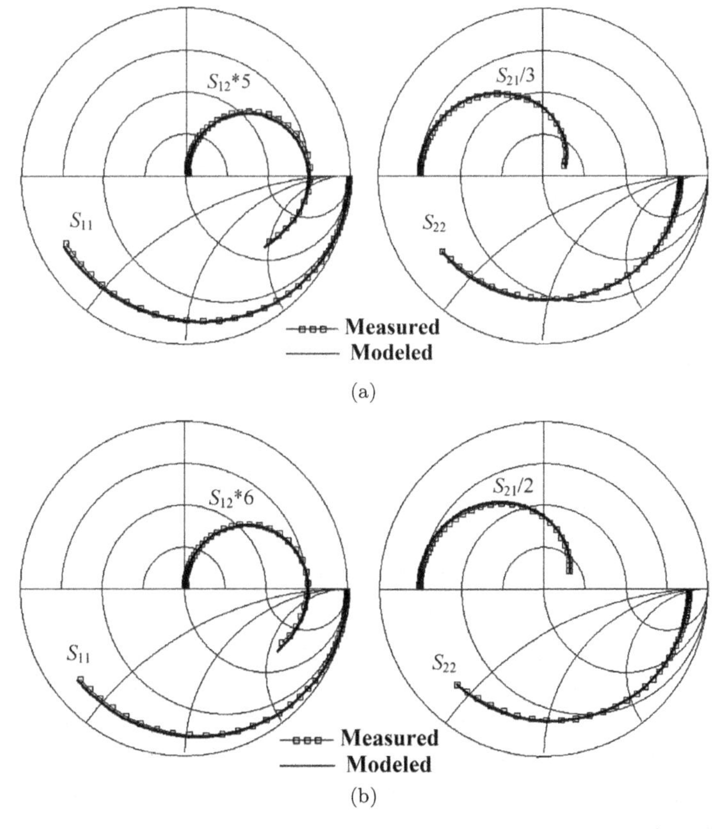

Figure 6.32. Comparison of S parameters between the measured and scaled MOSFET devices. Bias: $V_{\mathrm{gs}} = 0.6$ V, $V_{\mathrm{ds}} = 0.6$ V: (a) $8 \times 0.6 \times 12$ μm MOSFET; (b) $32 \times 0.6 \times 2$ μm MOSFET.

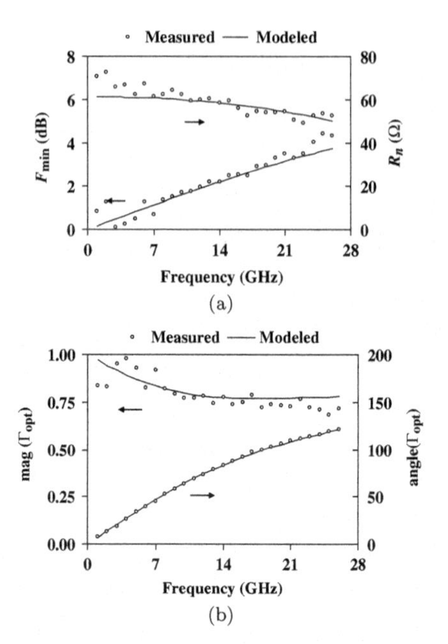

Figure 6.33. Comparison between the measured and modeled noise parameters for 4 finger × 0.6 μm × 18 cell MOSFET device. Bias: $V_{gs} = 0.6$ V, $V_{ds} = 0.6$ V. (a) F_{min} and R_n; (b) Γ_{opt}.

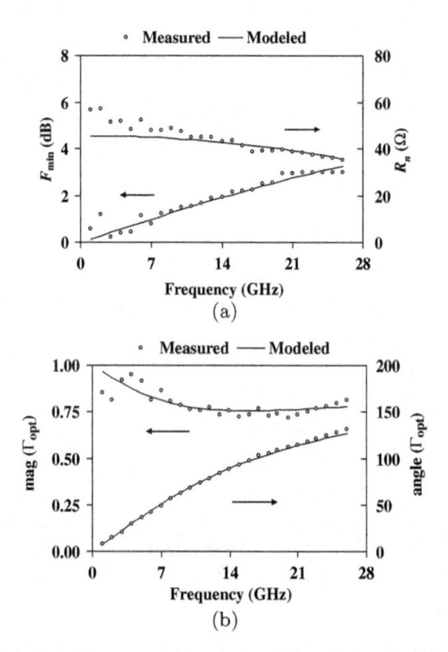

Figure 6.34. Comparison between measured and scaled noise parameters for 8 finger × 0.6 μm × 12 cell MOSFET device. Bias: $V_{gs} = 0.6$ V, $V_{ds} = 0.6$ V. (a) F_{min} and R_n; (b) Γ_{opt}.

6.6 Summary

The noise equivalent circuit model of MOSFET devices and the technology of signal and noise correlation matrix are described first. The expression of noise parameters based on the noise model is derived, and the extraction method of noise model parameters is given. Finally, the novel scalable noise and small-signal model for deep submicrometer MOSFET devices consist of multiple elementary cells. It allows exact modeling of all noise and small-signal model parameters from an elementary cell to a large size device. The scalable rules for noise and small-signal model parameters are given in detail.

References

[1] J. Gao, *Heterojunction Bipolar Transistor for Circuit Design — Microwave Modeling and Parameter Extraction.* Singapore: Wiley, 2015.

[2] J. Gao, *RF and Microwave Modeling and Measurement Techniques for Field Effect Transistors.* Raleigh, NC, USA: SciTech Publishing Inc., 2010.

[3] P. Sakalas, H.G. Zirath, A. Litwin, M. Schröter, and A. Matulionis, "Impact of pad and gate parasitics on small-signal and noise modeling of 0.35 μm gate length MOS transistors," *IEEE Transactions on Electron Devices*, 49(5): 871–880, 2002.

[4] Y. Cheng, C.-H. Chen, M. Matloubian, and M. Jamal Deen, "High-frequency small signal AC and noise modeling of MOSFETs for RF IC design," *IEEE Transactions on Electron Devices*, 49(3): 400–408, 2002.

[5] A. Pascht, M. Grözing, D. Wiegner, and M. Berroth, "Small-signal and temperature noise model for MOSFETs," *IEEE Transactions on Microwave Theory Techniques*, 50(8): 1927–1934, 2002.

[6] M. J. Deen, C. H. Chen, S. Asgaran, G. A. Rezvani, J. Tao, and Y. Kiyota, "High frequency noise of modern MOSFETs: Compact modeling and measurement issues," *IEEE Transactions on Electron Devices*, 53(9): 2062–2081, September 2006.

[7] M. W. Pospieszalski, "Modeling of noise parameters of MESFET's and MODFET's and their frequency and temperature dependence," *IEEE Transactions on Microwave Theory Technology*, 37: 1340–1350, September 1989.

[8] M. W. Pospieszalski, "Interpreting transistor noise," *IEEE Microwave Magazine*, 11(6): 61–69, October 2010.
[9] H. Hillbrand and P. Russer, "An efficient method for computer-aided noise analysis of linear amplifier networks," *IEEE Transactions on Circuits and Systems*, CAS-23(4): 235–238, April 1976.
[10] P. Yu–Microwave modeling and parameter extraction for 90-nm gate-length MOSFET devices, Ph.D Dissertation. East China Normal University, 2018.
[11] P. Yu, B. Chen, and J. Gao, "Microwave noise modeling for MOSFETs," *International Journal of Numerical Modeling: Electronic Networks, Devices and Fields*, 28(6): 639–648, 2015.
[12] R. Q. Lane, "The determination of noise parameters," *Proceedings of IEEE*, pp. 1461–1462, August 1969.
[13] J. Gao, "Direct parameter-extraction method for MOSFET noise model from microwave noise figure measurement," *Solid-State Electronics*, 63(1): 42–48, 2011.
[14] J. Gao, *RF and Microwave Modeling and Measurement Techniques for Field Effect Transistors.* Raleigh, NC, USA: SciTech Publishing Inc., 2010.
[15] J. Gao and A. Werthof, "Scalable small signal and noise modeling for deep submicron MOSFETs," *IEEE Transactions on Microwave Theory Techniques*, 57(4): 737–744, 2009.
[16] A. van der Ziel, "Gate noise in field effect transistors at moderately high frequencies," *Proceeding IRE*, 51: 461–467, March 1963.
[17] R. A. Pucel and H. A. Haus, "Signal and noise properties of gallium arsenide microwave field effect transistors," *Advances in Electronics and Electron Physics*, New York: Academic, 38: 195–265, 1975.
[18] A. Cappy, "Noise modeling and measurements techniques," *IEEE Transactions on Microwave Theory Technology*, 36: 1–10, January 1988.
[19] P. Heymann, M. Rudolph, H. Prinzler, R. Doerner, L. Klapproth, and G. Böck, "Experimental evaluation of microwave field-effect transistor noise models," *IEEE Transactions on Microwave Theory Technology*, 47: 156–162, February 1999.

Index

D

E

F

G

H

I

J

K

L

M

P

Q

R

S

W

Z

Printed in the United States
by Baker & Taylor Publisher Services